PLANT TRANSPOSABLE ELEMENTS

Biology and Biotechnology

PLANT TRANSPOSABLE ELEMENTS

Biology and Biotechnology

Edited by

Deepu Pandita

Anu Pandita

AAP | APPLE ACADEMIC PRESS

First edition published 2023

Apple Academic Press Inc.
1265 Goldenrod Circle, NE,
Palm Bay, FL 32905 USA

760 Laurentian Drive, Unit 19,
Burlington, ON L7N 0A4, CANADA

CRC Press
6000 Broken Sound Parkway NW,
Suite 300, Boca Raton, FL 33487-2742 USA

4 Park Square, Milton Park,
Abingdon, Oxon, OX14 4RN UK

Apple Academic Press exclusively co-publishes with CRC Press, an imprint of Taylor & Francis Group, LLC

Library and Archives Canada Cataloguing in Publication

Title: Plant transposable elements : biology and biotechnology / edited by Deepu Pandita, Anu Pandita.
Names: Pandita, Deepu, 1977- editor. | Pandita, Anu, 1977- editor.
Description: First edition. | Includes bibliographical references and index.
Identifiers: Canadiana (print) 20220490031 | Canadiana (ebook) 2022049004X | ISBN 9781774911815 (hardcover) | ISBN 9781774911822 (softcover) | ISBN 9781003315193 (ebook)
Subjects: LCSH: Transposons. | LCSH: Plant genetic transformation. | LCSH: Plant genetics. | LCSH: Plant biotechnology.
Classification: LCC QH452.3 .P53 2023 | DDC 572.8/69—dc23

Library of Congress Cataloging-in-Publication Data

CIP data on file with US Library of Congress

ISBN: 978-1-77491-181-5 (hbk)
ISBN: 978-1-77491-182-2 (pbk)
ISBN: 978-1-00331-519-3 (ebk)

About the Editors

Deepu Pandita
Senior Lecturer, Government Department of School Education, Jammu, Union Territory of Jammu and Kashmir, India

Deepu Pandita is working as a Senior Lecturer in the Government Department of School Education, Jammu, Union Territory of Jammu and Kashmir, India. Deepu Pandita has more than 20 years of teaching experience and has a master's degree in Botany from the University of Kashmir, Jammu and Kashmir, India, and a master's degree of Philosophy in Biotechnology from the University of Jammu, Jammu and Kashmir, India. Deepu Pandita has a number of international and national courses to her credit. Deepu Pandita has qualified for fellowships, such as JRF-NET and SRF from the Council of Scientific and Industrial Research (CSIR), New Delhi India; Biotechnology Fellowship, Government Department of Science and Technology, Jammu & Kashmir, India; and IAS-INSA-NASI Summer Research Teacher Fellowship, India. Deepu Pandita has presented her research work at both the national and international conferences and was awarded a Best Oral Presentation Award at an International Conference on Biotechnology for Better Tomorrow, a Women Researcher Award, and a Research Excellence Award from two professional associations in India. Deepu Pandita is a life member of various scientific societies. Deepu Pandita acts as a reviewer (18 journals), associate editor, and editor of a number of international journals. Deepu Pandita has published a number of editorials, book chapters (Springer, CRC, and Elsevier), reviews, and research articles in various journals of national and international repute, including *Cells, Frontiers in Plant Sciences, Journal of Fungi,* and *Frontiers in Physiology*. Deepu Pandia has currently several books in the process of production.

Anu Pandita
Senior Dietician, Vatsalya Clinic, New Delhi, India

Anu Pandita is working as a Senior Dietician at Vatsalya Clinic, E-5/11, Krishna Nagar, New Delhi, India. Previously she worked as a lecturer at Bee Enn College of Nursing, Talab Tillo, Jammu, India, and Dietician at Ahinsa

Dham Bhagwan Mahavir Charitable Health Centre, New Delhi, India. Anu Pandita has done her MSc internship and a course in the Dietetics Department of PGI, Chandigarh, India. Anu Pandita conducted a case study at the Pediatric Gastroenterology Ward at Nehru Hospital, PGI, Chandigarh, India, on a patient suffering from chronic liver disease. She has done a Certificate in Food & Nutrition as well. She has a number of trainings, refresher courses, and workshops to her credit. Anu Pandita is a life-time member of the Indian Dietetic Association and Indian Science Congress Association, Kolkata, India. Anu Pandita has published various book chapters in Springer journals and a number of reviews and research articles in various journals of national and international repute, including Cells, *Journal of Fungi*, and *Frontiers in Physiology*. Anu Pandita also has currently various books in process of production. Anu Pandita has presented her research work at both the national and international conferences.

Dedication

Deepu Pandita dedicates this book to her Research Guide, Professor (Dr.) Manoj Kumar Dhar, School of Biotechnology, University of Jammu, Jammu, India. Professor M K Dhar introduced DP to the World of Plant Transposons.

Contents

Contributors

Zarmeena Amjad
Institute of Plant Breeding and Biotechnology, MNS University of Agriculture Multan, Old Shujabad Road, Multan 60000, Pakistan

Tabasum Ashraf
Department of Biochemistry, University of Kashmir, Srinagar 190006, Jammu and Kashmir, India

Sr Assumi
Division of System Research and Engineering, ICAR Research Complex for NEH Region, Umiam, Meghalaya, India

Pankaj Barah
Department of Molecular Biology and Biotechnology, Tezpur University, Sonitpur 784028, India

Masarat Bashir
CoTS, Sheri-Kashmir University of Agricultural Sciences and Technology, Shalimar, Jammu and Kashmir, India

Suman Bakshi
Nuclear Agriculture and Biotechnology Division Bhabha Atomic Research Centre, Mumbai 400085, India

Nguyen Tran Hai Bang
Cuu Long Delta Rice Research Institute, Can Tho city, Vietnam

Pooja Bhadrecha
University Institute of Biotechnology, Chandigarh University, Mohali, Punjab, India

Shankar Bhujbal
Nuclear Agriculture and Biotechnology Division Bhabha Atomic Research Centre, Mumbai 400085, India

Braulio Edgar Herrera-Cabrera
Colegio de Postgraduados, Campus Puebla. Boulevard Forjadores No 205, Santiago Momoxpan, Puebla. C.P. 72760, México

H. V. Chaitra
Department of Crop Physiology, University of Agricultural Sciences, GKVK 560065 Bangalore, Karnataka, India

Swarnavo Chakraborty
Post Graduate Department of Biotechnology, St. Xavier's College (Autonomous), 30, Mother Teresa Sarani, Kolkata 700016, West Bengal, India

Ambika Chaturvedi
Dayalbagh Educational Institute, Department of Botany, Dayalbagh, Agra 282005, India

Divya Chauhan
ICAR-National Bureau of Plant Genetic Resource, New Delhi 110012, India

Gonzalo Guillermo Lucho-Constantino
Universidad Tecnológica de Gutiérrez Zamora, Veracruz. Prolongación Dr. Miguel Patiño S/N, Centro, 93556, Veracruz, México

Hebert Jair Barrales-Cureño
Instituto de Investigaciones Químico Biológicas, Universidad Michoacana de San Nicolás de Hidalgo, Edificio B-3, Ciudad Universitaria, 58030, Morelia, Michoacán, México

Ashraf Dar
Department of Biochemistry, University of Kashmir, Srinagar 190006, Jammu and Kashmir, India

Akash Das
Department of Molecular Biology and Biotechnology, Tezpur University, Sonitpur 784028, India

Lonishree Dutta
Assam Agricultural University, Assam 785013, India

Vincent Ezin
Faculty of Agricultural Sciences, University of Aboemy-calavi

Akash Fatima
Institute of Plant Breeding and Biotechnology, MNS University of Agriculture Multan, Old Shujabad Road, Multan 60000, Pakistan

Rafael Salgado-Garciglia
Instituto de Investigaciones Químico Biológicas, Universidad Michoacana de San Nicolás de Hidalgo, Edificio B-3, Ciudad Universitaria, 58030, Morelia, Michoacán, México

Younis Ahmad Hajam
Department of Biosciences, Division Zoology, Career Point University, Hamirpur 176041, Himachal Pradesh, India

Mainu Hazarika
Assam Agricultural University, Assam 785013, India

Sanjay J. Jambhulkar
Nuclear Agriculture and Biotechnology Division Bhabha Atomic Research Centre, Mumbai 400085, India

Zulqurnain Khan
Institute of Plant Breeding and Biotechnology, MNS University of Agriculture Multan, Old Shujabad Road, Multan 60000, Pakistan

Gagan Kumar
Centurion University of Technology and Management, Odisha 761211, India

Praveen Kumar
Assam Agricultural University, Assam 785013, India

José Lorenzo-Laureano
Universidad Intercultural del Estado de Puebla, Calle Principal a Lipuntahuaca S/N, 73475, Puebla, México

Rafiq Lone
Department of Botany, Central University of Kashmir, Jammu and Kashmir, India

Kartik Madankar
Department of Genetics and Plant Breeding, Institute of Agricultural Sciences, Banaras Hindu University, Varanasi 221005, India

Javid Ahmad Malik
Department of Zoology, Guru GhasidasVishwavidayalaya, Bilaspur 495001, Chhattisgarh, India

Fabiola Zaragoza-Martínez
Centro de Investigación y Estudios Avanzadosdel IPN. Av. Instituto Politécnico Nacional 2508, San Pedro Zacatenco, Gustavo A. Madero, 07360, Ciudad de México, México

Gita Mishra
Department of Zoology, Guru Ghasidas Vishwavidyalaya, Bilaspur, Chhattisgarh 495009, India

Udit Nandan Mishra
Faculty of Agriculture, Sri Sri University, Sri Sri Vihar, Godi Sahi, Cuttack 754006, Odisha, India

Jorge Montiel-Montoya
Centro Interdisciplinario de Investigaciónpara el Desarrollo Integral Regional, Instituto Politécnico Nacional Unidad Sinaloa, Boulevard Juan de Dios Bátiz 250, Col. San Joachín, Guasave, Sinaloa, México

Mahantesha B. N. Naika
Department of Biotechnology and Crop Improvement, Kittur Rani Channamma College of Horticulture, Arabhavi 591218, University of Horticultural Sciences, Bagalkot, Karnataka, India

Namrata
Agricultural Research Station, Mandore, Agriculture University, Jodhpur 352304, India

P. Lakshmi Padmavathi
Department of Biosciences and Sericulture, Sri Padmavati Mahila Visvavidyalayam (Women's University), Tirupati 517502, Andhra Pradesh, India

Jyothi Chaitanya Pagadala
Department of Plant Sciences, University of Hyderabad Gachibowli, Hyderabad, Telangana 500046, India

Himanshu Pandey
Dr. YSP University of Horticulture and Forestry, Solan 173230, Himachal Pradesh, India

Anu Pandita
Senior Dietician, Vatsalya Clinic, Krishna Nagar, New Delhi, India

Deepu Pandita
Government Department of School Education, Jammu, Jammu and Kashmir, India

H. T. Paramesha
Department of Biotechnology and Crop Improvement, Kittur Rani Channamma College of Horticulture, Arabhavi 591218, University of Horticultural Sciences, Bagalkot, Karnataka, India

Sameer Quazi
GenLab Biosolutions Private Limited, Bangalore 560043, Karnataka India

Archana Rai
Nuclear Agriculture and Biotechnology Division Bhabha Atomic Research Centre, Mumbai 400085, India

Rajiv Ranajan
Dayalbagh Educational Institute, Department Of Botany, Dayalbagh, Agra 282005, India

Aryadeep Roychoudhury
Post Graduate Department of Biotechnology, St. Xavier's College (Autonomous), 30, Mother Teresa Sarani, Kolkata 700016, West Bengal, India

Pooja Saraswat
Department of Botany, Dayalbagh Educational Institute, Dayalbagh, Agra 282005, India

Humaira Shah
Department of Biochemistry, University of Kashmir, Srinagar 190006, Jammu and Kashmir, India

Muhammad Shan
Institute of Plant Breeding and Biotechnology, MNS University of Agriculture Multan, Old Shujabad Road, Multan 60000 Pakistan

Ashok Singamsetti
Department of Genetics and Plant Breeding, Institute of Agricultural Sciences, Banaras Hindu University, Varanasi 221005, India

Rajesh Kumar Singhal
ICAR-Indian Grassland and Fodder Research Institute, Ministry of Agriculture and Farmer Welfare, Jhansi 284003, Uttar Pradesh, India

Ayushi Singh
Dayalbagh Educational Institute, Department of Botany, Dayalbagh Agra 282005, India

Shilpy Singh
Department of Biotechnology, Mangalmay Institute of management and Technology, Greater Noida, India

Lopamudra Sinha
National Bureau of Plant Genetic Recourses, New Delhi 110012, India

Rachael C. Symonds
School of Biological and Environmental Sciences, Liverpool John Moores University, UK

Rouf Maqbool
Department of Biochemistry, University of Kashmir, Srinagar 190006, Jammu and Kashmir, India

Parul Tyagi
Dayalbagh Educational Institute, Department of Botany, Dayalbagh, Agra 282005, India

Luis Germán López-Valdez
Laboratorio de Productos Naturales, Área de Química. AP74 Oficina de correos Chapingo, Universidad Autónoma Chapingo, Km 38.5 Carretera México Texcoco, 56230, Texcoco, Estado de México, México

Abbreviations

ABA	abscisic acid
Ac	activator
AGO	Argonaute
Al	aluminum
ALS	amytrophic lateral sclerosis
ART1	aluminium resistance transcription factor1
BAH	bromo-adjacent homology
CAF-1	chromatin assembly factor 1
CG	cytosine-guanine
CMT3	chromomethylase 3
Col	Columbia
CPDs	cyclobutane pyrimidine dimers
CREs	cis-regulatory-elements
DCL2 and DCL4	dicer-like nucleases
DCL3	dicer-like protein 3
ddm1	decrease in DNA methylation1
DEF	DEFICIENS
DME	demeter
DML1	demeter-like1
DRM2	domains rearranged methyl transferases
DNA	deoxyribonucleic acids
DRO1	DEEPER ROOTING 1
dsRNA	double-stranded RNA
Ds	dissociation
Ds	dissociate
DSBs	double strand breaks
Dt	dotted
ECs	epigenetic components
EDM2	Enhanced Downy Mildew 2
En/In	enhancer/inhibitor
Epi-d1	"Epigenetic silencing" of DWARF1
ERVs	endogenous retroviruses
ETE	exapted transposable elements
FIE1	fertilization-independent endosperm 1

FLC	FLOWERING LOCUS C
GRF	generic repeat finder
GSP	genome sequencing program
HMM	Hidden Markov model
HR	homologous recombination
HREs	heat response elements
HSPs	heat shock proteins
IDN2	Involved In De Novo 2
IDP	IDN2 paralogue
IN	integrase
IPCR	inverse polymerase chain reaction
IRs	insertion regions
IS	insertion sequences
KYP	Kryptonite
LCV	local combinational variable
lincRNAs	long intergenic noncoding RNAs
LINEs	long interspersed nuclear elements
LTRs	long terminal repeats
LTR-RT	long terminal repeat retrotransposons
MAS	marker-assisted selection
MET1	methyltransferase 1
MITEs	miniature inverted-repeat transposable elements
MLR	multiretrotransposon-like
mPing	miniature ping
MSA	multisequence alignment
NCBI	National Centre for Biotechnology Information
nDart1	non-autonomous DNA-based active rice transposon one
NGS	next-generation sequencing
NHEJ	non-homologous end joining
OA	organic anions
ORFs	open reading frames
PBSs	primer-binding sites
phyA	phytochrome A
PIKK	protein kinase
PLEs	Penelope-like elements
PolIV and Pol V	polymerases
PPTs	polypurine tracts
PTGS	posttranscriptional gene silencing
PW	powdery mildew

RC	rolling-circle
RdDM	RNA-directed DNA methylation
RDRP2 or RDR2	RNA-dependent RNA polymerase 2
Rep	replication initiator
RH	RNaseH
ROS1	repressor of silencing 1
RPA	replicon protein A
RRP6L1	RRP6-LIKE 1
RSI	ROSINA
RT	reverse transcriptase
SASP	secret-associated senescence of phenotype
SDR	split direct repeats
SDSA	synthesis-dependent strand annealing
SINEs	short interspersed nuclear element
siRNAs	small interfering RNAs
Spm	suppressor/mutator
SREs	stress-responsive elements
TDR	terminal direct repeats
TEs	transposable elements
TE-lincRNAs	TEs that associate with long intergenic noncoding RNAs
TF	transcription factor
TFBSs	transcription factor-binding sites
TGEs	transposable genetic elements
TGS	transcriptional gene silencing
TIRs	terminal inverted repeats
TPRT	target-site primed reverse transcription
TRIMs	terminal-repeat retrotransposons in miniature
TSDs	target site duplications
TSS	transcription start site
tub	tubular-rayed
turf	tubular ray flower
VIM	variant in methylation
WGD	whole-genome duplication
WGS	whole-genome shotgun

Preface

This book offers a framework and functions of various ubiquitous mobile transposable factors (TEs), or jumping genes, which translocate in diverse genomic locations inside the cell of a plant. Discovered in maize in the late 1940s by Barbara McClintock (before the discovery of DNA itself), TEs have been recognized with mystification since the beginning. The organismal genomes have chromosomes made of double-stranded DNA helix, which is the heart of stability and integrity of cells. In spite of the fact that the DNA repair allows the DNA chemical integrity and provides protection to the DNA from various mutagens of metabolic and environmental nature, the process of meiotic recombination and activity of transposable elements counteracts molecular guards of genome stability. The TEs frame the bulk of genomic DNA, often up to 50% of genomes. Transposition of the TEs may influence the function of genes at or adjacent to the TE insertion sites. Identification and characterization of TEs in genomes, primordial or recently inserted TEs by using high-tech protocols and approaches like next-generation sequencing (NGS), retrotransposon capture sequencing (RC-seq), computational approaches for prediction and analysis of TEs, annotation of TEs, genome engineering in plants via transposable elements, transposon tagging and mutagenesis are described in the chapter contents. In conclusion, this book on TEs also has chapters that present how TEs can be practiced in biotechnological applications of vital significance.

Previously known as "junk DNA" but currently JUNK is Just Unexplored Novel Knowhow which has contributed to the evolution of structure, function and regulation of genome in plants, epigenetic regulation, genome mutation and organismal speciation. The evolutionary significance of TEs in architecture and development of plants, stress responses, significance in rice plant, how host genomes regulate the activity of TEs, plant genome engineering by TEs and their role in plants in the upcoming climate change are also discussed in various individual chapters which emphasize on the below mentioned aspects:

1. The book provides up-to-date information on numerous classes of TEs.

2. This book is a systematic interpretation of protocols designed to characterize TEs and their biotechnological roles.
3. This unique and comprehensive text is the only book in the market that focuses exclusively on transposon biology and its role in plant development, architecture, epigenetic regulation, DNA repair, evolution and speciation, highlighting their importance in the approaching epoch of climate change.
4. This book functions as a valuable resource of facts for college students, academicians, researchers, and general public on the applications of the transposons in genome editing and their biology, classification, structure, function, and evolution in rice model plant.
5. This book will be valuable to the research centers around the globe working on the plant transposons that lead to a connecting link between the researchers to collaborate.

Part I
Introduction to Transposable Elements (TEs)

CHAPTER 1

Introduction, Classification, and Transposition of Transposable Elements (TEs)

JYOTHI CHAITANYA PAGADALA[1], DEEPU PANDITA[2*], and ANU PANDITA[3]

[1]*Department of Plant Sciences, University of Hyderabad Gachibowli, Hyderabad, Telangana 500046, India*

[2]*Government Department of School Education, Jammu, Jammu and Kashmir, India*

[3]*Vatsalya Clinic, Krishna Nagar, New Delhi, India*

[*]*Corresponding author. E-mail: deepupandita@gmail.com*

ABSTRACT

Tremendous improvements in molecular biology, genomics, and genetics have disclosed researchers to know rigorously about the various features and roles of transposable elements (TEs) in biology. Acquainted with most of the plant genome DNA, is composed of TEs. The description of their effect on the structure and expression of the genome substantially impacts toward understanding the phenotypic variations and worldwide genome regulation. Noteworthy findings at the corner level of TEs are mobile DNA sequences that transport from one chromosome to a new contained by the equivalent chromosome or into one more nonhomologous chromosome. Notable findings at the crossing point of transposition mechanisms allowed identifying novel TEs and extending their taxonomic distribution to reveal associations between TEs and plant genomes. Studying the TEs clears to recognize

Plant Transposable Elements: Biology and Biotechnology, Deepu Pandita, Anu Pandita, (Eds.)

variations in the gene expression levels between populations and natural variation patterns used in the genomics programme also paved the way to understand the fundamental progress in the stress levels of plants during the post-genomics phase for crop improvement. These fields have provided new perceptions that discovered the movement of TEs can be well-ordered epigenetically through their host mechanism. TEs impact the arrangement, dynamics, and structure of the genomes they occupy. This chapter highlights the history and classes of TEs predominantly on transposable mechanisms corresponding to "copy and paste" and "cut and paste". We begin by presenting basic structural features, classes, transposition, and significance of these elements, focusing on aspects that manage their maintenance and propagation within the genome and throughout plant eukaryotic hierarchy life cycles.

1.1 INTRODUCTION

Transposable elements (TEs) are "jumping genes," which can move from one genome location to another location. Firstly, geneticist Barbara McClintock recognized these elements in maize over 50 years ago (in 1944) at Cold Spring Harbor Laboratory in New York. She published an article on her findings in *Genetics* in November 1953. The discovery was titled "Induction of uncertainty at selected loci in Maize." She was honored with the Noble Prize in 1983 for her contribution to science by identifying TEs more than after 30 years of initial findings. She had named these elements as controlling elements. Most eukaryotic genomes contain many repetitive sequences. TEs can move from one chromosome into another, bacterial chromosome to bacteriophage chromosome or a plasmid in bacteria, and are found in many plant genomes (Bennetzen, 2000; Tenaillon et al., 2010; Pereira and Ryan, 2019). TEs can proliferate crossways and delete or edit from one genome, resulting in various mutations. Gene evolution and regulation caused the phenotypic diversity by simple insertional sequential polymorphism and polyploidy. Gene knockouts are also responsible for somatic mutations (Negi et al., 2016). TEs turn into a treasure trove intended for evolutionary genomics. Many researchers focus on genome evolution, progression, and adaptation and also consider new approaches to raise the yield of genome diversity (Le et al., 2007). TEs are the essential factors of plant genome size. TEs, exclusively retro transposons, establish the prominent part of giant genomes of plants species. In a broad sense, especially in cereal crops, such as maize and sugarcane, an increase in genome size is undoubtedly associated with the rise in TE content (Paterson et al., 2009; Schnable et al., 2009; Tomato Genome Consortium, 2012; de Setta et al., 2014). In conifers, the assemblage of Penelope-like components as a separate group is described by Gladyshev and Arkhipova (2007).

There are different tools developed in biology. Next-generation sequencing (NGS) tools reformed biology and provided unique opportunities to study giant and composite genomes, such as maize or sugarcane. Yet, NGS is also a trial for bioinformatics algorithms. There are many trends in biology to classify and categorize transposable components in thousands of genome sequences, but their applications in TE's are limited. Presently, new logarithmic techniques like Machine Learning (ML) or Dl and some traditional methods are established to overcome the difficulties of assembly and annotation processes of TEs for next-generation genomics. Some significant elements resemble the length of retrotransposon, LTR length, or ORFs. Also, motifs use data mining, such as TATA box, AATAAA, TDS, and poly-A tails, to design specific machine learning programs (Arango-Lopez et al., 2017). Though trends are developing in TEs, no single tool can be collectively helpful in identifying all TE types in all species worldwide. Hence, several researchers used diverse practices, techniques, and softwares found in the literature. Benjak et al. (2008) identified transduplicated improved cellular sequences, besides certain cellular functions have been domesticated and perhaps fulfilled besides providing facts about the contribution and capacity of the mobility of TEs for the genomic variability of grapevine. Composite antiquity of domestication also progresses in crop improvement of tomato, categorized by two sequential genetic bottlenecks. This program can be followed subsequently by recent breeding programs in numerous introgression trials through wild tomatoes and lineages. In the direction of replacing an inadequate pool of disease-resistance genes, DNA transposons, especially MuDR, hAT, and CACTA superfamilies, are useful. Quadrana (2020) explained that the model of gene expression changes triggered by TE insertions is expected to be more considerable than that created by SNPs. An active MITE "*mJing*" allied with the *high-tillering dwarf* mutant in rice recognized by Tang et al. (2019), also identified that the copy number of *mJing* intensely differs between *Oryza sativa*, *Oryza rufipogon*, and *Oryza glaberrima*. These signify their vital role in improving genome progression and species divergence for amplifying or reducing *mJing* elements in rice. Finding the part of genetic functions or alliance with the economic features will expedite breeding. The transposon reference mutation libraries suitably study functional genomics in economic crops like Rice (*Oryza sativa*) (Ram et al., 2019) and Corn (*Zea mays*) (Liang et al., 2019). A collective amount of superior quality plant genomes has been pronounced by the hasty improvement in genome sequencing tools. Investigators envisioned that studying the genetic mechanisms also intended to recognize critical genes associated with significant economic traits might not acquire genetic data used in plant breeding. However, genomic resources could become countless resources for breeding and genetic breeding in conifers. However, more data are available regarding the whole-genome sequences of several plants. But the sequences existing in conifers are significantly less in line with their tremendously bulky genomes, and the association is complex (Mielich et al., 2018; Zhao et al., 2017).

Recent trends in studying the function and dynamics of TE's in plant genome compartments, long-read sequencing can make available to widespread assembled genomes at a low price, which would be essential towards the understanding of TEs for local adaptation (Gao et al., 2012a, 2012b). There is a need to focus on private, abundant, or fixed insertions and their involvement in the arrangement of the expendable genomes. The influence of the organization of TE insertion in neighboring genes has a perspective of great assistance to researchers and helps to understand variations in phenotypes between the accessions of *Arabidopsis thaliana* species as genes take structural and functional annotations (Gazzani et al., 2003). Specifically, TE annotation would designate TFBS and promotors on respectively interleaved copy to forecast the probable active role of neighboring genes (Quesneville, 2020). The difference in proportions of crossing over across chromosome 4 of *Arabidopsis thaliana* disclosed affecting the existence regards meiotic recombination hot spots (Drouaud et al., 2006). TEs have important roles in human genetic disorders, such as cancer, autoimmune, and neurological diseases as well (Pandita and Pandita, 2016).

1.2 CLASSIFICATION OF TRANSPOSABLE ELEMENTS

TE are categorized through their potential to modify their sites in the genome or transposition (Hua-Van et al., 2011; Lisch, 2013). There are present well-known plant transposable elements conferring to the extensive taxonomy. Based on relevant transposing units, transposable elements to their coding regions involved in replicating the features are aligned into two classes (Chaparro et al., 2015). Broadly, TE are categorized into two groups: (1)

Class I RNA transposons/Retrotransposons mobilize through a "copy and paste" method. (2) **Class II DNA transposons**, transpose by a "cut and paste" mechanism (Fig. 1.1). TE classification depends on the mode of transposition, that is, classical or nonclassical, and the existence or nonexistence of LTR, and TIR sequences (Gao et al., 2012a, 2012b). Class II transposons use the cut and paste approach and propagate themselves by transposition as DNA copies, whereas RNA transposons mobilize themselves not by transposition but by making DNA copies of their RNA transcripts or by a copy and paste method. Class I retro transposons are again subclassified into LTR and non-LTR retrotransposons. Classical and nonclassical types of transposons belong to the DNA transposons (class 2) (Gladyshev and Arkhipova, 2007). These two categories of TEs vary among species. The abundance of LTR-containing retro-transposons occurs in plant genomes (Fedoroff, 2012). Several studies are available about the classes of transposable elements: DNA transposons vs. retrotransposons and superfamily as Copia vs. Gypsy level. These evaluations make available appreciated summaries of TE populations. And also discovered specific wide-ranging configurations, for instance, DNA transposons dispose to nearby genes relative to retrotransposons (Neumann et al., 2019). The Gypsy superfamily comprises more than 12 lineages that are reliable as long standing just as main divisions of plants (Feschotte and Pritham, 2005). Broad views about duplications, retrotransposons or Gypsy components stop leveling up biologically significant differences between TE lineages (Stritt et al., 2019).

In contrast, class 2 DNA transposons are subclassified into TIR and non-TIR transposons (Vitte et al., 2014). Supplementary approaches are established (RED, TEdna, transposome, REP de novo, LTR Classifier, Inpactor, etc.). Besides, TE sequences cause the progression of cellular functions and lead to the development of new proteins (Kidwell and Lisch, 1997; Jangam et al., 2018). A classic example is probably the perseverance of TE transposase, a protein involved in couple of transcription factors, such as FHY3 and FAR1 for regulating plant light signaling pathways (Lin et al., 2007).

1.2.1 RETROTRANSPOSONS/COPY AND PASTE TRANSPOSONS

Retrotransposons exist in almost all eukaryotes but not in prokaryotes (Mita and Boeke, 2016). On the other hand, a straight association is present in the proportions of the eukaryotic genome and the availability of retrotransposons. However, initially retro-transposons were considered as junk DNA

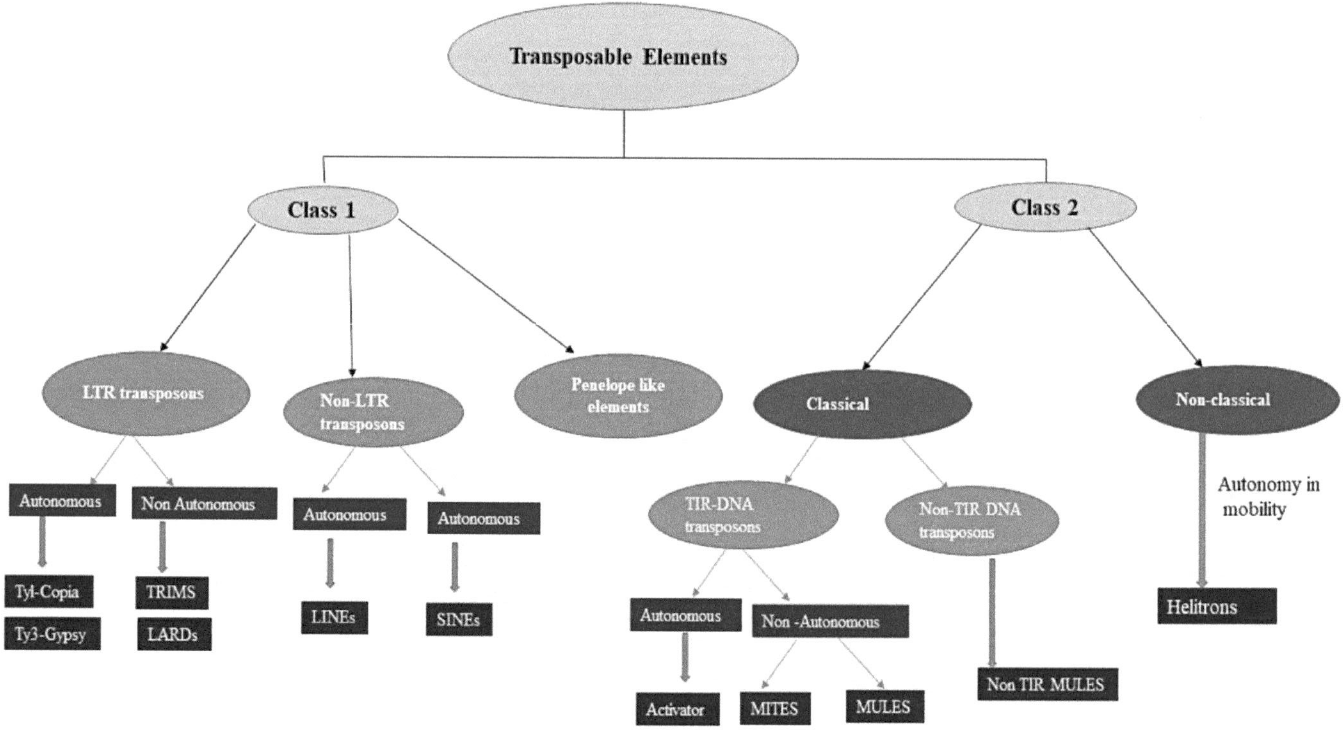

FIGURE 1.1 The overall classification of transposable elements.

(Bonchev, 2016). LTR retrotransposons and non-LTR retrotransposons are subcategories of retrotransposons (Fig. 1.2). Yeast genome comprises of 3% of LTR retrotransposons (Muszewska et al., 2011). The human genome has approximately 37% retrotransposons precisely LINEs and SINEs (Ou et al., 2018). Lastly, almost 85% of the maize genome is primarily composed of retrotransposons, specifically LTR retrotransposons (Keidar et al., 2018). The transposable element length and LTR length are essential to categorize LTR retrotransposons and anticipated that motifs privileged LTRs preserved superfamilies transversely by means of HMMs (hidden Markov models) current neural network (Fischer et al., 2018). Yuan and Wessler (2011) elucidated that class I TEs are considered as retrotransposons, which transpose through component encrypted mRNA intermediary contrived to commence a promoter in LTR retrotransposons. Initially, an interior promoter of non-LTR retrotransposons is reverse-transcribed into DNA and is incorporated somewhere else in the genome ("copy and paste"). Retrotransposons are present in the majority of plant genomes, mainly Ty1/Copia and Ty3/Gypsy families. Retrotransposons mainly determine the size of plant genomes, comprising monocots and dicots (Feschotte and Pritham, 2005). There is a configuration of amplification and variation in Brassicaceae family members about retrotransposons viz., Ty1/Copia, and Ty3/Gypsy (Fujimoto et al., 2008). Polymorphic amplifications in *Pinus massoniana* can be efficiently used to identify germplasm and genetic relationships. Conferring to the preservation of the reverse transcriptase sequences of Ty1-copia and Ty-gypsy-type retrotransposons, IRAP markers were established in *Pinus massoniana* on genomic LTR retrotransposon sequences (Cui et al., 2016). Commonly, the association of retrotransposons and the host genomes has long-lasting relations. The association varies in retrotransposons from the transposons in being dynamic and intended for only a short period in any genome. Among different species, it also relies on horizontal relocations for their long-standing survival. The significant vertical inheritance of retrotransposons is marked chiefly in LINEs (Petit et al., 2010). In plants, LTR retrotransposons mainly gather in the pericentromeric regions where small number of genes are commonly present. The long-standing association between the retrotransposons and the host genome demands whatever controls their copy number (Chu et al., 2016). It also raises the question of the positive and adverse properties of the genome (Belyayev et al., 2010).

The elements and structural features involved in the life cycle of retrotransposons were subcategorized into four orders. The four categories are LTR-RT, non-LTR retrotransposons, PLEs, and DIRS (Rahman et al.,

2013). Still, LTR-RT remains the utmost familiar one, also they can subsidize up and about to a significant part of plant genome size approximately 80%, for instance in wheat, barley and rubber tree (Gao et al., 2012a, 2012b). There are two superfamilies in plant LTR-RT order, Copia and Gypsy, which were classified based on the coding domain of internal association (Gao et al., 2012a, 2012b). Further, Copia and Gypsy superfamilies are subdivided through phylogenetic analysis into lineages and families depending on similarities in the coding region of enzymatic domain recognized as reverse transcriptase (Wiker et al., 2007). And also, Tork lineages fit into the Copia superfamily. Gypsy superfamily varies in plant genomes, such as Retrofit, Alesia, Angela, Bianca, Bryco, Lyco, Gymco, Oryco, Osser, SIRE, Tork. However, Athila, Clamyvir, Galadriel, Selgy, Tcn1, Reina, Tekay, Centromeric Retrotransposon, Phygy, and TAT are also present (Llorens et al., 2009; Llorens et al., 2010). Based on some phylogenetic studies, a chromodomain Gypsy superfamily is divided into diverse groups. Under the Chromovirus branch, Galadriel, Reina, Tekay, and CRM lineages were clustered (De Castro Nunes et al., 2018; Neumann et al., 2019). Tieman et al. (2017) and Labate and Robertson (2012) investigated the tomato mobilome composition, that is, TE families with modern deployment motion. Also analyzed 602 tomato accessions designed for short-read whole-genome resequencing. These families signify that the complete assortment of LTR and non-LTR retroelements is GYPSY, COPIA, and LINE superfamilies (Sultana et al., 2017). The variation in the copy number of transposable elements of *Triticum–Aegilops* recommends underlying evolutionary and revolutionary forces result in allopolyploidization (Yaakov et al., 2013).

1.2.2 DNA TRANSPOSONS/CUT AND PASTE TRANSPOSONS

DNA transposons (Class II) transpose by means of DNA that is removed and reinserted into the host genome through the mechanism of "cut and paste." Even if the genomes of blossoming plants have an abundant assemblage of a couple of TE classes, among Class II TEs, five superfamilies, such as CACTA, Mutator, PIF/Harbinger, hAT, and Tc1/mariner have been found in plant genomes (Girgis 2015). The abundant assortment of DNA transposons can be classified into three main subclasses (Pritham et al., 2007):

1. Classic transposons with the 10 most essential superfamilies. The cut and paste transposons.
2. Helitrons (Rolling-circle transposons).

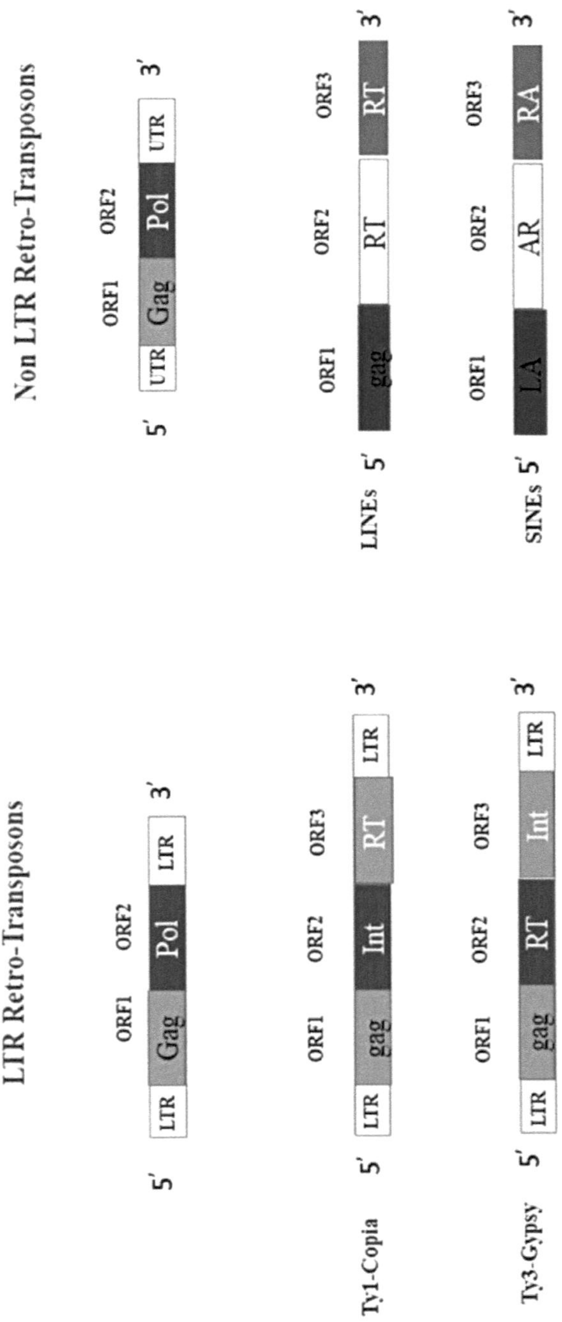

FIGURE 1.2 Structure of LTR retrotransposons and non-LTR retrotransposons.

3. Mavericks (Self-replicating transposons).

DNA-mediated transposons highlight their arrangement, extension dynamics, and genomic effect distinguished from other mobile elements. The systematic distribution and diversity of all dominant forms of eukaryotic DNA transposons have Helitrons and Mavericks. These evolutionary forces impact their conservation and variation in different genomic locations. DNA transposons have unique biological characteristics and provide shape to genome structural design besides direct to the beginning of genomic improvements in diverse eukaryotic lines (Feschotte and Pritham, 2005). Benjak et al. (2008) presented a worldwide and in-depth exploration of the class II transposons, particularly "cut and paste" transposons existing in the *Vitis vinifera* L genome. They characterized 1160 potentially widespread grapevine transposons and 2086 defective copies of transposons. The consequences of the study showed transduplication and amplified cellular sequences of TE elements. Some have probably missed their capacity to transpose as a conventional cellular gene to accomplish cellular functions. The mobility of TEs in grapevine display the high mutagenic capability and extraordinary genetic variability capacity to induce grapevine species. Furthermost, class II transposons excise from the donor site as double-stranded DNA and reinsert somewhere else in the genome by a process frequently known as "cut and paste" transposition. The transposition mechanism of Helitrons and Mavericks is yet not known. Class II components that transpose by diverse tools are Helitrons and associated elements, which transpose by rolling-circle replication, Mavericks, whose transposition process is unknown (Feschotte and Pritham, 2005). But in the prokaryotic family, IS200/605 of insertion sequences alter as a single-stranded transposon circle (Guynet et al., 2008; Barabas et al., 2008). Class II transposons typically comprise terminal inverted repeats (TIRs), which code a transposase protein that catalyzes its mobilization. In plants, protein transposase recognizes the TIRs sequences and is also helpful in affecting the classification of class II TEs into 10 various superfamilies, such as CACTA, hAT, Merlin, Mutator, P element, PIF, piggyBac, Tc1/Mariner (Feschotte and Pritham, 2005; Wicker et al., 2007; Feschotte et al., 2005). Velasco et al. (2007) explored the genomic sequence of the grapevine and found class II transposons and their five superfamilies shotgun sequences through BLAST searches. This sequence was later used by Jaillon et al. (2007) to find the genome sequence of grapevine that suggests angiosperm phylum ancestral hexaploidization. They characterized a Tc1-Mariner element grapevine sequence, and found few sequences with minimal similarity to these elements do exist, perhaps denoting ancient defective elements. These defective elements were identified through BLASTN analyses using representatives of broad TEs. Relatively, the hAT superfamily is quite old and widespread in eukaryotic genomes. The hAT is the most ubiquitous superfamily of transposons in the grapevine and also establishes 1459 hAT-related components in the grapevine genome. In terms of copy number, hATs are the most pervasive "cut and paste" transposon family in grapevine. It suggested that hATs can retain the capacity to transpose (Rubin et al., 2001). CACTA elements are profusely present as class II elements in *Brassica oleracea* and *Triticum* sp., and in less number in Arabidopsis (Rubin et al., 2001; Wicker et al., 2001; Zhang and Wessler, 2004). Grapevine prominently contains elements of MULE family known as MuDR, Jittery, and Hop. The Mutator superfamily named Mu present in maize, is a diverse superfamily of class II elements. Plant species, such as *Zea mays*, *Arabidopsis thaliana*, *Lotus japonicus*, and *Oryza sativa* have the high abundance of class II factors (Zhang and Wessler, 2004; Robertson, 1978; Lisch 2002; Holligan et al., 2006; Juretic et al., 2005; Turcotte et al., 2001). The Jittery family has been described in maize (Xu et al., 2004) and Hop elements in legumes (Chalvet et al., 2003). MULEs show vast variability, particularly in evolutionary dynamics of plant stress mechanisms. Despite the fact, existence of Jittery elements is in several plant genomes. Hop-like transposons originated in fungi and legumes. These elements appeared during the ancient horizontal gene transfer in both the legumes and fungus (Holligan et al., 2008). Hoen et al. (2006) distinguished all superfamilies that contain transduplicated cellular gene fragments of *Vitis* class II.

Transposons can mobilize and amplify host genome sequences together with their sequences by transduplication. But most of these apprehended gene fragments appear like nonfunctional pseudogenes (Hoen et al., 2006). In specific cases, transduplicated exons give rise to new host proteins through alternative splicing by incorporating within host transcripts (Zabala and Vodkin 2007). Transduplicated sequences can undergo transcription even after missing their capacity to code, and they also have a regulatory function. Yu et al. (2000) have reported that MULEs have transduplicated gene fragments in Arabidopsis, melon, rice, and *Lotus japonicas* (Van Leeuwen et al., 2007). In particular cases, transduplications are unable to inactivate the transposition of PIF elements shared by multiple copies. Transduplications are common in all the TEs of all superfamilies. Cultivated tomato has composite antiquity of domestication and crop yield improvement capability. And two consequently genetic bottlenecks can characterize, followed by modern breeding by several introgression events. Subsequently, wild tomatoes and relatives replenish the limited pool of disease-resistance genes. Class II DNA transposons of these superfamilies like MuDR, hAT, and CACTA pass by the mechanism of cut and paste manner. TE insertions mainly persisted for downstream analysis. Recent studies revealed that most TE insertions were present in a few tomato accessions only. However, cluster analysis based on the phylogenetic relationship between acquisitions previously determined using SNPs recapitulated 6906 TIPs (Baduel et al., 2019). DNA transposons are debated in detail in Chapter 2 of this book.

1.3 TRANSPOSITION OF TRANSPOSABLE ELEMENTS

The transposition of composite transposons is sponsored by one or any one of the two I.S. components located at their ends. Only the ends of transposons are needed for transposition. These 15 elements can transpose any DNA sequence between two 1S elements (Hua-Van et al., 2011; Lisch, 2013). The transposition mechanism established in plant transformation allows hereditarily altering plants without inserting bacterial DNA into their genomes. An earlier established way employs plant-derived transfer DNAs as substitutes for *Agrobacterium* T-DNAs (Rommens et al., 2004). The available events during the insertion of transposons are as follows: (1) staggered breaks are produced in the target DNA, (2) the transposon joint to the distended single-stranded ends, and (3) the leftover gaps are then occupied, producing the repeats of target DNA at the site of insertion. Transposition may be (1) conservative, (2) non-replicative, or (3) replicative (Fedoroff, 2000).

In conservative transposition, a transposon moves to a new site so that a double-strand break is not produced at the old site. The excision of transposons may be either precise or imprecise. In precise excision, which is rare, the transposon and a copy of the duplicated sequence are removed. Then in imprecise excision, a remnant of the transposon is left at the site, which may prevent the reactivation of the gene into which the transposon was located earlier. But precise excision always leads to activation of the affected gene. In non-replicative transposition, the transposon moves out of one site and becomes inserted at a new location. It leaves a double-strand break at the old location. The gap may or may not be repaired. A recent copy of the transposon is produced in replicative transposition, which becomes inserted at the new site. As an outcome, the original site remains unchanged (Machida et al., 1997).

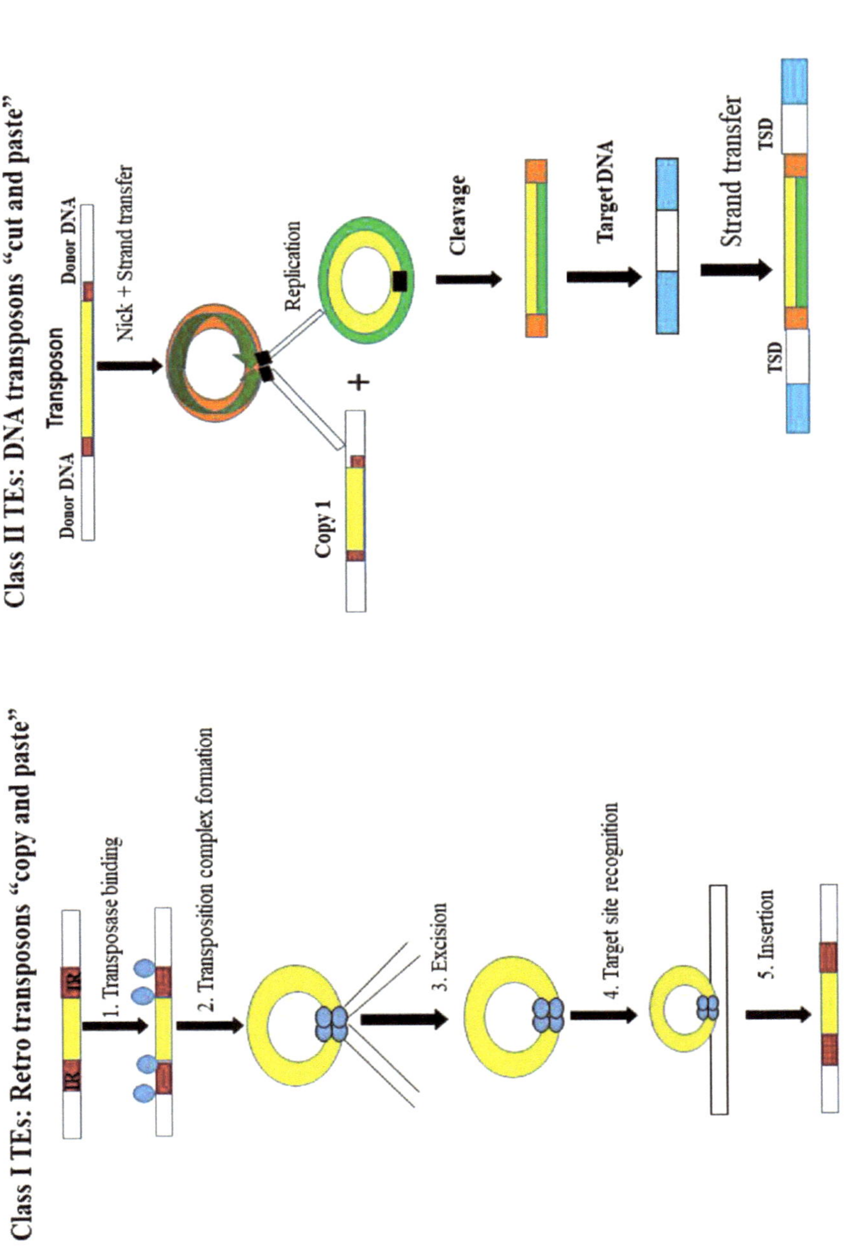

FIGURE 1.3 Transposition of retrotransposons and DNA transposons.

1.4 SIGNIFICANCE OF TRANSPOSABLE ELEMENTS

Transposons play an essential part in genome evolution and species divergence in tobacco, tomato, and other plants (Tang et al., 2019). Fascinating results were observed by Negi et al. (2016) that several TEs in the plant genome are precisely triggered and possibly transposed at the time of plants exposed to biotic and abiotic pressures. This can be achieved by regulating methylation or transposition themselves, that has been noticed in crops viz., tomato, tobacco, and other plants. Further, certain studies recommended the stimulation of TEs during stress might speed up genome recombination, which directly reflects on the plant's environmental adaptation process, which might impose on the specific long evolutionary history of species (Miura et al., 2001). TEs can play a potential role in genome evolution, gene regulation, and epigenetics of the plant through its life cycle (Henaff et al., 2014). TEs are also practical tools for plant breeding in genetics research. Due to their pronounced potential, TEs have been used in the breeding program in conifers. However, less allied findings have been described. There is a possibility for further studies that construct mutant libraries of conifers to recognize vital genes related to marked individuals, conferring to the genotype and phenotype differences in the library. Furthermore, molecular markers were developed based on TEs aimed at use in molecular breeding in conifers. Besides, it is essential to remark the substantial genome sizes plus comparatively long life cycles of conifers. Using TEs in conifer breeding added more excitement than in annual crop breeding, which needs additional initial work (Wang et al., 2020).

Mobile components have been recommended to source variation in the sequence that could allow hosts to progress quickly. In contrast, the extreme quantities of particular elements in various species for the welfare of the host genome advise a wanton disregard. Recent studies revealed that the eukaryotic genomes had progressed with intricate tools, such as epigenetic suppression by the RNAi silencing pathway (Vitte et al., 2014), several deactivating alterations to terminate the transcription and amplification events that avoid transposase production of these elements. Besides genetic association, gene function, and genome evolution studies, these elements have also been used to study evolutionary progressions (Hollister and Gaut, 2007). Transposons promote rearrangements, such as deletions, inversions, and duplicative inversions in host DNA. Some events result from recombination between multiple transposon copies, while others result from the transposition mechanisms (Freeling et al., 2008). When a transposon adds a copy of itself at a second site near its original location, recombination between the two copies may occur, generating either deletion or inversion of the host DNA depending mainly on the orientation of the repeats of 15 elements. If the repetitions are direct, that is, in the same direction, recombination between two elements would excise one copy of the element and the host DNA lying between the two elements in the form of a circle. That left behind a copy of the host DNA element will result in the deletion of the host DNA. However, most deletions that arise in the vicinity of transposons probably result from a variation in the pathway of the transposition event itself (Stuart et al., 2016). In contrast, recombining a pair of inverted repeats leads only to inversion of the host DNA located between the two repeats; there is no deletion of the transposon or the host sequence. Therefore, composite transposons whose I.S. modules are inverted become stable components of the genome. Some transposons promote duplicative inversions in which transposons lie in inverted orientation on either side of an inverted central region (Stuart et al., 2016). The role of transposable elements is deliberated extravagantly in different chapters of this book.

1.5 CONCLUSION

An incredible increase has taken place in improvements in the field of molecular biology, genomics, and genetics to study the whole genome levels of plants for crop improvement and disease resistance globally. To raise production, improvement of high-yielding varieties of plants with forbearance for biotic and abiotic stress is needed. Stress is a harmful threat to plant production and yield. Hitherto, there are plenty of practices and tools existing for crop production and disease resistance. Additionally prerequisite to studying the depiction of transposable elements and their effect on the structure and expression of the genome contributes to a significant influence headed on understanding the phenotypic variations corresponding to genome regulation. Noteworthy outcomes at the crossing point of transposition mechanisms attracted researchers to study and identify novel TEs and extending their taxonomic distribution revealed associations between TEs and plant genome.

Research on transposable elements provided novel insights that exposed the movement of TEs can be well-ordered epigenetically through their host mechanism. These elements influence the arrangement, dynamics, and structure of the genomes in which TEs transposed. This chapter highpoints the history, classes of TEs, transposition mechanisms of "copy and paste," and "cut and paste" mainly in plants. We instigate by giving basic structural features, classes, transposition, and significance of transposons and their transposition with respect to plants, also concentrating on features that manage their maintenance and propagation within the genome and all over plant eukaryotic hierarchy. In conclusion, the acquaintance extended from side to side study must lead to and established to the researchers, formerly helpful for further findings in the field of molecular approaches methods for studying the variations in the gene expression levels between populations and natural variation patterns used in the genomics program, and also to facilitate and comprehend the vital progress in the stress levels of plants during the post-genomics phase and complete the life cycle for crop improvement.

KEYWORDS

- **transposable elements**
- **copy and paste**
- **cut and paste**
- **transposition**
- **mobile DNA sequences**

REFERENCES

Arango-Lopez, J.; Orozco-Arias, S.; Salazar, J. A.; Guyot, R. Application of Data Mining Algorithms to Classify Biological Data: The Coffea Canephora Genome Case. In *Colombian Conference on Computing*; Springer: Cham, Switzerland, 2017; pp 156–170.

Baduel, P.; Quadrana, L.; Hunter, B.; Bomblies, K. Colot, V. Relaxed Purifying Selection in Autopolyploids Drives Transposable Element Overaccumulation Which Provides Variants for Local Adaptation. *Nat. Commun.* **2019,** *10*, 5818.

Barabas, O.; Ronning, D. R.; Guynet, C.; Hickman, A. B, Ton-Hoang, B.; Chandler, M.; Dyda, F. Mechanism of IS200/IS605 Family DNA Transposases: Activation and Transposon-Directed Target Site Selection. *Cell* **2008,** *132*, 208–220.

Belyayev, A.; Kalendar, R.; Brodsky, L.; Nevo, E.; Schulman, A.; Raskina, O. Transposable Elements in a Marginal Plant Population: Temporal Fluctuations Provide New Insights Into Genome Evolution of Wild Diploid Wheat. *Mobile DNA* **2010,** *1*, 6.

Benjak, A.; Forneck, A.; Casacuberta, J. M. Genome-Wide Analysis of the "Cut-and-Paste" Transposons of Grapevine. *PLoS ONE* **2008,** *3* (9), e3107.

Chalvet, F.; Grimaldi, C.; Kaper, F.; Langin, T.; Daboussi, M-J. Hop, an Active Mutator-Like Element in the Genome of the Fungus Fusarium oxysporum. *Mol. Biol. Evol.* **2003,** *20,* 1362–1375.

Chaparro, C.; Gayraud, T.; de Souza, R. F.; Domingues, D. S.; Akaou, S.; Laforga Vanzela, .L.; de Kochko, A.; Rigoreau, M.; Crouzillat, D.; Hamon, S.; Hamon, P.; Guyot, R. Terminal-Repeat Retrotransposons with GAG Domain in Plant Genomes: A New Testimony on the Complex World of Transposable Elements. *Genome Biol. Evol.* **2015,** *7,* 493–504.

Chu, C.; Nielsen, R.; Wu, Y. REPdenovo: Inferring de novo Repeat Motifs from Short Sequence Reads. *PLoS ONE* **2016,** *11,* 1–17.

Cui, B. W.; Fan, F. H.; Ding, G. J.; Yang, Z. Q.; Wen, X. P. Development of IRAP Markers Based on Genomic LTR Retrotransposon Sequences in Masson Pine (*Pinus massoniana*). *For Res.* **2016,** *29,* 348–353.

De Castro Nunes, R.; Orozco-Arias, S.; Crouzillat, D.; Mueller, L. A.; Strickler, S. R.; Descombes, P.; Fournier, C.; Moine, D.; de Kochko, A.; Yuyama, P. M.; Vanzela, A. L. L.; Guyot, R. Structure and Distribution of Centromeric Retrotransposons at Diploid and Allotetraploid Co_ea Centromeric and Pericentromeric Regions. *Front. Plant Sci.* **2018,** *9,* 175.

Fedoroff, N. V. Transposable Elements, Epigenetics, and Genome Evolution. *Science.* **2012,** *338,* 758–776.

Feschotte, C.; Osterlund, M. T.; Peeler, R.; Wessler, S. R. DNA-Binding Specificity of Rice Mariner-Like Transposases and Interactions with Stowaway MITEs. *Nucl. Acids Res.* **2005,** *33,* 2153–2165.

Feschotte, C.; Pritham, E. DNA Transposons and the Evolution of Eukaryotic Genomes. *Annu. Rev. Genet.* **2005,** *41,* 331–368.

Freeling, M.; Lyons, E.; Pedersen, B.; Alam, M.; Ming, R.; Lisch, D. Many or Most Genes in Arabidopsis Transposed After the Origin of the Order Brassicales. *Genome Res.* **2008,** *18,* 1924–1937.

Fujimoto, R.; Takuno, S.; Sasaki, T.; Nishio, T. The Pattern of Amplification and Differentiation of Ty1/Copia and Ty3/Gypsy Retrotransposons in Brassicaceae Species. *Genes Genet. Syst.* **2008,** *83,* 13–22.

Gao, D.; Chen, J.; Chen, M.; Meyers, B. C.; Jackson, S. A Highly Conserved, Small LTR Retrotransposon That Preferentially Targets Genes in Grass Genomes. *PLoS ONE* **2012a,** *7,* e32010.

Gao, D.; Jimenez-Lopez, J. C.; Iwata, A.; Gill, N.; Jackson, S. A. Functional and Structural Divergence of an Unusual LTR Retrotransposon Family in Plants. *PLoS ONE* **2012b,** *7,* e48595.

Gazzani, S.; Gendall, A. R.; Lister, C.; Dean, C. Analysis of the Molecular Basis of Flowering Time Variation in Arabidopsis Accessions. *Plant Physiol.* **2003,** *132* (2), 1107–1114.

Girgis, H. Z. Red: An Intelligent, Rapid, Accurate Tool for Detecting Repeats de-novo on the Genomic Scale. *BMC Bioinform.* **2015,** *16,* 1–19.

Grandbastien, M.-A. LTR Retrotransposons, Handy Hitchhikers of Plant Regulation and Stress Response. *Biochim. Biophys. Acta* **2015,** *1849,* 403–416.

Guynet, C.; Hickman, A. B.; Barabas, O.; Dyda, F.; Chandler, M. et al. In Vitro Reconstitution of a Single-Stranded Transposition Mechanism of IS608. *Mol. Cell* **2008,** *29,* 302–312.

Henaff, E.; Vives, C.; Desvoyes, B.; Chaurasia, A.; Payet, J.; Gutierrez, C.; Casacuberta, J. M. Extensive Amplification of the E2F Transcription Factor Binding Sites by Transposons During Evolution of Brassica Species. *Plant J.* **2014,** *77,* 852–862.

Holligan, D.; Zhang, X.; Jiang, N.; Pritham, E. J.; Wessler, S. R. The Transposable Element Landscape of the Model Legume Lotus japonicus. *Genetics* **2006,** *174*, 2215–2228.
Hollister, J. D.; Gaut, B. S. Population and Evolutionary Dynamics of Helitron Transposable Elements in *Arabidopsis thaliana*. *Mol. Biol. Evol.* **2007,** *24*, 2515–2524.
Jaillon, O.; Aury, J. M.; Noel, B.; Policriti, A.; Clepet, C. et al. The Grapevine Genome Sequence Suggests Ancestral Hexaploidization in Major Angiosperm Phyla. *Nature* **2007,** *449*, 463–467.
Jangam, D.; Feschotte, C.; Betrán, E. Transposable Element Domestication as an Adaptation to Evolutionary Conflicts. *Trends Genet.* **2018,** *33*, 817–831.
Juretic, N.; Hoen, D. R.; Huynh, M. L.; Harrison, P. M.; Bureau, T. E. The Evolutionary Fate of MULE-Mediated Duplications of Host Gene Fragments in Rice. *Genome Res.* **2005,** *15*, 1292–1297.
Keidar, D.; Doron, C.; Kashkush, K. Genome-Wide Analysis of a Recently Active Retrotransposon, Au SINE, in Wheat: Content, Distribution Within Subgenomes and Chromosomes, and Gene Associations. *Plant Cell Rep.* **2018,** *37*, 193–208.
Kidwell, M. G.; Lisch, D. Transposable Elements as Sources of Variation in Animals. *Proc. Natl. Acad. Sci. USA* **1997,** *94*, 7704–7711.
Labate, J. A.; Robertson, L. D. Evidence of Cryptic Introgression in Tomato (*Solanum lycopersicum* L.) Based on Wild Tomato Species Alleles. *BMC Plant Biol.* Aug 7 **2012,** *12*, 133.
Le, Q.; Melayah, D.; Bonnivard, E.; Petit, M.; Grandbastien, M. Distribution Dynamics of the Tnt1 Retrotransposon in Tobacco. *Mol. Genet. Genom.* **2007,** *278*, 1617–4615.
Le, Q. H.; Wright, S.; Yu, Z.; Bureau, T. Transposon Diversity in *Arabidopsis thaliana*. *Proc. Natl. Acad. Sci.* **2000,** *97*, 7376–7381.
Liang, L.; Zhou, L.; Tang, Y. et al. A Sequence-Indexed Mutator Insertional Library for Maize Functional Genomics Study. *Plant Physiol.* **2019,** *181*, 1404–1414.
Lin, R.; Ding, L.; Casola, C.; Ripoll, D. R.; Feschotte, C.; Wang, H. Transposase-Derived Transcription Factors Regulate Light Signaling in Arabidopsis. *Science* **2007,** *23*, 318 (5854).
Lisch, D. Mutator Transposons. *Trends Plant Sci.* **2002,** *7*, 498–504.
Llorens, C.; Futami, R.; Covelli, L.; Domínguez-Escribá, L.; Viu, J. M.; Tamarit, D.; Aguilar-Rodríguez, J.; Vicente-Ripolles, M.; Fuster, G.; Bernet, G. P.; Maumus, F.; Pomer, A. M.; Sempere, J. M.; Latorre, A.; Moya, A. The Gypsy Database (GyDB) of Mobile Genetic Elements: Release 2.0. *Nucl. Acids Res.* **2010,** *1*, 39.
Llorens, C.; Muñoz-Pomer, A.; Bernad, L.; Botella, H.; Moya, A. Network Dynamics of Eukaryotic LTR Retroelements Beyond Phylogenetic Trees. *Biol. Direct* **2009,** *4*, 41.
Loureiro, T.; Fonseca, N.; Camacho, R. Application of Machine Learning Techniques on the Discovery and Annotation of Transposons in Genomes. Master's Thesis, Faculdade de Engenharia, Universidade Do Porto, Porto, Portugal, 2012.
Machida, C.; Onouchi, H.; Koizumi, J.; Hamada, S.; Semiarti, E.; Torikai, S.; Machida, Y. Characterization of the Transposition Pattern of the AC Element in *Arabidopsis thaliana* Using Endonuclease I-SceI. *Proc. Natl. Acad. Sci.* **1997,** *94*, 8675–8680.
Mielich, K.; Shtifman-Segal, E.; Golz, J.; Zeng, G.; Wang, Y.; Berman, J.; Kunze, R. Maize Transposable Elements AC/DS as Insertion Mutagenesis Tools in Candida Albicans. *G3.* **2018,** *8*, 1139–1145.
Mita, P.; Boeke, J. D. How Retrotransposons Shape Genome Regulation. *Curr. Opin. Genet. Dev.* **2016,** *37*, 90–100.

Miura, A.; Yonebayashi, S.; Watanabe, K.; Toyama, T.; Shimada, H.; Kakutani, T. Mobilization of Transposons by a Mutation Abolishing Full DNA Methylation in Arabidopsis. *Nature* **2001,** *411*, 212–214.

Monat, C.; Tando, N.; Tranchant-Dubreuil, C.; Sabot, F. LTR Classifier: A Website for Fast Structural LTR Retrotransposons Classification in Plants. *Mob. Genet. Elem.* **2016,** *6*, e1241050.

Mustafin, R. N.; Khusnutdinova, E. K. The Role of Transposons in Epigenetic Regulation of Ontogenesis. *Russ. J. Dev. Biol.* **2018,** *49*, 61–78.

Muszewska, A.; Homan-Sommer, M.; Grynberg, M. LTR Retrotransposons in Fungi. *PLoS ONE* **2011,** *6*, 12.

Negi, P.; Rai, A. N.; Suprasanna, P. Moving Through the Stressed Genome: Emerging Regulatory Roles for Transposons in Plant Stress Response. *Front Plant Sci.* **2016,** *7*, 1448.

Neumann, P.; Novák, P.; Hoštáková, N.; MacAs, J. Systematic Survey of Plant LTR-Retrotransposons Elucidates Phylogenetic Relationships of Their Polyprotein Domains and Provides a Reference for Element Classification. *Mob. DNA* **2019,** *10*, 1–17.

Orozco-arias, S.; Liu, J.; Id, R. T.; Ceballos, D.; Silva, D.; Id, D.; Ming, R.; Guyot, R. Inpactor, Integrated and Parallel Analyzer and Classifier of LTR Retrotransposons and Its Application for Pineapple LTR Retrotransposons Diversity and Dynamics. *Biology* **2018.**

Ou, S.; Chen, J.; Jiang, N. Assessing Genome Assembly Quality Using the LTR Assembly Index (LAI). *Nucl. Acids Res.* **2018,** *46*, 1–11.

Pandita, D.; Pandita, A. Jumping Genes-"The Other Half of the Human Genome" and the Missing Heritability Conundrum of Human Genetic Disorders. *Biotechnol. J. Int.* **2016,** *11* (3), 1–18. https://doi.org/10.9734/BBJ/2016/13904.

Petit, M.; Guidat, C.; Daniel, J.; Denis, E.; Montoriol, E.; Bui, Q.; Lim, K.; Kovarik, A.; Leitch, A.; Grandbastien, M.; Mobilization of Retrotransposons in Synthetic Allotetraploid Tobacco. *New Phytol.* **2010,** *186*, 135–147.

Pritham, E. J.; Putliwala, T.; Feschotte, C. Mavericks, a Novel Class of Giant Transposable Elements Widespread in Eukaryotes and Related to DNA Viruses. *Gene.* **2007,** *390* (1–2), 3–17.

Quadrana, L. The Contribution of Transposable Elements to Transcriptional Novelty in Plants: *FLC Affair. Transcript.* **2020,** *11*, 192–198.

Quesneville, H. Twenty Years of Transposable Element Analysis in the *Arabidopsis thaliana* Genome. *Mobile DNA.* **2020,** *11*, 28.

Rahman, A. Y. A.; Usharraj, A. O.; Misra, B. B.; Thottathil, G. P.; Jayasekaran, K.; Feng, Y.; Hou, S.; Ong, S. Y.; Ng, F. L.; Lee, L. S. Tan, H. S.; Sakaff, M. K. L. M.; The, B. S.; Khoo, B. F.; Badai, S. S.; Aziz, N. A.; Yuryev, A.; Knudsen, B.; Laporte, A. D.; Mchunu, N. P.; Yu, Q.; Langston, B. J.; Freitas, T. A. K.; Young, A. G.; Chen, R.; Wang, L.; Najimudin, N.; Saito, J. A.; Alam, M. Draft Genome Sequence of the Rubber Tree Hevea Brasiliensis. *BMC Genom.* **2013,** *14*, 75.

Ram, H.; Soni, P.; Salvi, P.; Gandass, N.; Sharma, A.; Kaur, A.; Sharma, T. R. Insertional Mutagenesis Approaches and Their Use in Rice for Functional Genomics. *Plants.* **2019,** *8*, 310.

Robertson, D. Characterization of a Mutator System in Maize. *Mutant Res.* **1978,** *51*, 21–28. 24.

Rubin, E.; Lithwick, G.; Levy, A. A. Structure and Evolution of the hAT Transposon Superfamily. *Genetics* **2001,** *158*, 949–957.

Staton, S. E.; Burke, J. M. Transposome: A Toolkit for Annotation of Transposable Element Families from Unassembled Sequence Reads. *Bioinformatics* **2015,** *31*, 1827–1829.

Sultana, T.; Zamborlini, A.; Cristofari, G. et al. Integration Site Selection by Retroviruses and Transposable Elements in Eukaryotes. *Nat. Rev. Genet.* **2017,** *18*, 292–308.

Tang, Y.; Ma X, Zhao S, Zheng WXX, Sun, H.; Gu, P.; Zhu, Z.; Sun, C.; Liu, F.; Tan, L. Identification of an Active Miniature Inverted Repeat Transposable Element mJing in Rice. *Plant J.* **2019,** *98*, 639–653.

Tieman, D. et al. A Chemical Genetic Roadmap to Improved Tomato Flavor. *Science* **2017,** *355*, 391–394.

Turcotte, K.; Srinivasan, S.; Bureau, T. Survey of Transposable Elements from Rice Genomic Sequences. *Plant J.* **2001,** *25*, 169–179.

Van Leeuwen, H.; Monfort, A.; Puigdomenech, P. Mutator -Like Elements Identified in Melon, Arabidopsis and Rice Contain ULP1 Protease Domains. *Mol. Genet. Genom.* **2007,** *277*, 357–364.

Velasco, R.; Zharkikh, A.; Troggio, M.; Cartwright, D.; Cestaro, A. et al. A High Quality Draft Consensus Sequence of the Genome of a Heterozygous Grapevine Variety. *PLoS ONE* **2007,** *2*, e1326.

Vitte, C.; Fustier, M. A.; Alix, K.; Tenaillon, M. I. The Bright Side of Transposons in Crop Evolution. *Brief. Funct. Genom.* **2014,** *13*, 276–295.

Wang, J.; Lu, N.; Yi, F.; Xiao, Y. Identification of Transposable Elements in Conifer and Their Potential Application in Breeding. *Evol. Bioinform.* **2020,** *16*, 1–4.

Wicker, T.; Sabot, F.; Hua-Van, A.; Bennetzen, J. L.; Capy, P.; Chalhoub, B.; Flavell, A.; Leroy, P.; Morgante, M.; Panaud, O.; Paux, E.; Miguel, P. S.; Schulman, A. H. A Unified Classification System for Eukaryotic Transposable Elements. *Nat. Rev. Genet.* **2007,** *8*, 973–982.

Xu, Z.; Yan, X.; Maurais, S.; Fu, H.; O Brien, D. G.; Jittery, a Mutator Distant Relative with a Paradoxical Mobile Behavior: Excision Without Reinsertion. *Plant Cell* **2004,** *16*, 1105–1114.

Yaakov, B.; Meyer, K.; Ben-David, S.; Kashkush, K. Copy Number Variation of Transposable Elements in Triticum–Aegilops Genus Suggests Evolutionary and Revolutionary Dynamics Following Allopolyploidization. *Plant Cell Rep.* **2013,** *32*, 1615–1624.

Yu, Z.; Wright, S. I.; Bureau, T. E. Mutator-Like Elements in *Arabidopsis thaliana*: Structure, Diversity and Evolution. *Genetics* **2000,** *156*, 2019–2031.

Yuan, Y.; Wessler, S. The Catalytic Domain of All Eukaryotic Cut-and-Paste Transposase Superfamilies. *Proc. Natl. Acad. Sci. USA* **2011,** *108*, 7884–7889.

Zhang, X.; Wessler, S. R. Genome-Wide Comparative Analysis of the Transposable Elements in the Related Species *Arabidopsis thaliana* and *Brassica oleracea*. *Proc. Natl. Acad Sci.* **2004,** *101*, 5589–5594.

Zhao, G.; Zou, C.; Li, K.; Wang, K.; Li, T.; Gao, L.; Zhang, X.; Wang, H.; Yang, Z.; Liu, X.; Jiang, W.; Mao, L.; Kong, X.; Jiao, Y.; Jia, J. The Aegilops tauschii Genome Reveals Multiple Impacts of Transposons. *Nat. Plants* **2017,** *3*, 946–955.

Zytnicki, M.; Akhunov, E.; Quesneville, H. Tedna: A Transposable Element de novo Assembler. *Bioinformatics* **2014,** *30*, 2656–2658.

CHAPTER 2

DNA Transposons

JAVID AHMAD MALIK[1*], GITA MISHRA[1], YOUNIS AHMAD HAJAM[2], SAMEER QUAZI[3], and RAFIQ LONE[4]

[1]*Department of Zoology, Guru Ghasidas Vishwavidyalaya, Bilaspur, Chhattisgarh 495009, India*

[2]*Department of Biosciences, Division Zoology, School of Basic and Applied Sciences, Career Point University, Hamirpur, HP 176041, India*

[3]*GenLab Biosolutions Private Limited, Banglore, Karnataka 560043, India*

[4]*Department of Botany, Central University of Kashmir, Jammu and Kashmir, India*

**Corresponding author. E-mail: javidmalikr@gmail.com*

ABSTRACT

The genomes of most eukaryotes are larger and more complex as compared with those of prokaryotes which is not inherently surprising, since more genes are expected in organisms with more complexity. Plant genomes are unique among living species since they have a wide range of size variation. Although polyploidization is one of these factors, transposable elements (TEs) appear to be the most important in not only modifying the host's genome size, but also disrupting and generating new functions and regulatory networks through interactions with other genes.

These are genetic elements which have the ability to move from one place to another by means of transposition, thus are called as transposable elements. Transposable elements constitute a major proportion of the total plant genomes, so their influence on genome structure and expression is

Plant Transposable Elements: Biology and Biotechnology, Deepu Pandita, Anu Pandita, (Eds.)

quite evident in understanding global genome regulation and phenotype variation. There are two kinds of transposons depending upon their mechanism of transposition; transposons which migrate by using a "cut and paste" method, which involves removing a transposon copy from one location in the genome and re-inserting it in a different one. Because numerous daughter copies can be formed from a mother copy and inserted across the genome, transposons have become a significant part of huge genomes. Because of their prevalence, usefulness, and involvement in genome evolution, transposable elements have piqued researchers' curiosity. Transposable elements (TEs) are essential for their hosts' ability to respond to environmental stressors. As diverse as the TEs and host genomes, the processes by which these elements influence host adaptive capacity are numerous. Overall, transposons are a promising, but untapped area of research that should be put on the agendas of evolutionary geneticists, molecular ecologists, conservation biologists, and plant breeders. In this chapter, a comprehensive review regarding DNA transposons will be taken into consideration.

2.1 EUKARYOTIC GENOME

Proteins and eukaryotic DNA are inextricably linked throughout the cell. The combination of DNA and protein is known as chromatin. Euchromatin, which goes through the typical condensation/decondensation processes during the cell cycle, and heterochromatin, which is strongly condensed throughout the cell cycle, including interphase, are the two main chromatin types. Euchromatin is transcriptionally active and makes up the majority of chromosomes, whereas heterochromatin is found at the centromeres and telomeres of all chromosomes. Heterochromatin can also be found on chromosomes in other locations, such as the complete latent X chromosome in female mammals. Heterochromatin lacks transcription, crossing over, and replication that occurs late in the S phase. Histones are positively charged proteins that come in five different types: H1, H2A, H2B, H3, and H4. Histones are the most common proteins in chromatin. All histones include a considerable quantity of positively charged amino acids lysine and arginine, giving them a net positive charge. The positive charges on DNA's phosphates attract the negative charges on histones, keeping the DNA in contact with them. Nonhistone chromosomal proteins are also found in a wide variety of forms. In addition to being condensed throughout the cell cycle, heterochromatin is identified by lack of transcription, the absence of crossing over, and late replication in the S stage. Histones are tiny, positively charged proteins that are classified as H1, H2A, H2B, H3, or H4. The most frequent proteins in chromatin are histones.

A large amount of lysine and arginine, both positively charged amino acids, is found in all histones, giving them a net positive charge. The opposite charges on the phosphates of DNA attract the histones, which keeps the DNA in touch with the histones. Nonhistone chromosomal proteins are also found in a variety of forms. The nucleosome is a kind of DNA molecule. The structure of chromatin is extremely complicated having numerous levels of organization. The double-helical structure of DNA is the most basic level. The DNA is coupled with proteins and is heavily folded to generate a chromosome at a more complex level. When chromatin is extracted from a cell's nucleus and examined under electron microscope, it often resembles the beads on a string. When nuclease is added to this setup, the "string" connecting the "beads" is broken, leaving individual beads linked to around 200 bp of DNA. When more nuclease is introduced, the enzyme completely digests the DNA in the area between the strands. When you add more nuclease, the enzyme consumes entire DNA between the beads, leaving only a protein core connected to a DNA segment. Experiments like this showed that chromatin is more than just a jumble of proteins and DNA; it also has a repeating structure at its core. The nucleosome is the most fundamental level of chromatin structure, comprised of a protein and DNA repeating core produced by nuclease enzyme breakdown. The nucleosome is a crucial component of all living organisms. The nucleosome is a spool-like core particle made up of two copies of DNA around a histone octamer of eight proteins (two copies of H2A, H2B, H3, and H4). The nucleosome is a spool-like core particle made up of two copies of DNA wrapped around an octamer of eight histone proteins (two copies of H2A, H2B, H3, and H4). Each of these histone proteins has a flexible "tail" that extends out from the nucleosome and contains anywhere from 11 to 37 amino acids. The tails of histone have been positively charged. One nucleosome's tails may interact with the tails of nearby nucleosomes, making nucleosome compacting easier. Chromatin structure changes are essential for gene expression when histone tails are chemically altered. Despite the fact that H1, the fifth form of histone, is not a core particle, it is required for nucleosome assembly. H1 attaches to 20–22 bp of DNA where the octamer joins and departs, functioning as a clamp about the nucleosome octamer to help lock the DNA into place. The chromatosome is the next level of chromatin architecture, consisting of the core particle and its associated H1 histone. Each chromatosome contains around

167 base pairs of DNA. Linker DNA connects chromosomes and varies in size depending on the cell type; in most cells, linker DNA is between 30 and 40 bp long. This linker DNA may be coupled to nonhistone chromosomal proteins, and a few appear to bind directly to the core particle genome

2.2 TYPE OF DNA SEQUENCES IN EUKARYOTES

Unique-sequence DNA, moderately repetitive DNA, and highly repetitive DNA are all forms of sequences found in eukaryotic DNA. Sequences that appear only once or a few times throughout the genome are known as unique-sequence DNA. This DNA contains protein-encoding sequences as well as huge amount of DNA with no known function. In most multicellular eukaryotes, single-copy genes account for about 25–50% of the protein-encoding genes. Other genes present in distinct sequence DNA are found in many copies that are similar but not identical, collectively referred to as gene family. Most gene families are the result of duplication of an earlier existing gene. Sequences that appear in several copies are referred to as repetitive DNA. Repetitive DNA is found in abundance in a variety of eukaryotic organisms. For example, repetitive DNA accounts for more than half of the human genome. Moderately repetitive DNA is a type of repetitive DNA that typically consists of sequences that are repeated thousands of times and range in length from 150 to 300 base pairs. Some of these sequences are required for cell activity. Half of the cell's moderately repetitive DNA, on the other hand, has no recognized purpose. There are two sorts of repetitions in moderately repetitive DNA. Tandem repeat sequences emerge one after the other on the chromosomes and tend to cluster in certain places.

Repeat sequences with interspersions can be found all over the place. Tandem repeat sequences come one after the other on the chromosomes and tend to cluster at specific sites. Interspersed repeat sequences can be found all over the genome. The Alu sequence, a 200-bp sequence found over a million times and accounting for 11% of the human genome, is an example of an interspersed repeat. It does not seem to have any cellular function. SINEs are short repeats, similar to Alu sequences (short interspersed elements). LINEs are longer interspersed repeats containing tens of thousands of base pairs (long interspersed elements). LINE1 is a kind of LINE that makes up roughly 17% of the human genome. Transposable repeats make up the majority of interspersed repeats

Transposable repeats make up the majority of interspersed repeats. These small sequences, which are often less than 10 bp long, are found in abundance on the chromosome, where they are repeated in tandem and grouped in certain places, particularly around centromeres and telomeres. Satellite DNA is a type of extremely repetitive DNA that separates into a satellite fraction while centrifuged at high speeds due to differences in the percentages of the four nucleotides. Highly repetitive DNA is rarely translated into RNA. Although some highly repetitive DNA sequences may play a role in the operation of centromeres and telomeres, the vast majority of highly repetitive DNA has no known function. As we now know, gene density varies significantly between and within chromosomes. Human chromosome 19 has a high gene density, with around 26 genes per million bp. While as, chromosome 13 has just roughly 6.5 genes per million base pairs. Even within the same chromosome, gene density varies: some parts of chromosome 13's long arm have as few as 3 genes per million bp, while others have as many as 30 genes per million bp. The chromosome 13 short arm, which is entirely made up of heterochromatin, has almost no genes. DNA sequences that can move are known as TEs. All species' genomes contain TEs, which are mobile DNA sequences. Many genomes contain them, accounting for at least 45% of human DNA, for example. They insert into other genes and induce mutations and disrupting it, or by causing DNA rearrangements like chromosome deletions, duplications, and inversions.

2.3 TRANSPOSON FORMS AND SHAPES

Genetic sequences known as transposable elements can hop from one area of the genome to another. Due to their lengthy evolutionary history and variation, TEs come in a range of shapes and sizes. Based on their way of transposition, TEs can be split into two fundamental types, each of which can be further subdivided into subclasses on the basis of chromosomal intergration mechanism. Class I retrotransposons utilize an RNA intermediate that is reverse-transcribed into cDNA and integrated into the genome in a "copy-and-paste" process (Boeke et al, 1985). Long terminal repeat (LTR) retrotransposons are incorporated using an integrase-like enzyme that cleaves and exchanges strands. The method of target-primed reverse transcription is used to integrate non-LTR retrotransposons (LINEs and SINEs) into the chromosome (Luan et al, 1993). Class 2 DNA transposons use a DNA intermediate to transfer from one site in the genome to another (Rubin et al., 1982; Greenblatt et al., 1963). Each of these TE subclasses is further subdivided into subgroups found in a range of organisms but have a genetic organization in common. Two important LTR r elements are Ty1/copia and Ty3/gypsy, located in almost all eukaryotes (Malik et al., 2001). Similarly, the three superfamilies of DNA transposable elements hAT (hobo-Ac-Tam3), Tc1/mariner, and MULEs (Mutator-like elements) are found in eukaryotes (Feschotte et al., 2007). The descendants of a unit ancestral sequence can be traced back to the most comprehensive defined subgroups. This ancestor can be thought of as a consensus sequence that represents the entire (sub) family (Smit, 1999). Ts classification differs from taxonomics categorization in that it is subject to constant change due to the discovery of completely new TE types (Arkhipova et al., 2017).

2.4 TEs DISTRIBUTION IN THE GENOME

The genome is home to a multitude of TEs, which multiply and spread through unique interactions with one another and other cell components (Venner e al, 2009). TEs have a propensity for inserting themselves into specific areas or compartments of the genome. Natural selection as well as genetic drift also have a significant impact in TE distribution (Lynch M, 2007). Selective factors can influence whether certain genetic elements are deleted or preserved in specific genomic regions (Campos-Sánchez et al., 2016; Kvikstad and Makova, 2010). TEs make up a large part of the genome, including a significant amount of species-specific DNA. LTR retrotransposons, for example, account for 60–70% of the genome in maize, where Barbara McClintock made her seminal work (McClintock B, 1957). (Springer et al., 2018; Lazarow et al, 2013; Schnable et al, 2009). Most TE groups are still transposing, and the process is extremely mutagenic. In the laboratory, more than half of all recognized phenotypic mutants in *D. melanogaster* are due to spontaneous TE insertions (Eickbush et al., 2002). Mutagenic transposition activities have also been documented in laboratory mice, with continuous activity of numerous LTRs accounting for 10–15% of all mutant phenotypes inherited (Maksakova et al., 2006). Because transposons are more active when animals are stressed, regardless of their herbal environment (Lanciano and Mirouze, 2018, Horváth et al., 2017), their contribution to genetic diversity may be overlooked. Because TE insertions do not always give a fitness benefit to their hosts right away, those that do attain population fixation do so primarily through genetic drift before being degraded by point mutations

(Lynch, 2007). As a result of these modifications throughout time, TEs that can no longer express transposition enzymes and generate novel integration activities are formed. About 100 active L1 factors are expected to be present in each genome, with the majority of these being recent insertions that still segregate in the human population (Beck et al., 2010; Brouha et al., 2003). Transposition is a fascinating genetic growth mechanism that is slowed over time by deleting DNA. In eukaryotes, the stability of these two systems is a major driving force in the evolution of genome size (Schubert and Vu, 2016). A fundamental driving force in genome evolution has been identified to be the pace at which TEs transpose, which is partly controlled by the host (Freeling et al., 2015; Bennetzen and Wang, 2014).

TEs must maintain a precise balance between expression and suppression in order to continue evolving. The expression should be strong enough to encourage amplification, but not so powerful that it compromises TEs' ability to increase copy numbers. This type of balance correction explains why some TEs are naturally ideal for transposition (Mátés et al., 2009) and why others have acquired a self-regulatory system that regulates their own copy numbers (Saha et al., 2015). Small RNA, chromatin, and DNA modification mechanisms, as well as the KRAB Zincfinger Protein sequence of certain repressors, have all been studied recently (Ecco et al., 2017; Imbeault et al., 2017: Yang et al., 2017; Imbeault and Trono, 2014) During the early embryo's development, however, several of these mechanisms should be partially released to the permissible spectrum of guest gene expression programs. The loss of DNA methylation in the genome, for example, should restore the imprinted gene in primitive genetic cells (Miyoshi et al., 2016).

2.5 TEs AS INSERTIONAL MUTAGENS

Humans, like other species, are involved in a variety of actively active TEs in which the intrinsic stability of expression and repression continues to play a role (Huang et al., 2012). L1 is an example, as are other mobile factors relying on L1-mediated proteins for retrotransposition (Burns and Boeke, 2012; Beck et al., 2011). The same mechanisms are accountable for novel germline insertions that can proliferate to lead genetic illness, since it has been discovered that approximately 120 TE insertions are associated with human disease (Hancks and Kazazian., 2014). One out of every 21 births in Alu (Xing et al., 2009) and one out of every 95 births (Xing et al., 2009) results in de novo germline transposition. Somatic cell transposition and its implications have received little attention in the past because it can be considered an evolutionary dead-end for transposons with no long-term ramifications for the host. According to some evidence, TEs may be active in somatic cells in a variety of species (Kazazian HH, 2011). L1 expression and transposition were witnessed in diverse human somatic processes, including stem cells and early embryos (Klawitter et al., 2016; Garcia-Perez et al., 2007).

Transposon expression in the mammalian brain is also a hot topic, with L1 transposition being proposed as a way to diversify neuronal populations (Erwin et al., 2014; Baillie et al., 2011; Muotri et al., 2005). Several researchers (Schauer et al., 2018; Tang et al., 2017; Scott et al., 2016; Tubio et al., 2014; Iskow et al., 2010) have also discovered somatic activity in human cancers, where tumors can accumulate a large number of recent L1 insertions. Somatic activity in human malignancies is due to a small number of so-called "hot" L1 loci, just as it is for human polymorphisms (Scott et al., 2016). Individual (Lee et al., 2012), tumor type (Lee et al., 2012), and time frame in the clonal evolution of the tumor (Rodic et al., 2015; Tubio et al., 2014) factors, all influence the actions of those copies. Even while the quite majority seem to be "passenger" mutations, some of these mobile element insertions impair critical tumor suppressors and oncogenes, forcing the genesis of most malignancies (Burns KH., 2017). To keep TEs in check, host cells have developed a variety of strategies. However, because natural selection pressure declines with age and eventually disappears, TEs can potentially become more active (Gorbunova et al., 2014).

2.6 TEs IN ACTIVITIES BESIDES TRANSPOSITION

While transposition-related DNA breaking and insertion is a definite cell damage source, it is not the only route for TEs to affect their hosts. Reactivated transposons are harmful to the host in a multiple ways. To begin with, transposon locus de-repression, including their own transcription, can affect host mRNA production and processing in a variety of ways (Elbarbary et al., 2016; Daniel et al., 2015). Human cells and a range of animal organs, such as muscle, brain, and liver, go through replicative senescence (De Cecco et al., 2013; Van Meter et al., 2014). TEs have had their transcriptional inhibition lifted across the genome. In cancer, de-repression of the LTR and L1 promoters can lead to oncogene activation (Babaian and

Mager, 2016). Second, TE encoded proteins can cause genomic instability, such as L1ORF2p's endonuclease activity (Hedges and Deininger, 2007). Furthermore, an innate immune response triggered by the buildup of extra-chromosomal DNA and RNA transcript copies obtained from TEs might result in autoimmune disorders and sterile inflammation. The role of the aforementioned processes in diseases is still unknown. Following TE transcription, the encoded proteins are translated, while retroelements undergo reverse transcription into cDNA substrates appropriate for transposition. The reverse-transcriptase protein that produced was subsequently used to make cytosolic DNAs. The reverse-transcriptase protein that results, as well as the cytosolic nucleic acids: DNA hybrids that follow, can drive inflammatory pathways.

Due to anomalies in systems that normally inhibit TE processing or destabilize TE-derived DNA, patients with Aicardi–Goutières syndrome, for example, have a buildup of TE-derived cytosolic DNA (Crow et al., 2015; Stetson et al., 2008). Some TEs, such as endogenous retroviruses that produce Gag, Pol, or Env proteins, do encode functional proteins (Vargiu et al., 2016). Some TEs code for some functional proteins, including retroviruses capable of expressing Gag, Pol, or envelope (Env) proteins (Vargiu et al., 2016). Multiple sclerosis has been connected to overexpression of specific Env proteins (Perron et al., 2001). and amytrophic lateral sclerosis (ALS) (Li et al., 2015).

2.7 TEs MODIFY TRANSCRIPTIONAL NETWORKS

Cis-regulatory networks control the transcription of a diverse genes that coordinate together to control entire pathways and biological processes. There is currently accumulating evidence that TEs have provided a plentiful supply of material for the control of gene expression, confirming Barbara McClintock's prediction (McClintock B, 1956). Indeed, TEs have the ability to spread a large number of promoters, enhancers, and protein binding sites. Indeed, TEs have the ability to disseminate a vast number of promoters and enhancers (Bejerano et al., 2006; Chuong et al., 2016; Trizzino et al., 2017; Thompson et al., 2016), transcription factor binding sites (Sun et al., 2018; Ito et al., 2017; Sundaram et al., 2014; Kunarso et al., 2010; Wang et al., 2007), insulator sequences (Schmidt et al., et al., 2012; Lunyak et al., 2007; Wang et al., 2015), and repressive elements (Lippman et al., 2004; Rebollo et al., 2011). A good example of gene that affects coat color and whose expression can be affected by the level of methylation of TE upstream of its promoter is the coat color of agouti mice. In case of oil palm, the methylation status of a TE inside a flowering gene controls whether the plants produce oil-rich fruit (Ong-Abdullah et al., 2015). Because TE families frequently cover a genome with similar copies, it has long been assumed that they can use the same cis-regulatory module to "wire" batteries of genes scattered all through the genome (Davidson and Britten, 1979)

TEs appear to have laid the foundation for the cis-regulatory networks evolution, including pathways underlying pregnancy (Lynch et al., 2015; Lynch et al., 2011), stem cell pluripotency (Lu et al., 2014; Wang et al., 2014; Kunarso et al., 2010), neocortex development (Notwell et al., 2015), and innate immunity in mammals (Notwell (Makarevitch et al., 2015). TE sequences, in truth, have all of the characteristics of a "conventional" gene regulation network (Feschotte C., 2008; Elbarbary et al., 2016). According to Sun et al., transcription factors bind to them, they integrate numerous inputs (activation/repression), they respond to both cis and trans signals, and they can coordinate gene expression (2018). In this setting, TEs are excellent agents for altering biological processes by building new cis-regulatory circuits and fine-tuning existing networks.

KEYWORDS

- **genes**
- **transposable elements**
- **plant genome**
- **genetic elements**
- **genome regulation**

REFERENCES

Ade, C.; Roy-Engel, A. M.; Deininger, P. L. Alu Elements: An Intrinsic Source of Human Genome Instability. *Curr. Opin. Virol.* **2013,** *3*, 639–645.

Arkhipova, I. R. Using Bioinformatic and Phylogenetic Approaches to Classify Transposable Elements and Understand Their Complex Evolutionary Histories. *Mob. DNA* **2017,** *8*, 19.

Babaian, A.; Mager, D. L. Endogenous Retroviral Promoter Exaptation in Human Cancer. *Mob. DNA* **2016,** *7*, 24.

Baillie, J. K.; Barnett, M. W.; Upton, K. R.; Gerhardt, D. J.; Richmond, T. A.; De Sapio, F. et al. Somatic Retrotransposition Alters the Genetic Landscape of the Human Brain. Nature. **2011,** *479*, 534–537.

Beck, C. R.; Collier, P.; Macfarlane, C.; Malig, M.; Kidd, J. M.; Eichler, E. E. et al. LINE-1 Retrotransposition Activity in Human Genomes. *Cell* **2010,** *141*, 1159–1170.

Beck, C. R.; Garcia-Perez, J. L.; Badge, R. M.; Moran, J. V. LINE-1 Elements in Structural Variation and Disease. *Annu. Rev. Genom. Hum. Genet.* **2011,** *12*, 187–215.

Bejerano, G.; Lowe, C. B.; Ahituv, N.; King, B.; Siepel, A.; Salama, S. R. et al. A Distal Enhancer and an Ultraconserved Exon Are Derived from a Novel Retroposon. *Nature* **2006,** *441*, 87–90.

Belfort, M.; Curcio, M. J.; Lue, N. F. Telomerase and Retrotransposons: Reverse Transcriptases That Shaped Genomes. *Proc. Natl. Acad. Sci. USA* **2011,** 108, 20304–20310.

Bennetzen, J. L.; Wang, H. The Contributions of Transposable Elements to the Structure, Function, and Evolution of Plant Genomes. *Annu. Rev. Plant Biol.* **2014,** *65*, 505–30.

Boeke, J. D.; Garfinkel, D. J.; Styles, C. A.; Fink, G. R. Ty Elements Transpose Through an RNA Intermediate. *Cell* **1985,** *40*, 491–500.

Brouha, B.; Schustak, J.; Badge, R. M.; Lutz-Prigge, S.; Farley, A. H.; Moran, J. V. et al. Hot L1s Account for the Bulk of Retrotransposition in the Human Population. *Proc. Natl. Acad. Sci. USA* **2003,** *100*, 5280–5285.

Brown, P. O.; Bowerman, B.; Varmus, H. E.; Bishop, J. M. Correct Integration of Retroviral DNA In Vitro. *Cell* **1987,** *49*, 347–356.

Burns, K. H.; Boeke, J. D. Human Transposon Tectonics. *Cell*. **2012,** *149*, 740–752.

Burns, K. H. Transposable Elements in Cancer. *Nat. Rev. Cancer*. **2017,** *17*, 415–424.

Campos-Sánchez, R.; Cremona, M. A.; Pini, A.; Chiaromonte, F.; Makova, K. D. Integration and Fixation Preferences of Human and Mouse Endogenous Retroviruses Uncovered with Functional Data Analysis. *PLOS Comput. Biol*. **2016,** *12*, e1004956.

Carvalho, C. M.; Lupski, J. R. Mechanisms Underlying Structural Variant Formation in Genomic Disorders. *Nat. Rev. Genet*. **2016,** *17*, 224–238.

Casola, C.; Hucks, D.; Feschotte C. Convergent Domestication of Pogo-Like Transposases Into Centromere-Binding Proteins in Fission Yeast and Mammals. *Mol. Biol. Evol.* **2007,** *25*, 29–41.

Cecco, M.; Criscione, S. W.; Peckham, E. J.; Hillenmeyer, S.; Hamm, E. A.; Manivannan, J. et al. Genomes of Replicatively Senescent Cells Undergo Global Epigenetic Changes Leading to Gene Silencing and Activation of Transposable Elements. *Aging Cell* **2013,** *12*, 247–256.

Chuong, E. B.; Elde, N. C.; Feschotte C. Regulatory Evolution of Innate Immunity Through Co-Option of Endogenous Retroviruses. *Science* **2016,** *351*, 1083–1087.

Crow, Y. J.; Manel N. Aicardi–Goutières Syndrome and the Type I Interferonopathies. *Nat. Rev. Immunol.* **2015,** *15*, 429–440.

Daniel, C.; Behm, M.; Öhman M. The Role of Alu Elements in the Cis-Regulation of RNA Processing. *Cell Mol. Life Sci*. **2015,** *72*, 4063–40676.

Davidson, E. H.; Britten, R. J. Regulation of Gene Expression: Possible Role of Repetitive Sequences. *Science* **1979,** 2*04*, 1052–1059.

De Cecco, M.; Criscione, S. W.; Peterson, A. L.; Neretti, N.; Sedivy, J. M.; Kreiling, J. A. Transposable Elements Become Active and Mobile in the Genomes of Aging Mammalian Somatic Tissues. *Aging* **2013,** *5*, 867–883.

Ecco, G.; Imbeault, M.; Trono D. KRAB Zinc Finger Proteins. *Development*. **2017,** *144*, 2719–2729.

Eickbush, T. H.; Furano, A. V. Fruit Flies and Humans Respond Differently to Retrotransposons. *Curr. Opin. Genet. Dev.* **2002,** *12*, 669–674.

Elbarbary, R. A.; Lucas, B. A.; Maquat, L. E. Retrotransposons as Regulators of Gene Expression. *Science* **2016,** *351*, aac7247.

Erwin, J. A.; Marchetto, M. C.; Gage, F. H. Mobile DNA Elements in the Generation of Diversity and Complexity in the Brain. *Nat Rev Neurosci.* **2014,** *15*, 497–506.

Feschotte, C.; Pritham, E. J. DNA Transposons and the Evolution of Eukaryotic Genomes. *Annu. Rev. Genet.* **2007,** *41*, 331–68.

Feschotte C. Transposable Elements and the Evolution of Regulatory Networks. *Nat. Rev. Genet.* **2008,** *9*, 397–405.

Freeling, M.; Xu, J.; Woodhouse, M.; Lisch D. A Solution to the C-Value Paradox and the Function of Junk DNA: The Genome Balance Hypothesis. *Mol. Plant.* **2015,** *8*, 899–910.

Fulcher, N.; Derboven, E.; Valuchova, S.; Riha, K. If the Cap Fits, Wear It: An Overview of Telomeric Structures Over Evolution. *Cell Mol. Life Sci.* **2014,** *71*, 847–865.

Garcia-Perez, J. L.; Marchetto, M. C.; Muotri, A. R.; Coufal, N. G.; Gage, F. H.; O'shea, K. S. et al. LINE-1 Retrotransposition in Human Embryonic Stem Cells. *Hum. Mol. Genet.* **2007,** *16*, 1569–1577.

Gorbunova, V.; Boeke, J. D.; Helfand, S. L.; Sedivy, J. M. Sleeping Dogs of the Genome. *Science* **2014,** *346*, 1187–1188.

Greenblatt, I. M.; Brink, R. A. Transpositions of Modulator in Maize Into Divided and Undivided Chromosome Segments. *Nature* **1963,** *197*, 412–413.

Han, K.; Lee, J.; Meyer, T. J.; Remedios, P.; Goodwin, L.; Batzer, M. A. L1 Recombination-Associated Deletions Generate Human Genomic Variation. *Proc. Natl. Acad. Sci. USA* **2008,** *105*, 19366–19371.

Hancks, D. C.; Kazazian, H. H. Roles for Retrotransposon Insertions in Human Disease. *Mob. DNA* **2016,** *7*, 9.

Hedges, D. J.; Deininger, P. L. Inviting Instability: Transposable Elements, Doublestrand Breaks, and the Maintenance of Genome Integrity. *Mutat. Res. Mol. Mech. Mutagen.* **2007,** *616*, 46–59.

Horváth, V.; Merenciano, M.; González J. Revisiting the Relationship Between Transposable Elements and the Eukaryotic Stress Response. *Trends Genet.* **2017,** *33*, 832–841.

Imbeault, M.; Helleboid P-Y, Trono D. KRAB Zinc-Finger Proteins Contribute to the Evolution of Gene Regulatory Networks. *Nature*. **2017,** *543*, 550–554.

Imbeault, M.; Trono D. As Time Goes By: KRABs Evolve to KAP Endogenous Retroelements. *Dev. Cell* **2014,** *31*, 257–258.

Iskow, R. C.; McCabe, M. T.; Mills, R. E.; Torene, S.; Pittard, W. S.; Neuwald, A. F. et al. Natural Mutagenesis of Human Genomes by Endogenous Retrotransposons. *Cell* **2010,** *141*, 1253–1261.

Ito, J.; Sugimoto, R.; Nakaoka, H.; Yamada, S.; Kimura, T.; Hayano, T. et al. Systematic Identification and Characterization of Regulatory Elements Derived from Human Endogenous Retroviruses. *PLOS Genet.* **2017,** *13*, e1006883.

Jangam, D.; Feschotte, C.; Betrán E. Transposable Element Domestication as an Adaptation to Evolutionary Conflicts. *Trends Genet.* **2017,** *33*, 917–931.

Kazazian, H. H. Mobile DNA transposition in Somatic Cells. *BMC Biol.* **2011,** *9*, 62.

Klawitter, S.; Fuchs, N. V.; Upton, K. R.; Munoz-Lopez, M.; Shukla, R.; Wang, J. et al. Reprogramming Triggers Endogenous L1 and Alu Retrotransposition in Human Induced Pluripotent Stem Cells. *Nat. Commun.* **2016,** *7*, 10286.

Kunarso, G.; Chia, N. Y.; Jeyakani, J.; Hwang, C.; Lu, X.; Chan, Y. S. et al. Transposable Elements Have Rewired the Core Regulatory Network of Human Embryonic Stem Cells. *Nat Genet.* **2010,** *42*, 631–634.

Kursel, L. E.; Malik, H. S. Centromeres. *Curr. Biol.* **2016,** *26*, R487–R490.

Kvikstad, E. M.; Makova, K. D. The (R)evolution of SINE versus LINE Distributions in Primate Genomes: Sex Chromosomes Are Important. *Genome Res.* **2010,** *20*, 600–613.

Lanciano, S.; Mirouze M. Transposable Elements: All Mobile, All Different, Some Stress Responsive, Some Adaptive? *Curr. Opin. Genet. Dev.* **2018,** *49*, 106–114.

Lazarow, K.; Doll M-L, Kunze R. Molecular Biology of Maize AC/DS Elements: An Overview. In *Plant Transposable Elements*; Peterson, T., Ed.; Humana Press: Totowa; 2013; pp 59–82.

Lee, E.; Iskow, R.; Yang, L.; Gokcumen, O.; Haseley, P.; Luquette, L. J. et al. Landscape of Somatic Retrotransposition in Human Cancers. *Science* **2012,** *337*, 967–971.

Lee, J. A.; Carvalho, C. M.; Lupski, J. R. A DNA Replication Mechanism for Generating Nonrecurrent Rearrangements Associated with Genomic Disorders. *Cell* **2007,** *131*, 1235–1247.

Li, W.; Lee M-H, Henderson, L.; Tyagi, R.; Bachani, M.; Steiner, J. et al. Human Endogenous Retrovirus-K Contributes to Motor Neuron Disease. *Sci. Transl. Med.* **2015,** *7*, 307ra153.

Lippman, Z.; Gendrel A-V, Black, M.; Vaughn, M. W.; Dedhia, N.; McCombie, W. R. et al. Role of Transposable Elements in Heterochromatin and Epigenetic Control. *Nature* **2004,** *430*, 471–476.

Lu, X.; Sachs, F.; Ramsay, L.; Jacques, P. E.; Goke, J.; Bourque, G. et al. The Retrovirus HERVH Is a Long Noncoding RNA Required For Human Embryonic Stem Cell Identity. *Nat. Struct. Mol. Biol.* **2014,** *21*, 423–425.

Luan, D. D.; Korman, M. H.; Jakubczak, J. L.; Eickbush, T. H. Reverse Transcription of R2Bm RNA Is Primed By a Nick at the Chromosomal Target Site: A Mechanism for Non-LTR Retrotransposition. *Cell.* **1993,** *72*, 595–605.

Lunyak, V. V.; Prefontaine, G. G.; Nunez, E.; Cramer, T.; Ju, B. G.; Ohgi, K. A. et al. Developmentally Regulated Activation of a SINE B2 Repeat as a Domain Boundary in Organogenesis. *Science* **2007,** *317*, 248–251.

Lynch M. *The Origins of Genome Architecture*, 1st ed.; Sinauer Associates: Sunderland, 2007.

Lynch, V. J.; Leclerc, R. D.; May, G.; Wagner, G. P. Transposon-Mediated Rewiring of Gene Regulatory Networks Contributed to the Evolution of Pregnancy in Mammals. *Nat. Genet.* **2011,** *43*, 1154–1159.

Lynch, V. J.; Nnamani, M. C.; Kapusta, A.; Brayer, K.; Plaza, S. L.; Mazur, E. C. et al. Ancient Transposable Elements Transformed the Uterine Regulatory Landscape and Transcriptome During the Evolution of Mammalian Pregnancy. *Cell Rep.* **2015,** *10*, 551–561.

Makarevitch, I.; Waters, A. J.; West, P. T.; Stitzer, M.; Hirsch, C. N.; Ross-Ibarra, J. et al. Transposable Elements Contribute to Activation of Maize Genes in Response to Abiotic Stress. *PLoS Genet.* **2015,** *11*, e1004915.

Maksakova, I. A.; Romanish, M. T.; Gagnier, L.; Dunn, C. A.; Van de Lagemaat, L. N.; Mager, D. L. Retroviral Elements and Their Hosts: Insertional Mutagenesis in the Mouse Germ Line. *PLoS Genet.* **2006,** *2*, e2.

Malik, H. S.; Eickbush, T. H. Phylogenetic Analysis of Ribonuclease H Domains Suggests a Late, Chimeric Origin of LTR Retrotransposable Elements and Retroviruses. *Genome Res.* **2001,** *11*, 1187–1197.

Mátés, L.; Chuah MKL, Belay, E.; Jerchow, B.; Manoj, N.; Acosta-Sanchez, A. et al. Molecular Evolution of a Novel Hyperactive Sleeping Beauty Transposase Enables Robust Stable Gene Transfer in Vertebrates. *Nat Genet.* **2009,** *41*, 753–761.

McClintock, B. *Controlling Elements and the Gene: Cold Spring Harb Symp Quant Biol*; Cold Spring Harbor Laboratory Press, 1956; pp 197–216.

Miyoshi, N.; Stel, J. M.; Shioda, K.; Qu, N.; Odahima, J.; Mitsunaga, S. et al. Erasure of DNA Methylation, Genomic Imprints, and Epimutations in a Primordial Germ-Cell Model Derived from Mouse Pluripotent Stem Cells. *Proc. Natl. Acad. Sci. USA* **2016,** *113*, 9545–9550.

Muotri, A. R.; Chu, V. T.; Marchetto, M. C.; Deng, W.; Moran, J. V.; Gage, F. H. Somatic Mosaicism in Neuronal Precursor Cells Mediated by L1 Retrotransposition. *Nature* **2005,** *435*, 903–910.

Notwell, J. H.; Chung, T.; Heavner, W.; Bejerano, G. A Family of Transposable Elements Co-Opted Into Developmental Enhancers in the Mouse Neocortex. *Nat. Commun.* **2015,** *6*, 6644.

Ong-Abdullah, M.; Ordway, J. M.; Jiang, N.; Ooi S-E, Kok S-Y, Sarpan, N. et al. Loss of Karma Transposon Methylation Underlies the Mantled Somaclonal Variant of Oil Palm. *Nature* **2015,** *525*, 533–537.

Perron, H.; Jouvin-Marche, E.; Michel, M.; Ounanian-Paraz, A.; Camelo, S.; Dumon, A. et al. Multiple Sclerosis Retrovirus Particles and Recombinant Envelope Trigger an Abnormal Immune Response In Vitro, by Inducing Polyclonal Vβ16 T-Lymphocyte Activation. *Virology* **2001,** *287*, 321–332.

Rebollo, R.; Karimi, M. M.; Bilenky, M.; Gagnier, L.; Miceli-Royer, K.; Zhang, Y. et al. Retrotransposon-Induced Heterochromatin Spreading in the Mouse Revealed by Insertional Polymorphisms. *PLOS Genet.* **2011,** *7*, e1002301.

Rodić, N.; Steranka, J. P.; Makohon-Moore, A.; Moyer, A.; Shen, P.; Sharma, R. et al. Retrotransposon Insertions in the Clonal Evolution of Pancreatic Ductal Adenocarcinoma. *Nat Med.* **2015,** *21*, 1060–1064.

Rubin, G. M.; Kidwell, M. G.; Bingham, P. M. The Molecular Basis of P-M Hybrid Dysgenesis: The Nature of Induced Mutations. *Cell.* **1982,** *29*, 987–994.

Saha, A.; Mitchell, J. A.; Nishida, Y.; Hildreth, J. E.; Ariberre, J. A.; Gilbert, W. V. et al. A Trans-Dominant Form of Gag Restricts Ty1 Retrotransposition and Mediates Copy Number Control. *J Virol.* **2015,** *89*, 3922–3938.

Schauer, S. N.; Carreira, P. E.; Shukla, R.; Gerhardt, D. J.; Gerdes, P.; Sanchez-Luque, F. J. et al. L1 Retrotransposition Is a Common Feature of Mammalian Hepatocarcinogenesis. *Genome Res.* **2018,** *28*, 639–653.

Schmidt, D.; Schwalie, P. C.; Wilson, M. D.; Ballester, B.; Goncalves, A.; Kutter, C. et al. Waves of Retrotransposon Expansion Remodel Genome Organization and CTCF Binding in Multiple Mammalian Lineages. *Cell.* **2012,** *148*, 335–348.

Schnable, P. S.; Ware, D.; Fulton, R. S.; Stein, J. C.; Wei, F.; Pasternak, S. et al. The B73 Maize Genome: Complexity, Diversity, and Dynamics. *Science* **2009,** *326*, 1112–1115.

Schubert, I.; Vu, G. T. H. Genome Stability and Evolution: Attempting a Holistic View. *Trends Plant Sci.* **2016,** *21*, 749–757.

Scott, E. C.; Gardner, E. J.; Masood, A.; Chuang, N. T.; Vertino, P. M.; Devine, S. E. A Hot L1 Retrotransposon Evades Somatic Repression and Initiates Human Colorectal Cancer. *Genome Res.* **2016,** *26*, 745–755.

Smit, A. F. Interspersed Repeats and Other Mementos of Transposable Elements in Mammalian Genomes. *Curr. Opin. Genet. Dev.* **1999,** *9*, 657–663.

Springer, N. M.; Anderson, S. N.; Andorf, C. M.; Ahern, K. R.; Bai, F.; Barad, O. et al. The Maize W22 Genome Provides a Foundation for Functional Genomics and Transposon Biology. *Nat Genet.* **2018,** *50*, 1282–1288.

Stetson, D. B.; Ko, J. S.; Heidmann, T.; Medzhitov R. Trex1 Prevents Cell-Intrinsic Initiation of Autoimmunity. *Cell.* **2008,** *134*, 587–598.

Sun, X.; Wang, X.; Tang, Z.; Grivainis, M.; Kahler, D.; Yun, C. et al. Transcription Factor Profiling Reveals Molecular Choreography and Key Regulators of Human Retrotransposon Expression. *Proc. Natl. Acad. Sci. USA* **2018,** *115*, E5526–E5535.

Sundaram, V.; Cheng, Y.; Ma, Z.; Li, D.; Xing, X.; Edge, P. et al. Widespread Contribution of Transposable Elements to the Innovation of Gene Regulatory Networks. *Genome Res.* **2014,** *24*, 1963–1976.

Tang, Z.; Steranka, J. P.; Ma, S.; Grivainis, M.; Rodić, N.; Huang CRL, et al. Human Transposon Insertion Profiling: Analysis, Visualization and Identification of Somatic LINE-1 Insertions in Ovarian Cancer. *Proc. Natl. Acad. Sci. USA* **2017,** *114*, E733–E740.

Trizzino, M.; Park, Y.; Holsbach-Beltrame, M.; Aracena, K.; Mika, K.; Caliskan, M. et al. Transposable Elements Are the Primary Source of Novelty in Primate Gene Regulation. *Genome Res.* **2017,** *27*, 1623–1633.

Tubio, J. M.; Li, Y.; Ju, Y. S.; Martincorena, I.; Cooke, S. L.; Tojo, M. et al. Extensive Transduction of Nonrepetitive DNA Mediated by L1 Retrotransposition in Cancer Genomes. *Science* **2014,** *345*, 1251343.

Van Meter, M.; Kashyap, M.; Rezazadeh, S.; Geneva, A. J.; Morello, T. D.; Seluanov, A. et al. SIRT6 Represses LINE1 Retrotransposons by Ribosylating KAP1 But This Repression Fails with Stress and Age. *Nat. Commun.* **2014,** *5*, 5011.

Vargiu, L.; Rodriguez-Tomé, P.; Sperber, G. O.; Cadeddu, M.; Grandi, N.; Blikstad, V. et al. Classification and Characterization of Human Endogenous Retroviruses; Mosaic Forms Are Common. *Retrovirology* **2016,** *13*, 7.

Venner, S.; Feschotte, C.; Biémont C. Dynamics of Transposable Elements: Towards a Community Ecology of the Genome. *Trends Genet.* **2009,** *25*, 317–323.

Wang, J.; Vicente-García, C.; Seruggia, D.; Moltó, E.; Fernandez-Miñán, A.; Neto, A. et al. MIR Retrotransposon Sequences Provide Insulators to the Human Genome. *Proc. Natl. Acad. Sci. USA* **2015,** *112*, E4428–E4437.

Wang, T.; Zeng, J.; Lowe, C. B.; Sellers, R. G.; Salama, S. R.; Yang, M. et al. Speciesspecific Endogenous Retroviruses Shape the Transcriptional Network of the Human Tumor Suppressor Protein p53. *Proc. Natl. Acad. Sci. USA* **2007,** 10*4*, 18613–18618.

Xing, J.; Zhang, Y.; Han, K.; Salem, A. H.; Sen, S. K.; Huff, C. D. et al. Mobile Elements Create Structural Variation Analysis of a Complete Human Genome. *Genome Res.* **2009,** *19*, 1516–1526.

Yang, P.; Wang, Y.; Macfarlan, T. S. The role of KRAB-ZFPs in Transposable Element Repression and Mammalian Evolution. *Trends Genet.* **2017,** *33*, 871–881.a

Part II

Protocols Designed to Characterize TEs and Biotechnological Applications

CHAPTER 3

Computational Approaches for Prediction and Analysis of Transposable Elements in Plants

MAHANTESHA B. N. NAIKA[1*], AKASH DAS[2], H. T. PARAMESHA[1], PANKAJ BARAH[2], and H. V. CHAITRA [3]

[1]*Department of Biotechnology and Crop Improvement, Kittur Rani Channamma College of Horticulture, Arabhavi, University of Horticultural Sciences, Bagalkot, Karnataka 591218, India*

[2]*Department of Molecular Biology and Biotechnology, Tezpur University, Sonitpur 784028, India*

[3]*Department of Crop Physiology, University of Agricultural Sciences, GKVK, Bangalore, Karnataka 560065, India*

**Corresponding author. E-mail: mahantesha.bnn@uhsbagalkot.edu.in*

ABSTRACT

Transposable elements (TEs) are DNA sequences and a well-known class of repetitive sequences and also known as jumping genes that have a natural potential to travel from one position to another within a genome. They are known to involve in various cellular processes and play a role in genetic research. These are commonly found in prokaryotes and eukaryotes. The next-generation sequencing platforms are generating big data from model and non-model organisms. Experimental approaches for identification and characterization of TEs are expensive and time-consuming. Advances in computational approaches lead to the development of databases and tools to identify and characterize TEs in the omics era. This chapter covers the overview of TEs, classification, and present about the available computational

Plant Transposable Elements: Biology and Biotechnology, Deepu Pandita, Anu Pandita, (Eds.)

approaches to predict and analyze the role of TEs in biotic and abiotic stress response.

3.1 INTRODUCTION

A genome can be defined as the entire genetic repository of an organism (Goldman and Landweber, 2016). Since the introduction of next-generation sequencing (NGS) technologies, researchers have witnessed a revolution in DNA sequencing technologies that are widely accepted in the field of Biology, enabling them to sequence millions of bases in model and non-model organisms. As a result of NGS, sequencing has become more throughput and speedy while also reducing costs over time (Fig. 3.1). In plant biology applications, NGS can be used in various fields, such as genome sequencing and resequencing (genomics), RNA sequencing (transcriptomics), metabolite sequencing (metagenomics), exon and genome capturing, evolutionary studies, explore large genebank collections, identification and development of molecular markers resources (genic and genomic SSR, SNP, transposons), polyploid genetics, phylogenetic and ecological studies, and also knowing the mechanisms involved in gene expression and secondary metabolism (Egan et al., 2012; Unamba et al., 2015; Goldman and Landweber, 2016; Sharma et al., 2018).

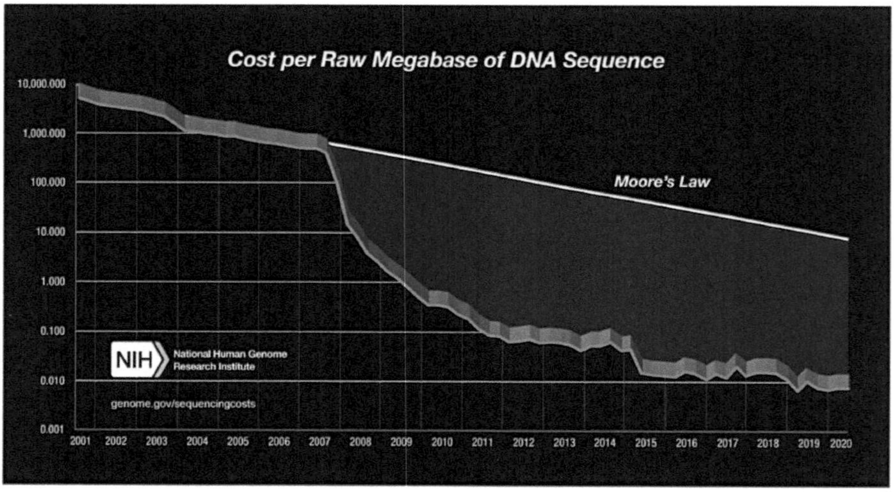

FIGURE 3.1 NGS sequence cost per raw mega base.

Source: Image courtesy: Wetterstrand KA. DNA Sequencing Costs: Data from the NHGRI Genome Sequencing Program (GSP) www.genome.gov/sequencingcostsdata.

Till June 2021, the following genomes are publicly available at National Centre for Biotechnology Information (NCBI): Eukaryotes (18792), Prokaryotes (358326), Viruses (44764), Plasmids (31618), and Organelles (19328). (Information retrieved on August 20, 2021 from NCBI (https://www.ncbi.nlm.nih.gov/genome/browse/#!/overview/). In spite of the availability of this magnitude of data, it remains a challenge to analyze and interpret it in order to determine its biological significance. In the early 1980s, TE sequencing began and was mainly analyzed manually. During 1990s, there was a surge in eukaryotic DNA sequence availability and at the same time, usage of computers was widely used in biological labs (Janicki et al., 2011). In most eukaryotes, noncoding sequence regions make up the majority of the genome. Intergenic DNA, repetitive elements including TEs and introns are among these sequences (Sahebi et al., 2018).

Transposable elements (TEs) are defined as DNA sequences that are able to move, or duplicate from one location to another in the genome. Prokaryote and eukaryotes occupy a large proportion of TEs (Munoz-Lopez and Garcia-Perez, 2010). The TEs are also known as jumping genes or mobile sequence or transposons or translocatable elements or insertion elements or parasitic elements or selfish DNA (Tripathi et al., 2014). Most flowering plants have a high proportion of TEs in their genomes (Oliver et al., 2013).

3.2 HISTORICAL PROSPECTIVE OF TEs

The activator (*Ac*)/dissociator (*Ds*) transposons in maize were first identified by Barbara McClintock (McClintock, 1948, 1950). For the first time, Finnegan 1989 proposed the TEs classification system and notable two broad classes were characterized based on transposition intermediate: RNA (class I or retrotransposons) also called "copy and paste" mechanism and DNA (class II or DNA transposons) also called "cut and paste" mechanism. In 2007, Wicker et al. proposed a unified and refined classification based on the mechanisms of transposition, sequence similarities, and structural relationship. Kapitonov and Jurka (2008) reported a universal hierarchical classification system similar to Wicker et al. (2007) in eukaryotic transposable elements. The classification was based on transposition mechanism, sequence similarities, and structural relationships, and which was made available as the Repbase database (https://www.girinst.org/repbase/). In 2010, Biemont provided a context of transposable elements during subsequent times. The historical milestone of TE is depicted in Figure 3.2 as reported recently by Makałowski et al. (2019). The classification of eukaryotic TEs was updated with the addition of the SVA and Retrogenes super families to the order SINE (Fig. 3.3).

The increasing interest and impact of NGS on genome or transcriptome sequencing, assembly, functional characterization of transcripts, and alignment of model and non-model plants had led to developing novel computational tools (Fig. 3.4). A genome-level TE analysis and annotation can provide information on TEs evolution and their population dynamics, changes in genome size and structure, TE regulation by the host (Janicki et al., 2011).

Bergman and Quesneville (2007) reviewed on post-genomic era to identify, annotate, discover new TE families and TE copies detection using computational approaches including *de novo*, homology-based, structure-based, and comparative genomic methods. The TEs contribute to the organization, structure, and evolution of genomes. Further, this has created interest in sequencing of the whole genome followed by assembly, alignments, and functional characterization of transcripts. This has paid much attention in the field of bioinformatics to develop new methods and tools to analyze such large-scale data in order to decipher its biological relevance (Bergman and Quesneville, 2007; Makałowski et al., 2019). The sequence length of TEs ranges from less than 100-bp to more than 20-kb. The transposition of TEs is largely dependent on the diverse TE-encoded enzymatic machinery. So far, indentified enzymes are reverse transcriptase, endonuclease, DD[E/D]-transposase (Tpase), Tyrosine-recombinase, and Rep/Helicase1 (https://www.girinst.org/education/index.html). The technology of genomic sequencing has enhanced current research advancements and paved the way for new initiatives. Bioinformatics communities face an increasing challenge in developing algorithms and implementations that efficiently handle large sequence data sets (Janicki et al., 2011). Now, a good number of model and non-model plant genomes have been sequenced and identified the percentage of TEs classes as shown in Figure 3.5. Among the selected genomes, *Medicago truncatula* contains the lowest, that is, 18.30 % of TEs, whereas *Zea mays* contains 84.20%. These data indicate the abundance and distribution of TEs across plant genomes.

In the post-genomic era, bioinformatics has undergone significant changes. The data sets were handled, mined, compared, extracted, clustered, analyzed, and visualized using novel computational methods, machine learning techniques, and web-based server tools. A global effort is underway to develop resources for regional hosting, organized and structured access, and to assess and analyze the ever-growing amounts of data generated each day.

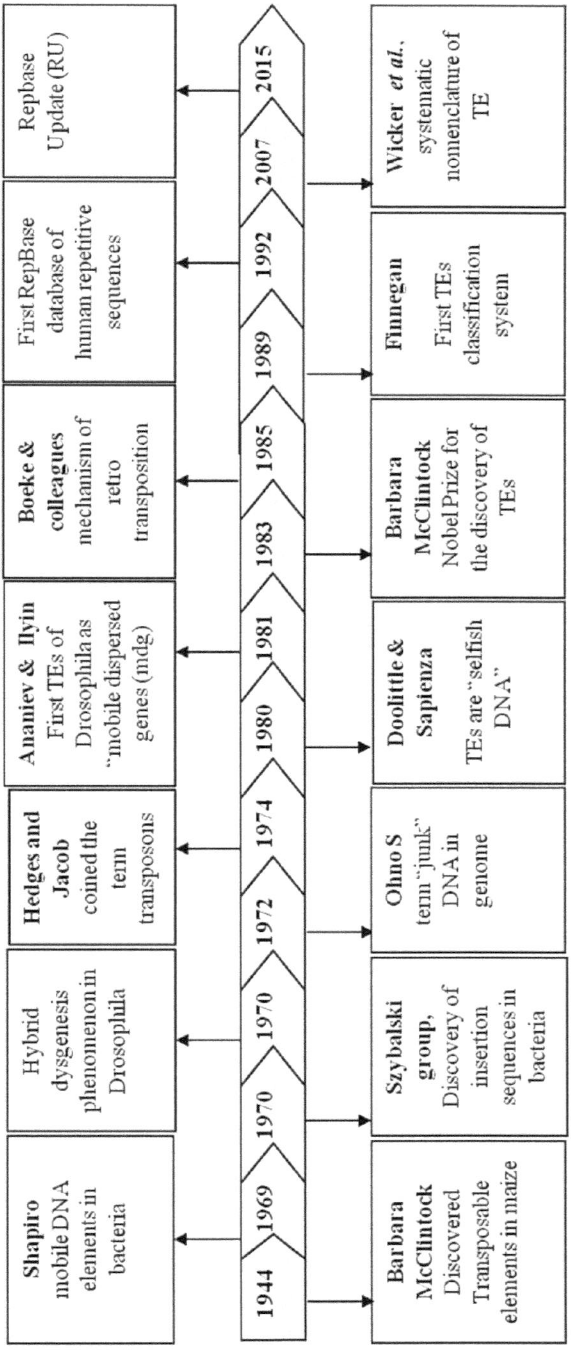

FIGURE 3.2 Milestones of transposable elements.

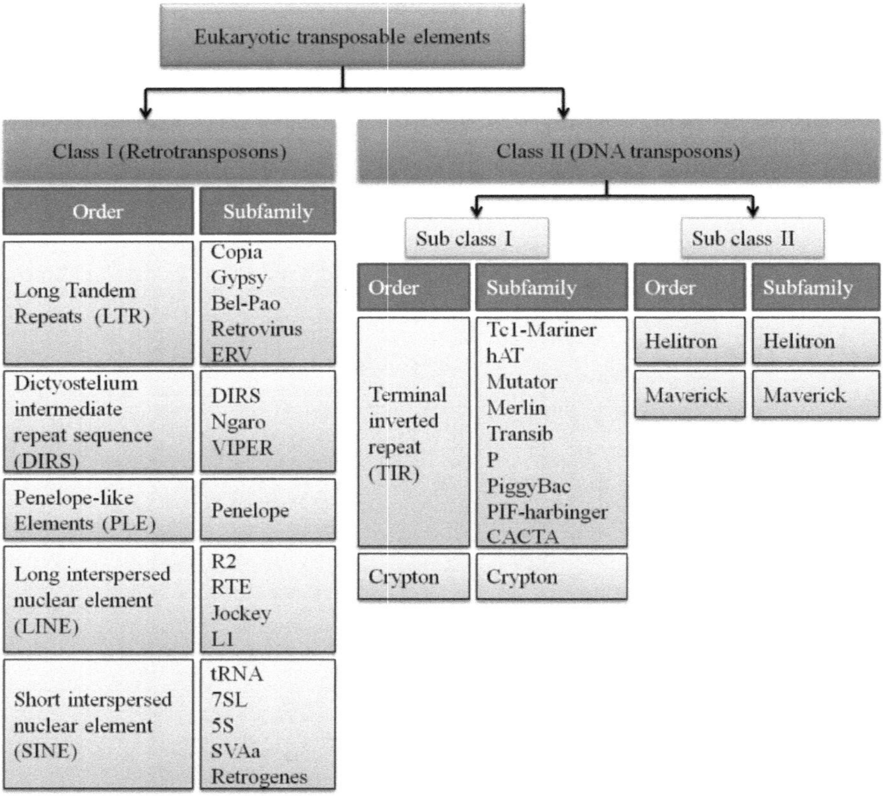

FIGURE 3.3 Classification of eukaryotic transposable elements as proposed by Wicker et al. (2007) and updated by Makałowski et al. (2019).

3.3 BIOINFORMATICS APPROACHES FOR IDENTIFICATION OF TEs

High throughput NGS has provided large data on different organisms. However, the biocuration approach is time-consuming and infeasible for handling such big data. Hence, the use of Bioinformatics approaches to analyze and decipher biological functions from big data is gaining importance. In contrast, manual detection of TE would be time-consuming and unpractical.

It presents a continuous challenge for TE discovery, which is why many tools have been developed specifically for TE discovery (Janicki et al., 2011). It has been necessary for scientists to develop automated methods to discover and classify the TE types present in sequenced genomes. Since TEs have computationally observable structure signatures, they are ideal for large-scale

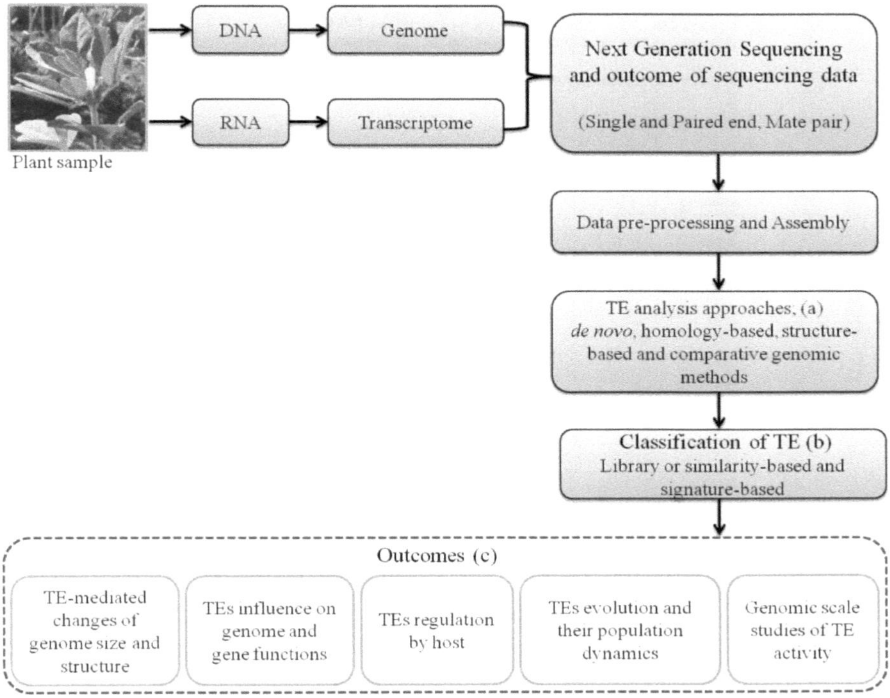

FIGURE 3.4 General outline of NGS followed by computational approaches to analysis and classification of TEs in plants ((a) Bergman and Quesneville (2007), (b) Makałowski et al. (2019), and (c) Janicki et al. (2011)).

bioinformatics approaches with varying degrees of success at detecting TE types. Computer programming languages like Perl and Python as well as modules, such as Bioperl and Biopython aided the progress of bioinformatics, which included computational analyses of TEs (Janicki et al., 2011).

Lerat (2009) reviewed comprehensively presented programs/methods/approaches Library based, Signature based, de novo, self-comparison, k-mer, and spaced seed, Pipelines of programs, Classification programs and Programs intended to detect particular repeats other than TEs in genomes. TE sequences are often discovered, retrieved, compared, aligned, and phylogenetically analyzed computationally. BLAST, ClustalW, PHYLIP, and PAUP are commonly used bioinformatics tools. Janicki et al. (2011) presented about the flow to TE discovery, classification, masking and annotation, analysis, visualization, and databases. In 2015, Biscotti et al., reported that *de novo*, homology-based, structure-based, and comparative genomic methods are used to identify and functionally characterize the TEs using genomics data.

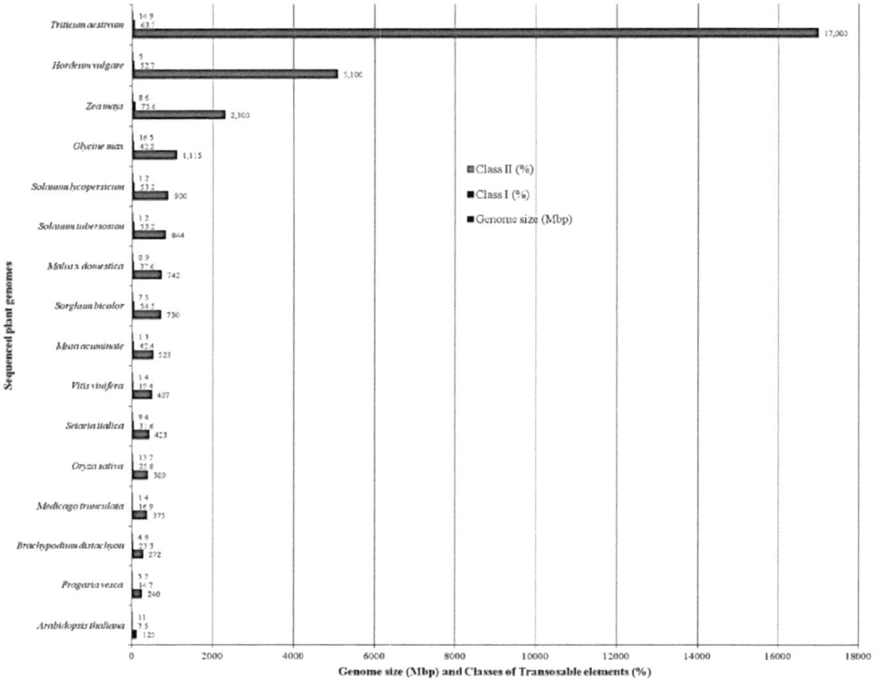

FIGURE 3.5 Selected model and non-model plant genome size and TE classes contents.
Source: The figure was created using the data presented by Oliver et al. (2013).

Makałowski et al. (2019) in their paper, discussed about TE analyses methods and tools and presented two scenarios.

1. Already available species genomic or transcriptome sequences with the transposon repertoire, previously characterized genome, or earlier studies on TEs.
2. Dealing with a totally unknown genome or a genome with little information on TEs.

Presented on De Novo Approaches, Determination in NGS Data, Population-Level Analyses, Comparative Genomics of TE Insertions, Classification of Transposable Elements, Pipelines, and Meta-analyses.

In addition to low selection pressure, TEs evolve quickly and exist in rapidly evolving environments, and because of this fact, it is more difficult to identify, characterize, and annotate TEs. To identify and annotate TEs in plant genomes, four categories of methods have been developed:

de novo, structure-based, comparative genomics, and homology-based (Orozco -Arias et al., 2019). Plant-based TE databases are depicted in Table 3.1.

TABLE 3.1 Plant Specific Transposable Elements Databases.

Crop/plants	Database	References/ PMID
Rice	RiTE	Copetti et al. (2015)
Brassica spp	BrassicaTED	Murukarthick et al. (2014)
Soybean	SoyTEdb	Du et al. (2010)
Mulberry	MnTEdb	Ma et al. (2015)
Plants and fungi	TREP	Wicker et al. (2002)
Plants	P- MITE	Chen et al. (2014)
Plants	MASiVEdb	Bousios et al. (2012)
Dioecious plants	DPTEdb	Li et al. (2016)

Computational tools that are specialized in the analysis and characterization of TE are available. The bioinformatics community has developed multiple bioinformatics tools, and many of them can identify specific classes of TEs (Table 3.2).

TABLE 3.2 Selected Tools for Transposable Elements Identification and Analyses.

Software/Tools	Approach	References
Identification of TEs		
FORRepeats	Homology-based	Lefebvre et al. (2003)
LTR_MINER	Structure-based	Pereira (2004)
LTR_retriever	Structure-based	Ou and Jiang (2017)
LTRType	Homology-based	Zeng et al. (2017)
PILER	Structure-based *De novo*	Edgar and Myers (2005)
Classification and analysis		
Inpactor	Structure-based, Homology-based	Orozco-arias et al. (2018)
LTRClassifier	Structure-based	Monat et al. (2016)

3.4 ROLE OF TRANSPOSABLE ELEMENTS IN PLANT STRESS RESPONSE

Plants are sessile organisms and need to constantly adapt themselves to changing environmental conditions by reprogramming their genome and transcriptome (Song et al., 2021). Our understanding of plant stress biology and global food production could be improved through advances in bioinformatics, plant genomics, and data science (Shameer et al., 2019). TEs are one such modulator which can alter the expression of a gene in order to bring about desired responses (Fig. 3.6). Upon exposure to stress, certain regulatory sequences known as stress-responsive elements (SREs) present in these TEs get activated leading to enhanced TE activity known as "transpositional burst." These TEs then participate in genomic reprogramming which helps organisms cope with challenging situations (Bucher et al., 2012; McClintock 1984; Negi et al., 2016).

But how do these TEs modulate stress response? It is known that TEs greatly enhance resistance to stressors in plants (Mao et al., 2015). They do it essentially by getting inserted into different regions of the neighboring protein-coding genes either by cut-paste or copy-paste mechanisms from their original location. These regional insertions can either activate or deactivate adjacent genes bringing about changes in overall transcriptional response of the host plant to various biotic and abiotic stress conditions (Negi et al., 2016). For example, insertion of a retrotransposon known as Tcs1 in the promoter region of Ruby gene activates it under cold conditions for producing anthocyanin pigment, responsible for the deep red color of pulp in blood red oranges (Butelli et al., 2012). Contrary to this, insertion of a LTR retrotransposon into powdery mildew (PW) susceptible gene CsaMLO8-induced loss-of-function mutation and led to increased hypocotyl-mediated resistance in cucumber (Berg et al., 2015). Retrotransposons are also known to be activated in rice plants upon iron stress condition. These retrotransposons shared similar Cis-Regulatory-Elements (CREs) with promoter of overexpressed rice genes suggesting its crucial role in rice iron stress tolerance (Finatto et al., 2015). Similarly, the insertion of a novel miniature inverted-repeat transposable elements (MITE) TE known as miniature ping (mPing) in the promoter of adjacent genes activated it only under cold stresses as compared with normal conditions in rice. Hence, such TEs can also be used as a "genetic tool" to manipulate gene expression for a given condition (Yasuda et al., 2013). A recent WGAS study showed the insertion of an 82 bases long MITE in the promoter region of a NAC-transcription factor (TF)—ZmNAC111 suppressed its expression using RNA-directed DNA methylation and lead to increased susceptibility of maize plants to drought stress (Mao et al., 2015). Similarly, insertion of MITE in the 3′ UTR sequence of TaHSP16.9–3A gene in wheat resulted in increased tolerance to heat stress (Li et al., 2014). ONSEN is a retrotransposon which is expressed in the Col ecotype of *Arabidopsis thaliana* under heat stress conditions (Ito et al., 2011). A number of heat shock proteins (HSPs) bind to its

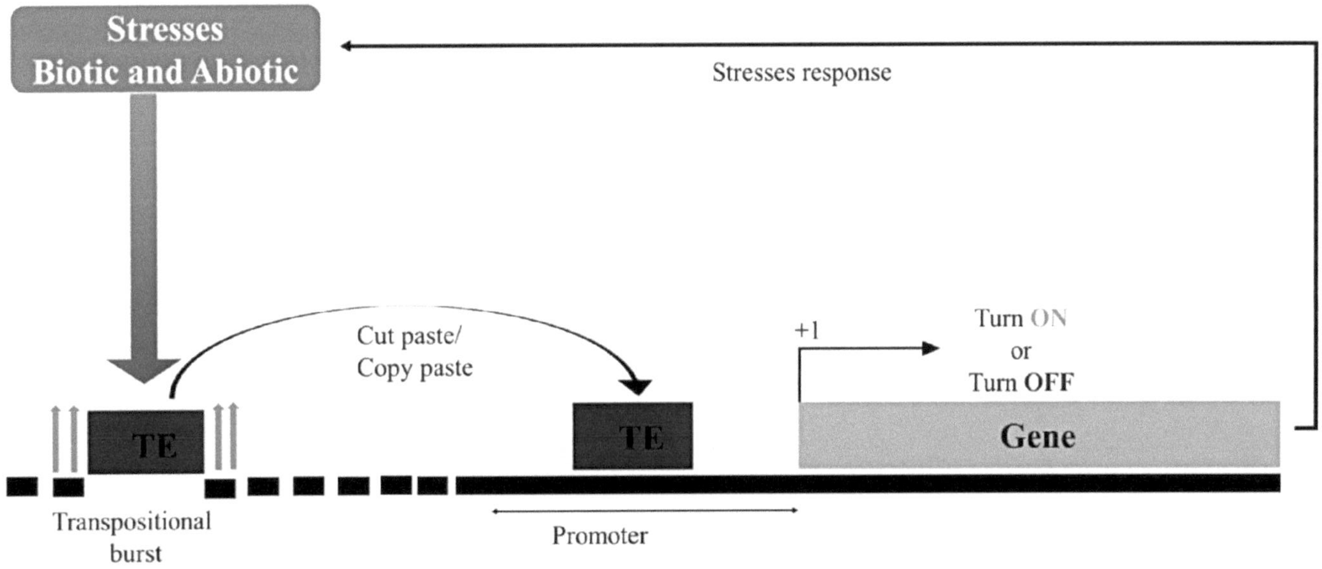

FIGURE 3.6 Schema showing how transposable elements (TEs) modulate stress response by reprogramming host plant genome and transcriptome.

promoter region bringing about stress-induced activation of ONSEN (Lim et al., 2006; Cavrak et al., 2014). Epigenetic changes like DNA-methylation can also activate TEs. For example, demethylation of TEs in promoter sequence of certain defense genes conferred resistance in *Arabidopsis* to fungus *Fusarium oxysporum* (Le et al., 2014). Recently, it was also found that stress-induced TE families undergo "preferential cluttering", that is, certain families of TEs are more likely to stay near a set of genes as compared with others. For example, 20 out of 576 TE families preferentially cluttered only near the upregulated genes, whereas three families were found to be preferentially cluttered only around down-regulated genes in maize under different abiotic stress (Makarevitch et al., 2015).

It is now proven that TEs can no longer be considered as "JUNK." Their role in shaping the transcriptional landscape of plants to various stressors is indispensable. TEs constitute a large portion of the nuclear genome of major crops, such as rice (20–40%) and maize (85%) (Li et al., 2017; Schnable et al., 2009). The cross-talk between TEs and stress response is complex, multifaceted, and is yet to be fully explored (Negi et al., 2016). Studies dealing with TEs in plant stress response have surfaced recently. With the advent of state-of-the-art technologies like RNA-Seq, it is now possible to easily capture and establish a link between a large number of TEs with both coding and noncoding genes (Wang et al., 2017). The field of TE role in plant stress response is still in its nascency and requires further investigation. However, the clue that they are major role players in plant stress seem to be well established (Table 3.3).

TABLE 3.3 Abiotic and Biotic Stress-Responsive Transposable Elements Found in Plants.

Abiotic stress			
TE	**Stress**	**Plant**	**References**
ONSEN	Prolonged heat stress	*Arabidopsis thaliana*	Pecinka et al. (2010)
ONSEN	Heat stress	*Arabidopsis thaliana*	Matsunaga et al. (2012)
ATCopeg1	Salt stress, cytokine	*Arabidopsis thaliana*	Duan et al. (2008)
CLCoy1	Wounding and salt stress	*Citrus lemon*	Felice et al. 2009)
mPing	Cold stress	*Oryza sativa*	Yasuda et al. (2013)
mPing	Gamma radiation	*Oryza sativa*	Nakazaki et al. (2003)
Reme1	UV light	*Cucumis melo*	Ramallo et al. (2008)
TLC1	High salt	*Lycopersicon chilense*	Tapia et al. (2005)
BARE1	Adaptation to moisture	*Hordeum vulgare*	Kalendar et al. (2000)
HRET1	Wounding	*Hibiscus syriacus*	Jeung et al. (2005)

TABLE 3.3 *(Continued)*

Biotic stress			
TE	**Stress**	**Plant**	**References**
AtGP1	Bacterial flagellin	*Arabidopsis thaliana*	Yu et al. (2013)
Queenti	Fungal elicitor	*Nicotiana tabacum*	Anca et al. (2014)
Morgane	Fungal infection	*Triticum aestivum*	Sabot et al. (2006)
Tos17	Bacterial blight	*Oryza sativa*	Sha et al. (2005)
Erika	*Fusarium graminearum*	*Triticum aestivum*	Ansari et al. (2007)
Athila	Geminivirus	*Arabidopsis thaliana*	Buchmann et al. (2009)

3.5 CONCLUSION

Since advances in NGS keep on increasing and producing large amounts of data and advances in computational research are putting more efforts into developing tools to analyze and store such data. An increased interest and improvement in computational tools specialized in TE analyses resulted from genome-level analysis of TE. Innovative tools and technologies enable biologists to explore new research avenues that may provide novel, fascinating insights into the biology of mobile elements. TEs that respond to stress could be used in molecular breeding. The availability of such a large number of TE sequences facilitates research on TEs in many areas, addressing both fundamental and applied questions.

KEYWORDS

- **genomics**
- **transposable elements**
- **bioinformatics**
- **database**
- **tools**
- **stress-responsive genes**

REFERENCES

Chen, J.; Hu, Q.; Zhang, Y.; Lu, C.; Kuang, H. P-MITE: A Database for Plant Miniature Inverted-Repeat Transposable Elements. *Nucl. Acids Res.* **2014,** *42* (D1), D1176–D1181.

Copetti, D.; Zhang, J.; El Baidouri, M.; Gao, D.; Wang, J.; Barghini, E.; Cossu, R. M.; Angelova, A.; Roffler, S.; Ohyanagi, H.; Wicker, T. RiTE Database: A Resource Database for Genus-Wide Rice Genomics and Evolutionary Biology. *BMC Genom.* **2015,** *16* (1), 1–10.

Du, J.; Grant, D.; Tian, Z.; Nelson, R. T.; Zhu, L.; Shoemaker, R. C.; Ma, J. SoyTEdb: A Comprehensive Database of Transposable Elements in the Soybean Genome. *BMC Genom.* **2010,** *11* (1), 1–7.

Duan, K.; Yi, K.; Dang, L.; Huang, H.; Wu, W; Wu, P. Characterization of a Sub- Family of Arabidopsis Genes with the SPX Domain Reveals Their Diverse Functions in Plant Tolerance to Phosphorus Starvation. *Plant J.* **2008,** *54*, 965–975.

Edgar, R. C.; Myers, E. W. PILER: Identification and Classification of Genomic Repeats. *Bioinformatics* **2005,** *21* (1), 152–158.

Egan, A. N.; Schlueter, J.; Spooner, D. M. Applications of Next-Generation Sequencing in Plant Biology. *Am. J. Bot.* **2012,** *99* (2), 175–185.

Felice, B.; Wilson, R. R.; Argenziano, C.; Kafantaris, I.; Conicella, C. A. Transcriptionally Active Copia-Like Retroelement in *Citrus limon. Cell. Mol. Biol. Lett.* **2009,** *14*, 289–304.

Finatto, T.; de Oliveira, A. C.; Chaparro, C.; Da Maia, L. C.; Farias, D. R.; Woyann, L. G.; Picault, N. Abiotic Stress and Genome Dynamics: Specific Genes and Transposable Elements Response to Iron Excess in Rice. *Rice* **2015,** *8*, 1–18.

Finnegan, D. J. Eukaryotic transposable elements and genome evolution. *Trends in genetics.* **1989,** *5*, 103–107.

Goldman, A. D.; Landweber, L. F. What Is a Genome? *PLoS Genet.* **2016**. *12* (7), e1006181.

https://www.girinst.org/education/index.html

https://www.girinst.org/repbase

https://www.ncbi.nlm.nih.gov/genome/browse/#!/overview

Ito, H.; Gaubert, H.; Bucher, E.; Mirouze, M.; Vaillant, I.; Paszkowski, J. An siRNA Pathway Prevents Transgenerational Retrotransposition in Plants Subjected to Stress. *Nature* **2011,** *472*, 115–119.

Janicki, M.; Rooke, R. and Yang, G. Bioinformatics and Genomic Analysis of Transposable Elements in Eukaryotic Genomes. *Chromosome Res.* **2011,** *19* (6), 787–808.

Jeung, J. U.; Cho, S. K.; Lee, S. J.; Shin, J. S. Characterization of Ty3-gypsy-like Elements in *Hibiscus syriacus. Mol. Cells (Springer Science & Business Media BV).* **2005,** *19.*

Kalendar, R.; Tanskanen, J.; Immonen, S.; Nevo, E.; Schulman, A. H. Genome Evolution of Wild Barley (*Hordeum spontaneum*) by BARE-1 Retrotransposon Dynamics in Response to Sharp Microclimatic Divergence. *Proc. Natl. Acad. Sci.* **2000,** *97*, 6603–6607.

Kapitonov, V. V.; Jurka, J. A Universal Classification of eukaryotic Transposable Elements Implemented in Repbase. *Nat. Rev. Genet.* **2008,** *9* (5), 411–412.

Lefebvre, A.; Lecroq, T.; Dauchel, H.; Alexandre, J. FORRepeats: Detects Repeats on Entire Chromosomes and Between Genomes. *Bioinformatics* **2003,** *19* (3), 319–326.

Lerat, E. Identifying Repeats and Transposable Elements in Sequenced Genomes: How to Find Your Way Through the Dense Forest of Programs. *Heredity* **2010,** *104* (6), 520–533.

Li, J.; Wang, Z.; Peng, H.; Liu, Z. A MITE Insertion Into the 3′-UTR Regulates the Transcription of TaHSP16.9 in Common Wheat. *Crop J.* **2014,** *2*, 381–387.

Li, S. F.; Zhang, G. J.; Zhang, X. J.; Yuan, J. H.; Deng, C. L.; Gu, L. F.; Gao, W. J. DPTEdb, an Integrative Database of Transposable Elements in Dioecious Plants. *Database (Oxford)* **2016,** *12*, 2016.

Li, X.; Guo, K.; Zhu, X.; Chen, P.; Li, Y.; Xie, G.; Peng, L. Domestication of Rice Has Reduced the Occurrence of Transposable Elements Within Gene Coding Regions. *BMC Genom.* **2017,** *18*, 1–12.

Lim, C. J.; Yang, K. A.; Hong, J. K.; Choi, J. S.; Yun, D. J.; Hong, J. C.; Lim, C. O. Gene Expression Profiles During Heat Acclimation in *Arabidopsis thaliana* Suspension- Culture Cells. *J. Plant Res.* **2006,** *119*, 373–383.

Loureiro, T. D. S. D. C. Application of Machine Learning Techniques on the Discovery and Annotation of Transposons in Genomes, Master's Thesis; Faculdade de Engenharia, Universidade Do Porto, Porto, Portugal, 2012.

Ma, B.; Li, T.; Xiang, Z.; He, N. MnTEdb, a Collective Resource for Mulberry Transposable Elements. *Database (Oxford).* **2015,** *27*, 2015.

Makałowski, W.; Gotea, V.; Pande, A.; Makałowska, I. Transposable Elements: Classification, Identification, and Their Use as a Tool for Comparative Genomics. *Evol. Biol.* **2019,** 177–207.

Makarevitch, I.; Waters, A. J.; West, P. T.; Stitzer, M.; Hirsch, C. N.; Ross-Ibarra, J; Springer, N. M. Transposable Elements Contribute to Activation of Maize Genes in Response to Abiotic Stress. *PLoS Genet.* **2015,** *11*, e1004915.

Mao, H.; Wang, H.; Liu, S.; Li, Z.; Yang, X.; Yan, J.; Qin, F. A Transposable Element in a NAC Gene Is Associated with Drought Tolerance in Maize Seedlings. *Nat. Commun.* **2015,** *6*, 1–13.

Matsunaga, S.; Iwata, E.; Ikeda, S.; Abe, N.; Kobayashi, A.; Kurata, M.; Yoshioka, Y.; Criqui, M. C.; Genschik, P.; Ito, M. Roles of GIG1 and UVI4 in Genome Duplication in *Arabidopsis thaliana*. *Plant Signal. Behav.* **2012,** *7*, 1079–1081.

McClintock, B. The Origin and Behavior of Mutable Loci in Maize. *Proc. Natl. Acad. Sci.* **1950,** *36* (6), 344–355.

McClintock, B. The Significance of Responses of the Genome to Challenge. *Science* **1984,** *226* (4676), 792–801.

Monat, C. Tando, N.; Tranchant-Dubreuil, C.; Sab Monat, C.; Tando, N.; Tranchant-Dubreuil, C.; Sabot, F. LTRclassifier: A Website for Fast Structural LTR Retrotransposons Classification in Plants. *Mob. Genet. Elements* **2016,** *6* (6), 1241050.

Munoz-Lopez, M. and Garcia-Perez, J. L. DNA Transposons: Nature and Applications in Genomics. *Curr. Genom.* **2010,** *11* (2), 115–128.

Murukarthick, J.; Sampath, P.; Lee, S. C.; Choi, B. S.; Senthil, N.; Liu, S.; Yang, T. J. BrassicaTED-a Public Database for Utilization of Miniature Transposable Elements in Brassica Species. *BMC Res. Notes* **2014,** *7* (1), 1–11.

Nakazaki, T.; Okumoto, Y.; Horibata, A.; Yamahira, S.; Teraishi, M.; Nishida, H.; Inoue, H.; Tanisaka, T. Mobilization of a Transposon in the Rice Genome. *Nature* **2003,** *421*,170–172.

Negi, P.; Rai, A. N.; Suprasanna, P. Moving Through the Stressed Genome: Emerging Regulatory Roles for Transposons in Plant Stress Response. *Front. Plant Sci.* **2016,** *7*, 1448.

Oliver, K. R.; McComb, J. A.; Greene, W. K. Transposable Elements: Powerful Contributors to Angiosperm Evolution and Diversity. *Genome Biol. Evol.* **2013,** *5* (10), 1886–1901.

Orozco-Arias, S.; Isaza, G.; Guyot, R. Retrotransposons in Plant Genomes: Structure, Identification, and Classification Through Bioinformatics and Machine Learning. *Int. J. Mol. Sci.* **2019,** *20* (15), 3837.

Orozco-Arias, S.; Liu, J.; Tabares-Soto, R.; Ceballos, D.; Silva Domingues, D.; Garavito, A. Ming, R.; Guyot, R. Inpactor, Integrated and Parallel Analyzer and Classifier of LTR

Retrotransposons and Its Application for Pineapple LTR Retrotransposons Diversity and Dynamics. *Biology* **2018,** *7* (2), 32.

Ou, S.; Jiang, N. LTR_Retriever: A Highly Accurate and Sensitive Program for Identification of Long Terminal Repeat Retrotransposons. *Plant Physiol.* **2018,** *176* (2), 1410–1422.

Pecinka, A.; Dinh, H. Q.; Baubec, T.; Rosa, M.; Lettner, N.; Scheid, O. M. Epigenetic Regulation of Repetitive Elements Is Attenuated by Prolonged Heat Stress in *Arabidopsis*. *Plant Cell* **2010,** *22*, 3118–3129.

Pereira, V. Insertion Bias and Purifying Selection of Retrotransposons in the *Arabidopsis thaliana* Genome. *Genome Boil.* **2004,** *5* (10), 1–10.

Ramallo, E.; Kalendar, R.; Schulman, A. H.; Martínez-Izquierdo, J. A. Reme1, a Copia Retrotransposon in Melon, Is Transcriptionally Induced by UV Light. *Plant Mol. Biol.* **2008,** *66*, 137–150.

Sabot, F.; Sourdille, P.; Chantret, N.; Bernard, M. Morgane, a New LTR Retrotransposon Group, and Its Subfamilies in Wheats. *Genetica* **2006,** *128*, 439–447.

Sahebi, M.; Hanafi, M. M.; van Wijnen, A. J.; Rice, D.; Rafii, M. Y.; Azizi, P.; Osman, M.; Taheri, S.; Bakar.; M. F. A.; Isa, M. N. M.; Noor, Y. M. Contribution of Transposable Elements in the Plant's Genome. *Gene*. **2018,** *665*, 155–166.

Schnable, P. S.; Ware, D.; Fulton, R. S.; Stein, J. C. Wei.; F. Pasternak S.; Presting G. G. The B73 Maize Genome: Complexity, Diversity, and Dynamics. *Science* **2009,** *326*, 1112–1115.

Sha, A. H.; Huang, J. B.; Zhang, D. P. Relationship of Activation of Tos17 and Rice Adult Plant Resistance to Bacterial Blight. *Yi Chuan*. **2005,** *27* (2), 181–184.

Shameer, K.; Naika, M. B. N.; Shafi, K. M.; Sowdhamini, R. Decoding Systems Biology of Plant Stress for Sustainable Agriculture Development and Optimized Food Production. *Prog. Biophys. Mol. Biol.* **2019,** *145*, 19–39.

Sharma, T. R.; Devanna, B. N.; Kiran, K.; Singh, P. K.; Arora, K.; Jain, P.; Tiwari, I. M.; Dubey, H.; Saklani, B.; Kumari, M.; Singh, J.; Jaswal, R.; Kapoor, R.; Pawar, D. V.; Sinha, S.; Bisht, D. S.; Solanke, A. U.; Mondal, T. K. Status and Prospects of Next Generation Sequencing Technologies in Crop Plants. *Curr. Iss. Mol. Biol.* **2018,** *27*, 1–36.

Song, Z. T.; Liu, J. X.; Han, J. J. Chromatin Remodelling Factors Regulate Environmental Stress Responses in Plants. *J. Integr. Plant Biol.* **2021,** *63*, 438–450.

Tapia, G.; Verdugo, I.; Yañez, M.; Ahumada, I.; Theoduloz, C.; Cordero, C.; Poblete, F.; González, E.; Ruiz-Lara, S. Involvement of Ethylene in Stress-Induced Expression of the TLC1. 1 Retrotransposon from Lycopersicon chilense Dun. *Plant Physiol.* **2005,** *138*, 2075–2086.

Tripathi, G.; Sharma, B. M.; Shasmal, J.; Tomar, U. K. Transposable Elements, 2014.

Unamba, C. I.; Nag, A.; Sharma, R. K. Next Generation Sequencing Technologies: The Doorway to the Unexplored Genomics of Non-Model Plants. *Front. Plant Sci.* **2015,** *6*, 1074.

Wang, D.; Qu, Z.; Yang, L.; Zhang, Q.; Liu, Z. H.; Do, T.; Zhu, J. K. Transposable Elements (TEs) Contribute to Stress-Related Long Intergenic Noncoding RNAs in Plants. *Plant J.* **2017,** *90*, 133–146.

Wetterstrand KA. DNA Sequencing Costs: Data from the NHGRI Genome Sequencing Program (GSP). www.genome.gov/sequencingcostsdata

Wicker, T.; Matthews, D. E.; Keller, B. TREP: A Database for Triticeae Repetitive Elements. *Trends Plant Sci.* **2002,** *7* (12), 561–562.

Wicker, T.; Sabot, F.; Hua-Van, A. et al. A Unified Classification System for Eukaryotic Transposable Elements. *Nat. Rev. Genet.* **2007,** *8*, 973–982.

Yasuda, K.; Ito, M.; Sugita, T.; Tsukiyama, T.; Saito, H.; Naito, K.; Okumoto, Y. Utilization of Transposable Element mPing as a Novel Genetic Tool for Modification of the Stress Response in Rice. *Mol. Breed.* **2013,** *32*, 505–516.

Yu, A.; Lepère, G.; Jay, F.; Wang, J.; Bapaume, L.; Wang, Y.; Abraham, A. L.; Penterman, J.; Fischer, R. L.; Voinnet, O.; Navarro, L. Dynamics and Biological Relevance of DNA Demethylation in Arabidopsis Antibacterial Defense. *Proc. Natl. Acad. Sci.* **2013,** *110*, 2389–2394.

Zeng, F. C.; Zhao, Y. J.; Zhang, Q. J.; Gao, L. Z. LTRtype, an Efficient Tool to Characterize Structurally Complex LTR Retrotransposons and Nested Insertions on Genomes. *Front. Plant Sci.* **2017,** *8*, 402.

CHAPTER 4

Annotation of Transposable Elements in the Evolution of Plant Genome Biology

ASHOK SINGAMSETTI[1,3], KARTIK MADANKAR[1], and
P. LAKSHMI PADMAVATHI[2*]

[1]Department of Genetics and Plant Breeding, Institute of Agricultural Sciences, Banaras Hindu University, Varanasi 221005, India

[2]Department of Biosciences and Sericulture, Sri Padmavati Mahila Visvavidyalayam (Women's University), Tirupati 517502, Andhra Pradesh, India

[3]Department of Genetics and Plant Breeding, M. S. Swaminathan School of Agriculture, Centurion University of Technology and Management, Paralakhemundi 761211, Odisha, India

[*]Corresponding author. E-mail: mlpadmavathi67@gmail.com

ABSTRACT

Transposable elements (TEs) is a mobile genetic material made up of repetitive DNA sequences considered as the main drives for genome evolution in most of the plants. Unraveling of multifaceted activities of TEs is the major breakthrough for the rise in functional genomics and other large-scale omics analyses. In general, the plant genomes are significantly dynamic, mainly due to the nature, types, and behavior of TEs. The class and forms of TEs and their pattern of movement in a plant genome, association with unique chromosome features and rearrangements in the genome and intrinsic equilibrium between expression and repression were major and fundamental aspects to study the effect of TEs on genome evaluation. Even a minimal

Plant Transposable Elements: Biology and Biotechnology, Deepu Pandita, Anu Pandita, (Eds.)

quantity of the heredity material (DNA) is enough for detecting the TEs with high copy number and to create specific primers for transposition experiments by which active TEs from different plant species could be isolated and make possible for TEs to continue to co-evolve with their hosts. The new technologies and methods are emerging in plant genomics and bioinformatics. There are various genome annotation methods available, however, because of their large copy number and the complicated structures induced by the insertions of new TE sequences into old TE sequences, identification of TEs is rather difficult. At present, there are some computational tools available for transposome annotation from family identification to repetitive TE copies in the plant genome. Based on the technique involved in the identification of TE sequences, the annotation methods were categorized into four groups viz., de novo, structure-based, homology-based, and comparative genomics. In this chapter, we tried to make an attempt to elucidate the available TE detection methodologies, identification of class and type or form of TEs, computational tools or combination of tools and developing models by machine learning techniques to identify TE sequence copies in the whole genome more precisely.

4.1 INTRODUCTION

Transposons/transposable elements (TEs) share a major portion of abundant non-genic component of plant genomes. Barbara McClintock identified the genes those prone to mobility during late 1940s, later called them as "transposable elements" and added to the existing concepts of genome structure and functions (McClintock, 1950). It took nearly a decade to explore and understand the importance of transposons and its applications. The identification of mobility nature of transposons leads to discovery of "selfish" or "parasitic DNA" hypothesis. The hypothesis cleared the suspense of ability to self-reproduce and to induce mutations in many of sequences within the genome organization. Although these self-replicated genome sequences have no role in the plant cells, these were maintained due to replication's capacity to disseminate copies inside and across genomes (Orgel and Crick, 1980; Doolittle and Sapienza, 1980). In such a manner, spreading copies themselves ("selfishness") stabilize and occasionally enhance their own viability at the disbursement of the gene of host plants (Werren, 2011; Burt and Trivers, 2006). Apart from the selfish and deleterious nature of TEs, there are several proven theories that strengthen the involvement of TEs, and their activities may occasionally gain a fitness benefit to their plant hosts.

Although the TEs are behaving selfishly, the majority of scientists are able to recognize their significant role in maintaining the genome integrity, diversification as well as evolution of the genetic makeup catalogue of plants. It was made possible by the advanced genomic technologies like next-generation sequencing (NGS) and other high-dimensional machine learning approaches.

In order to gain a better knowledge of genome functions and the role of TEs in the evolution of plant genome organization, a highly reliable and precise annotation of these elements is very crucial. Transposon origin, biogenesis, regulatory mechanisms, and functions in genomic integrity, gene architecture, and gene expression control require complete and precise annotation. Till date, several bioinformatics software packages have been released to address the annotations in TE, but these vary in few important aspects, such as speed of the analyses, precision, and reliability. The tools have limitations and miss appropriate annotations of transposons in plant genomes (Sreeskandarajan et al., 2014; Ye et al., 2014, 2016). The chapter primly focused on the significance of TEs as the enhancers of plant genome dynamics and evolution, and to further disseminate the research issue on the annotations of TEs contributed toward the evolution of plant genome biology. The first half of the chapter concentrated on basic information about TEs, types of TEs, and annotations that modify genome structure, while the second half focused on bioinformatics-based tools and algorithms that can make automatic predictions and judgments to discover TE annotations.

4.1.1 STRUCTURE AND TYPES OF TES

Transposable elements are nothing but the nucleotide sequences which were mobile throughout the genome (mostly within) through the mechanism of transposition. The major portion of plant genome is consisting of repetitive sequences including tandem, inverted, and interspersed repeats. TEs are a common subject of research among repeated elements because numerous evidences suggest that they were engaged with the epigenetic components as they provide flexibility to plants in response to the altering environments (Gao et al., 2016; Rey et al., 2016). They generated mutations by moving within the genome and noticeably increase the number of their replicas with the ability of replication. TEs are self-mobilizing elements that had a significant influence on plant evolution at various periods of evolutionary decline (Smit, 1999; Lander et al., 2001; Jurka et al., 2007;

Huang et al., 2012; Bourque et al., 2018). Barbara McClintock first discovered TEs in the maize plant genome, describing them as "mutable genes" or "unstable genes," and the discovery enabled her to receive the Nobel Prize (McClintock, 1950). TEs are commonly known as "jumping genes." The lengths of TE may vary from 10 nucleotides to thousand nucleotides. TEs are usually non-mutagenic, but as genomics progresses, their ability to get mutated is becoming more apparent through some processes viz., genetic and epigenetic. TEs play a key role in the corresponding processes, such as gene regulatory networks, phenotypic modification, and the creation of adaptive genetic variants (Wei and Cao, 2016). The TEs transformation into a newer genes or promoter region is in accordance with their function as an "engines of plant genome evolution" (Galindo-Gonzalez et al., 2017; Vicient and Casacuberta, 2017).

TEs were categorized in hierarchical manner (depicted in Fig. 4.1) based on its position and nature (Wicker, 2007). According to replicative processes and coding regions entangled in element replication, there are two primary groups of TEs, Class I and Class II (Fig. 4.1) and both of them comprise the majority of TEs discovered in plant genomes based on the way of their mobilization. Both of these classes replicate via RNA and DNA intermediates. Based on structural topographies and life cycle, Class I retrotransposons elements are subdivided into short interspersed nuclear element (SINEs), long interspersed nuclear element (LINEs) and long terminal repeats (LTRs), whereas, Class II DNA transposons are subdivided into terminal inverted repeats elements (TIR), Helitron, Crypton, and Maverik. TIR transpose through a "cut-and-paste" mechanism (Feschotte and Pritham, 2007), while other elements of class II, that is, Crypton, Helitron, and Maverik usage show replicative mechanisms (Grabundzija et al., 2018; Thomas and Pritham, 2015; Kapitonov and Jurka, 2001). This categorization approach had been applied successfully in numerous plant species (Wicker et al, 2007; Llorens et al., 2011; Domingues et al., 2012; Xu et al., 2017; Beule et al., 2015; Ming et al., 2016). Among the retrotransposons, LTR is the most common and abundant (Grandbastien, 2015; Gao et al., 2012), and these may contribute above 80% in plant genomes, as in barley and wheat genome (Rahman et al., 2013).

Annotation or genome annotation, a process of identifying the functional elements on a genome's sequence can be processed further. It is a necessary task to track the sequences of unknown function in DNA sequencing. It is an in silico approach that comprises of explaining the function of a predicted gene. The technique of genome annotation is led

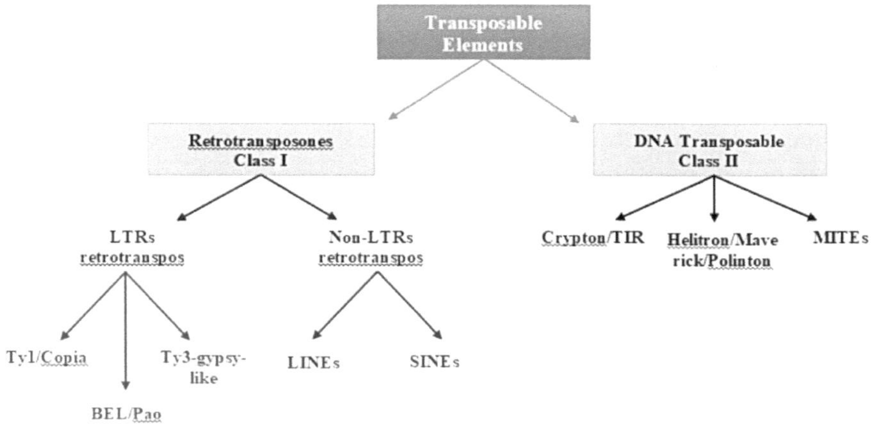

FIGURE 4.1 Classification of transposable elements.

by genome assembly using either a *de novo* technique or a reference genome-based method. Broadly, the genome annotation is divided into three categories, such as nucleotide, protein, and process-level annotation. Annotation at the nucleotide level aids in determining the physical position of DNA sequences along with its component positions viz., genes, RNAs and transposons, whereas the second step is to figure out what genes might do. The third aims to detect the different genes interactions and an efficient functional annotation assembling through the identifying the pathways involved (Pablo et al., 2018).

4.2 WHY ANNOTATION OF TES IS DIFFICULT?

The DNA sequences of TEs are extremely diverse within each class and they evolve rapidly (Bourque et al., 2018; Wicker, 2007; Arkhipova, 2017). Each transposon class has distinct sequences with distinct characteristics. Furthermore, TEs were detected in a wide range of abundances and activity levels, as well as levels of sequence degradation (Hua-Van et al., 2005; Smit, 2012). Once the TEs are inserted into plant genomes, each region is open to mutations and rearrangements, which include deletions, insertions, and truncations. Therefore, annotation of TEs is challenging and time-consuming practice due to its diverse natures and sequences repeats which are extremely diverse across the plant genomes (Bourque et al., 2018; Huang et al., 2012). The precise identification of TEs is an essential for its annotation (Yandell and Ence, 2012). With the advancement and development of whole-genome sequencing, the biological effect of TEs has gradually increased. The momentum of genome annotation can be accelerated by annotating the TEs. Comprehensive and precise annotation of transposons is necessary for complete understanding of its origin, biogenesis, regulatory mechanisms, and functional roles in the genome.

Identification of TEs is an epic task to annotate TEs in a plant genome for the researchers. The creation of a well-defined reference database with the required TE sequences is important in annotation. The family is the most basic level of TE categorization, which assigns TEs based on a consensus sequence from the original progenitor (Wicker, 2007). A multiple alignment of genomic copies, known as seeds, may be used to recreate this consensus sequence. Furthermore, multiple alignments of genomic copies can be employed to create a family-specific hidden Markov model (HMM). Many algorithms use consensus TE sequences and HMMs to annotation of TEs. Databases like Repbase (Bao et al., 2015) and Dfam (Wheeler et al., 2013) are usually used to store consensus TE sequences. Seed alignments and precise sequence generation models are essential for regeneration of evolutionary history of TEs. Years of manual curation had gone into creating the high-quality consensus libraries (Stitzer et al., 2019; Hubley et al., 2016; Lerat et al., 2003). Improvements in the genome-sequencing era will open up the possibility of producing thousands more at a rapid rate (Koepfli et al., 2015; Lewin et al., 2018). With the improvement of next-generation sequencing (NGS) technologies, the quality of genome assemblies, particularly in the case of TEs, is improving (Chang and Larracuente, 2019). This advancement is a need for the development of algorithm in the form of software that can correctly recognize the diverse panel of TEs present in plant genome sequences and furthermore annotate them.

4.3 METHODS OF ANNOTATIONS ON TEs

The attention toward developing new methodologies for computational analyses of TEs was gradually enhancing as the evident of TEs on genome sequencing, assembly, alignment, and annotations in many studies, as well as their contribution in genome organization and its evolution. The three important methods viz., *de novo*, homology-based, and structure-based methods are available for TE identification and their annotation (Han and Wessler, 2004; Bergman and Quesneville, 2007).

4.3.1 DE NOVO METHODS

De novo methods can be used to discover new TEs by using the inherent nature of repetition of hidden mobile DNA sequences in the genome. These methods are carried out without the involvement of any prior knowledge of the structure or similarity of TEs in sequences with known TEs. Although these *de novo* approaches are capable of detecting both novel and known TEs in the genome, detection results generally contain a combination of TEs and non-TE repetitions, that necessitate further categorization and fine filtering. These techniques are often highly reliant on both the sequencing and assembly processes as they rely on the quality of sequence data that has been assembled. However, the recently developed new methods that attempted to bypass these complexities through the usage of sequencing reads. The most standard technique in the *de novo* approaches is to detect similarly matched sequences at various positions in a self-genome followed by clustering the pairings to generate repeat families. Several processes, such as tandem repeats, satellites, and segmental duplication, as well as TE sequences generate repeats which can be recognized by *de novo* methods, as these are not specific to TEs. The primary problems in *de novo* techniques are sorting TEs among many repetition classes and identifying the unique TE families.

The fragmented nature of TEs, interspersion of the TEs into other repeat classes, and resolving closely related TE families are the probable complexities for the difficulty in the annotations of TEs. Large insertions or deletions interrupt the nested and partial matches among fragmented repeat instances. Formation of degenerated copies of the genome and aggregation of nested repeats into "meta-families" with distinct repeat families restricts the accurate annotation of TEs in these methods.

Basically, many of the *de novo* TE methods use classical computational approaches to identify repeat instances in the plant genome, such as pairwise similarity searches. Reputer, RepeatMatch, and RepeatFinder has been used earlier for detecting the exact repeats. Fewer, quicker techniques based on k-mer methods, such as BLAST to anchor pairwise similarity searches, were available. Multiple alignments of all the repeat copies in a cluster (RECON software; Bao and Eddy, 2002) and obtaining complete copies of the repeats by identifying families from pairs of repeats are the next steps in de novo TE detection. This phase involved finding the longest TE sequences in a cluster and creating a filter based on their presence. The outstanding example of this technique is the GROUPER program from the BLASTER suite (Quesneville et al., 2003, 2005).

4.3.2 STRUCTURE-BASED METHODS

These techniques look for specified structural sequences in the plant genomes of existing TEs. They can locate specific forms of TEs, however, they have an issue with low copy members and/or non-TE repetitions in their detection outputs, which leads to more TEs being detected. These methods differ from de novo repeat finding strategies in that they detect specific TE model architecture rather than just offering the transposition results.

Several structure-based techniques have recently been used to identify LTR retrotransposons by looking for common structural characteristics in this subclass of TE, such as primer-binding sites (PBSs), LTRs, target site duplications (TSDs), and polypurine tracts (PPTs). The tool LTR STRUC used alignment and locating local repetitions within a user-specified length using a heuristic seed-and-extend approach. The major limitations of this method are (1) this method detects only the TEs within the same contig and (2) this method is unable to detect the incomplete LTR elements like most of the other newly developed methods for discovering TE. Later, SMaRTFinder platform for efficient searches of structured sets of motifs in DNA was developed by Morgante et al. (2005). This enables for motifs to be edited and motifs to be removed from the overall compound pattern at a user-specified level. This technique detected the mutation that happened as a result of allowing variants on the structured motif.

4.3.3 HOMOLOGY-BASED METHODS

The most popular methods for identifying homology with known TE protein-coding sequences in order to discover new TE families are the homology-based methods. The researchers were able to capture genuine TEs and proceed to classify novel TEs, even the ones that only exist in a single copy in the plant genome. They examined the sequence similarities using tools, such as BLAST (Altschul et al., 1990), RepeatMasker (Smit et al., 2013), and HUMMER3 (Eddy, 2011) among putative and known TEs to find TEs hidden in plant genomes. They are capable of identifying authentic TEs in plant genomes, even if they just have a single copy. However, they are unable to detect novel TEs, and their detection findings frequently include sequences that do not contain a full-length copy or the whole structure of well-defined TEs. Homology-based protein sequence techniques outperform de novo approaches in terms of capturing a high number of TE sequences.

Protein homology-based techniques offer a particular advantage over de novo repeat finding approaches in that they take advantage of existing information obtained from a large number of previously known TE sequences. These techniques have certain drawbacks, such as bias in favor of identifying already identified families and significant protein homology-based TEs. These techniques are inapplicable to a few TE classes which are completely made up of noncoding sequences, such as SINEs (short inverted repeat transposable elements) and MITEs (miniature inverted repeat transposable elements). Methods for identifying TEs are frequently applied to fully assembled genomic sequences, however, preliminary genome resources, such as BAC sequences have also been discovered (Mao et al., 2000).

The software HMMER (Durbin et al., 1999) is an alternative to these techniques that predicts open-read frames (ORFs) from the PFAM database using profile hidden Markov models (HMMs) (Bateman et al., 2002; Berezikov et al., 2000; Rho et al., 2007). In general, these detection techniques need more research into structural features in order to get a complete reference sequence. Caspi and Pachter (2006) proposed a novel approach for identifying new TE families and occurrences that does not rely on structural or homology characteristics. These studies focus on the fact that transposition results in massive insertions, which may be seen in various sequence alignments. This technique identifies insertion regions (IRs) in orthologous genome sequences whenever multiple alignments are broken by a substantial insertion in one or more organisms.

Since 1994, a variety of computational techniques for detecting TE have been introduced. Several of these tools are discussed in detail in this chapter. A slightly modified form of the categorization of computational techniques for finding TE is included in Table 4.1.

TABLE 4.1 Classification of Computational Methods for Locating TEs.

S.No.	Methods	Classification criteria
1	Library-based methods	A library-based program looks for duplicates of known repeats that are already in a database and determines a sequence for them
		RepBase, most comprehensive database, which can be noted as a stock of manually annotated repeats
2	Learning-based methods	Is based on the idea that the nucleotide sequences of TEs are different from the rest of the genome
		Machine learning techniques are used to create models that differentiate TE from the rest of the genome in the development of tools based on these approaches
3	Signature-based methods	A signature-based technique examines a set of distinct features of each TE class
		A poly-A tail, target site duplication, long terminal repeats, terminal inverted repeats, and/or a hairpin loop are some of distinct features
4	Comparative genomics-based methods	Methods based on comparative genomics take use of TE's species-specific features
		When two genomes from closely related plant species are examined, TE should be present in one but lacking in the other
5	De novo methods	It is a new technique based on the repeating nature of TE.
		The majority of de novo techniques depend on either self-comparison or k-mer counts
6	Consensus methods	This approach is made up of a number of different tools.
		A pipeline of library-based approaches, learning-based techniques, and de novo approaches make up computational tools

4.4 DATABASES FOR ANNOTATIONS OF TEs

Databases are the repository of different types of TEs present in plant kingdom. There are many types of databases ranging from general databases having information about all the classes of TEs to database having specific class of TEs information. More specific, there are also some databases representing specific plant TEs. Here, we are focusing on two important general databases, RepBase and Dfam, while remaining other specific databases are covering in next section of this chapter.

RepBase (Bao et al., 2015) is a useful resource for TEs and other forms of repeats in eukaryotic genomes. For practically all eukaryotic genome sequence studies, it serves as a well-curated reference database. It has been used to compare the performance of TEs in terms of annotation to various curated databases. Currently, the Genetic Information Research Institute (GIRI) maintains it, and its official website is at http://www.girinst.org. RepBase now has over 38,000 sequences from various families and subfamilies of eukaryotic genomes. Abou 70% of them are full consensus sequences that have never been published before, and 30% are sample sequences derived from specific loci, 90% of families or subfamilies are derived from 134 species, with the remaining 10% made up of repeats from another 700 species. The process of identifying TEs and reconstructing family consensus sequences takes a long time. Furthermore, identifying TEs is a difficult process that changes depending on genome size. Several scholars have identified tools to annotate TE, including RECON, RepeatScout, PILER, RepeatModeler, REPCLASS, REPET, and PASTEC. These tools use RepBase background to annotate TEs in plant genome.

Dfam (Wheeler et al., 2013) is a set of multiple sequence alignments, each comprising a set of representative species in a given transposable element family. Its primary goal is to enhance the homology-based annotation step through hidden Markov models (profile HMMs). This allowed for better discovery of distant homologs of recognized families. Sensitivity is even more important in this technique since copies of earlier TEs with an accumulation of mutations might be exceedingly difficult to identify. Profile HMMs (Krogh, 1998; Durbin et al., 1998) are machine learning techniques and complex probabilistic models that capture information of position-specific conservation. The Dfam website (https://dfam.org) provides information on each family as well as genome annotations for a set of core genomes. Dfam maintained data of 347 species with 6915 curated families and 273,655 total families. During November 2020, Dfam released its new version 3.3 with a lot of upgradations, a significant step in the expansion of database.

TE discovery and annotation methods that detect TE repeat instances can be used based on low sensitivity (Bao and Eddy, 2002; Quesneville et al., 2005). The TE identification process is critical for completely annotating TE occurrence in plant genomes. TE detection begins with the assembly of de novo or a reference set of TE sequences detected using the process outlined above, and is continued by the resolution of the consensus sequence, categorization of the TE subfamilies, and some manual curation. The growing importance of TEs in plant genomic sequences has resulted in the development of a plethora of sophisticated tools and methodologies for TE detection and annotation in TEs. Some of the computational tools and algorithm were briefly explained hereunder:

4.5 GENERAL REPEAT ANNOTATORS

A significant percentage of plant genomes contain repetitive sequences. These sequences play a major role in evolution (McClintock, 1984; Orgel and Crick, 1980; Doolittle and Sapienza, 1980). There are three types of repetitive sequences: (1) local repeats, which comprise simple sequence repeats and tandem repeats, (2) families of scattered repeats, which include predominantly transposable elements, and (3) segmental duplications. With the growing number of plant genome sequencing efforts, identifying repeat families in plant genomes is becoming increasingly important. Annotation of repetitive sequences is difficult, because it is covering a large portion of genome and causes confusion during large-scale gene annotation.

RECON (Bao and Eddy, 2002)—an automized process for identifying the de novo repeats and the classification of its sequences basing on the multiple sequence alignment information as it is relatively easy. The RECON algorithm is a set of C programs, and Perl scripts are available at http://www.genetics. wustl.edu/eddy/recon including a demo and more related materials. The RECON algorithm had become the dominant tool in the past decades for *de novo* identification of repeat family in sequenced plant genomes. But, its main disadvantages include difficulty in defining element boundaries and computation for large genome size.

RepeatScout (Price et al., 2005) is a k-mer method for detecting repeating sequences. This technique creates a set of repeat families by selecting high frequency/k-mers as seeds and extending each seed to a significantly greater consensus sequence, which is then aligned with its locations in the plant genome. The algorithm can be easily downloaded from website, http://www.cse.ucsd.edu/groups/bioinformatics/software.html. RECON is slower than RepeatScout. As a result, RepeatScout is ideally suited for plant genomes.

RepeatModeler (Hubley and Smit, 2008): It is a repeat detection and classification algorithm that assists in the refinement of TE boundaries, its classification, and building of nonredundant TE libraries. RepeatModeler, on the other hand, is unable to generate an entire, no redundant library of full-length consensus sequences. RepeatModeler was unable to find a single contiguous consensus sequence for a specific TE family. Flynn et al. (2020) propose RepeatModeler2, a convenient update of this program to resolve this problem. This open-source package is available at https://github.com/Dfam-consortium/RepeatModeler. It is a multithreaded program that uses the NCBI BLAST engine.

Red (Girgis, 2015) is a C++ program capable of rapid detection of repetitive sequences. This software is not dependent on other programs and is easily available at http://toolsmith.ens.utulsa.edu. The input file comprises repetitive sequences. Both assembled and unassembled sequences can be handled by Red. This program uses a hidden Markov model (HMM) to scan the whole plant genome for repetitions. Red searches the genome for k-mers that appear at least three times. The detected repeated areas and probable nonrepetitive sections are utilized to train an HMM, which searches the whole genome for repeats. The HMM is trained using supervised learning, which eliminates the need for manually annotated data. Across each genome, the labeling and training operations are carried out automatically. It is the first repetition detection algorithm that can label its own training data and train itself on each genome independently (Girgis, 2015).

Generic Repeat Finder (GRF) (Shi and Liang, 2019) is an open-source C++ program that can detect multiple types of repeats including terminal direct repeats, terminal inverted repeats, and interspersed repeats. GRF is freely available at Github. The GRF method is based on quick, comprehensive numerical calculations in combination with optimized dynamic programming methodologies (Shi and Liang, 2019).

4.6 LTR RETROTRANSPOSONS

LTR retrotransposons are Class I TEs that is signalized by long terminal repeats (LTRs) ranging from 100 bp to 25 kb in size directly next to the internal coding site. They replicated via reverse transcription mechanism and deployed to other parts of the plant genome using freshly generated cDNA. It accounted for roughly 75% of the maize genome (Baucom et al., 2009). LTR retrotransposons are divided into three types: Ty1-copia-like, Ty3-gypsy-like, and BEL-Pao-like (Kumar and Bennetzen, 1999; Neumann et al., 2019). The first two types are found in plant, animal, fungus, and protista genomes, whereas the third is found solely in animal genomes (Copeland et al., 2005; Wicker et al., 2007).

For detecting the LTR retrotransposons, alignment of known elements in a database of the desired plant genome is generally used which helps in the identification of the elements in the database correctly, but they cannot find items that are not associated with or are not in the database. On the other side of the coin, some common structural features (signals) present in these elements, such as polypurine tract (PPT), long terminal repeats (LTRs), target site repeats (TSRs), primer-binding sites (PBSs), and TG-CA box, as well as reverse transcriptase (RT), RNaseH (RH), and Integrase (IN) sites. These findings enabled the discovery of LTR components. However, this technique for detecting LTR retrotransposons are currently relatively limited. LTR STRUC, LTR FINDER, and LTR par are a few examples. The traditional approach of locating LTR retrotransposons basing on the comparison of input nucleotide sequences to existing reverse transcriptases has three major inherent disadvantages:

1. biased search for reverse transcriptases sequences,
2. it overlooks LTR retrotransposon elements that lack a reverse transcriptase, and
3. identifying putative sequences is time-consuming. Although many bioinformatics methods have been developed for de novo LTR identification from plant genome sequences, for comparative research on automated and standardized software tool for both LTR identification and annotation would be beneficial and desirable.

LTR_STRUC (McCarthy and McDonald, 2003) is an earliest data-mining platform to detect the LTR retrotransposons. It is a Windows-based visual C++ program that detects and analyzes LTR retrotransposons within the genome databases automatically for structural feature of these elements having biological interest. LTR STRUC's search algorithm consists of four primary steps: (1) locating an initial pair of matches; (2) alignment of genomic areas near to the first matches; (3) detection of putative LTR end sequences; and (4) identification of exact end points. The following information is included in the analytic output file: Source contig's name; Score for the most recent hit; Its position; element lengths, LTR lengths, Contig lengths, and ORF lengths; TSRs, complete transposon nucleotide sequences; PPT, dinucleotide terminating the LTRs; LTRs, PBS, the transposon's orientation inside the contig; All ORF sequences, intra-element percent identification of LTRs, and alignment of putative LTR.

LTR par (Kalyanaraman and Aluru, 2006) may be executed in either serial or multiprocessor mode. LTR par is more accurate and faster than LTR STRUC, while LTR STRUC is more specific than LTR par. The program supports a strong parameter set that includes structural limitations as well as quality controls. Furthermore, LTR par offers greater versatility by giving the user more control. The presence/absence of TSRs and TG...CA motifs can be weighted by the user, and the software can output its predictions at varying confidence levels based on the user's weights. If a user is looking for LTR retro elements in a newly sequenced genome, they can experiment with different weights and scoring parameter values and see how the predictions vary before settling on an appropriate set of parameters. On genomes where the two LTRs of each retrotransposon are expected to be highly conserved, the algorithm provides a *de novo* full-length LTR retrotransposon prediction. Other structural characteristics typical of an LTR retrotransposon can be added specificity to the software in its current state: Special-purpose sequences, such as PPT, PBS, gag, pol, and env are commonly found in the genomic area between two LTR sequences, and recognizing these patterns is critical for establishing the biological identification of each prediction.

LTR_FINDER (Xu and Wang, 2007) LTR FINDER uses common structural patterns in DNA sequences to accurately predict the sites and organization of full-length LTR retrotransposons. This technology is the first web server for detecting LTR retrotransposons and is capable of scanning large-scale sequences quickly. Through a multirefinement process, LTR FINDER uses quick algorithms to create reliable LTRs and anticipate accurate element boundaries. It also detects essential enzyme domains, which improve the accuracy of autonomous element predictions. LTR_FINDER is a free tool available at http://tlife.fudan.edu.cn/ltr_finder/. However, LTR_FINDER allows only one-CPU for all jobs, limiting the scalability of this program. To overcome this limitation, Ou and Jiang, 2019 developed "LTR_FINDER_parallel.pl," which will chop the genome into small pieces to run LTR_FINDER in parallel.

LTR harvest (Ellinghaus et al., 2008) is another popular platform for *de novo* LTR detection. It comes with precompiled binary code makes it user-friendly. LTR harvest employs a different set of methods and characteristics than LTR par to construct the same LTR model. It enables the user to include biological variables, such as LTR length and distance, TSD length, and motifs into big data sets quickly and with customizable parameter settings. LTR harvest can deal with whole-genome shotgun (WGS) sequencing data, which is typically in the form of many unordered contigs, because it takes sequences in several FASTA formats. It is an open-source software available at http://www.zbh.uni-hamburg.de/ LTR harvest.

MGEScan3 (Lee et al., 2016) is a Galaxy-based program that combines MGEScan_LTR (Rho et al., 2010) and MGEScan-non-LTR (Rho and Tang, 2008) available at http://mgescan.readthedocs.org. Thus, it detects non-LTR retrotransposons along with both LTRs. MGEScan-long terminal repeats and

MGEScan-non-LTR have been effectively employed in eukaryotic genome sequences to identify LTRs and non-LTR retrotransposons. These tools, on the other hand, do not have user-friendly interfaces and are not well suited for data visualization in general data formats. MGEScan is a user-friendly solution that integrates these two applications with a Galaxy workflow system that is accelerated on compute clusters using MPI and Python threading. Researchers can use MGEScan and Galaxy to find transposable elements through a graphical user interface with ready-to-use procedures. In public genome browsers, MGEScan also visualizes the custom annotation tracks for mobile genetic elements. MGEScan can be used in four different ways: as a command-line tool, a Galaxy Toolshed, a Galaxy-based web server, and an Amazon cloud virtual cluster.

LtrDetector (Valencia and Girgis, 2019) is a recently developed tool for *de novo* LTR annotation https://github.com/TulsaBioinformatics Toolsmith/ LtrDetector. LtrDetector employs a revolutionary technology based on k-mer distributions that enables it to get high-quality results with relatively simple methods. It is simple to set up and utilize. It is not species-specific, and it works well on genomes of various sizes and repetition content using its default values. It operates efficiently on a standard personal computer and is automatically prepared for parallel processing. It offers a visualization tool for k-mer scores to make manual review of the highlighted items easier. These properties make LtrDetector an appealing tool for future long terminal repeat retrotransposons annotation studies.

LTR retriever (Ou and Jiang, 2018) is an open-source software, easily available from GitHub at https://github.com/oushujun/LTR retriever. It is a command-line Perl software having multithreading ability that can retrieve LTR-RT candidates from LTR finder, LTR harvest, and MGEScan-LTR to construct high-quality with compact LTR libraries for plant genome annotations or its exploration. LTR retriever accepts FASTA-formatted input plant genomic sequences file. This software can manage both fragmented and gapped regions of genome, which is useful for annotating draft genomes of selected plants. LTR retriever has been specially design for plant genomes, but can vary for other species. This tool generates an output file that contains a library of high-quality, comprehensive yet nonredundant LTR exemplars that are being used to detect using RepeatMasker software. A redundant library also becomes available for additional research. A summary table with LTR-RT coordinates, TSDs, length, motifs, insertion time, and LTR superfamilies is also generated. The application also outputs in gff3 format, which is useful for further examination.

GRF (Shi and Liang, 2019) detects terminal direct repeats (TDR), which provides a slightly modified version of LTR FINDER to find LTR retrotransposons. The algorithm called generic repeat finder (GRF) is already explained in generic repeat annotators section.

4.7 NON-LTR (SINE AND LINE)

Long interspersed nuclear elements (LINEs) and short interspersed nuclear elements (SINEs) belong to a class of non-retrotransposons (SINEs) whose retrotransposition is dependent on proteins encoded by a partner LINE (Dewannieux et al., 2003; Kajikawa and Okada, 2002; Boeke, 1997; Weiner et al., 1986; Singer, 1982). SINEs are nonautonomous retroelements that have a composite structure and can even be up to 500 nucleotides long. Like LINEs, they are frequently terminated at the 3′ end by a poly (A) stretch, or simple sequence motifs and flanked by TSDs (Sun et al., 2007; Okada, 1991a, 1991b; Weiner et al., 1986; Galli et al., 1981). The majority of SINEs are obtained from tRNAs. Thus, they are transcribed from degraded internal promoters by RNA polymerase III. LINEs, on the other hand, are self-contained, but SINEs are noncoding and depends on the reverse transcriptional machinery of LINEs. SINE retrotransposition necessitates active LINEs and the LINE reverse transcriptase's sequence-dependent recognition of the SINE 3′ terminus.

SINEs are made up of two or more modules, often the head, body, and tail. The cellular RNA generated by pol III forms the 5′-terminal head. In reverse transcription, SINEs containing such a region imitate LINE RNA (Okada, 1997). The body's origin is either unknown or descended from a companion LINE. The 3′-terminal tail is a variable-length simple sequence repeats usually degenerate. Furthermore, two SINEs can unite to form a dimeric SINE, resulting in the formation of a new SINE family. Simple SINEs are those that simply have the head and tail, whereas complex SINEs are those that are dimeric, trimeric, and so on. Ohshima and Okada (2005), Kramerov and Vassetzky (2011) investigated various characteristics of SINE structure, biology, and evolution. SINEs are mobile genetic components that infiltrate higher eukaryotes' genomes (exceeding 10 percent of some genomes). Despite the fact that they may be damaging to the cell, SINEs are an important component of genetic diversity, including regulatory elements for gene expression, alternative splice sites, polyadenylation signals, and sometimes even functioning RNA genes (Makalowski, 2000; Belancio et al., 2008; Okada et al., 2010; Kramerov and Vassetzky, 2011).

SINEBase (Vassetzky and Kramerov, 2013) is a database (http://sines.eimb.ru) of SINEs. Its website may be used in two ways: (i) to browse the database of SINE families and (ii) to utilize specially built tools to analyze possible SINE sequences. Its purpose is to be a resource for scientists working on transposable elements and biologists exploring nucleic acid sequences. Individual SINE sequences access to identify SINE families and/or analyzed using SINEBase's toolset, which may be thought of as a compendium of SINEs.

SINE-Finder (Wenke et al., 2011) is a Python program designed to identify tRNA-derived SINEs. There is no installation required for this program and parameter settings are straightforward. However, only the forward strand of the input sequences could be searched without errors, which requires manually generating and providing the reverse complement strand for the complete search. SINE-Finder only identifies tRNA-type SINE while unable to identify rest of the SINE-type's, that is, 7SLRNA and 5SRNA. The SINE-Finder tool can be obtained from http://tudresden.de/die_tu_dresden/fakultaeten/fakultaet_mathematik_und_naturwissenschaften/fachrichtung_biologie/botanik/zellmolbiopflanzen.

MGEScan3 (Lee et al., 2016) is a Galaxy-based program used to detect *de novo* non-LTR retrotransposons. The details of MGEScan3 are already mentioned in LTR section.

SINE_Scan (Mao and Wang, 2017) is a Perl program representing the latest development of *de novo* SINE identification methods which is based on SINE-Finder. SINE_Scan can identify all three known types of SINEs, which are tRNA, 7SLRNA, and 5SRNA. SINE_Scan allows the users to start their analyses from different steps and generates multisequence alignment (MSA) files for each candidate for manual curation. Its code is available freely at http://github.com/maohlzj/SINE_Scan.

4.8 TIR/MITE

MITEs (miniature inverted repeat transposable elements) belong to the subcategory of DNA transposable elements which share similar characteristics of DNA transposons. MITE, like retrotransposons, has short conserved terminal inverted repeats (TIRs) but high copy numbers in plant genomes (Fattash et al., 2013; Wessler et al., 1995). As shown in Figure 4.2, a typical MITE consists of an internal sequence and a TIR pair of 10 nucleotide sequences. The inner sequence is flanked by the TIR pair, and the entire MITE is then surrounded by a TSD of length 2–10 nucleotides. MITEs can be anywhere between 50 and 800 nucleotides long. Because MITEs do not encode proteins and have no coding potential for transposition, they are nonautonomous TEs. They do, however, frequently appear in genic regions' introns or at gene ends in intergenic sections (Lu et al., 2012; Wright et al., 2003).

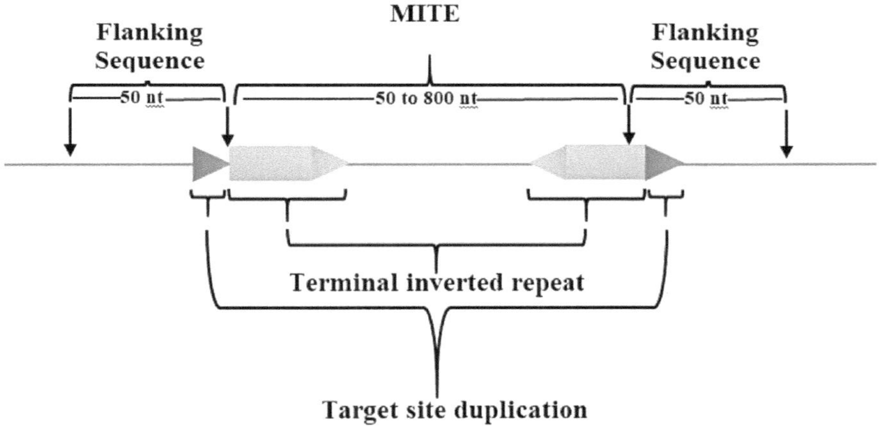

FIGURE 4.2 Structure of miniature inverted repeat transposable element.

MITEs are found in abundance in the genomes of eukaryotic organisms, including plants. Their movement in plant genomes can alter gene architecture and functions. Stowaway, an MITE discovered in potatoes, was reported to produce phenotypic diversity in tuber skin color by inserting the flavonoid 3′, 5′-hydroxylase gene into the first exon (Momose et al., 2010). MITEs contribute to genomic diversity, emergence of new gene and variants of mRNA transcripts in rice, according to a genome-wide MITE investigation (Oki et al., 2008). Genes far from MITEs were discovered to have greater expression than those of adjacent or including MITEs in *Oryza sativa* (Lu et al., 2012). They are known to have a dynamic role in genome evolution of *Brassica* according to a comparative investigation of MITEs in *Brassica oleracea, B. rapa, and Arabidopsis thaliana* (Sampath et al., 2014).

Detection of MITE encounters three major challenges genome-wide, in general: (1) quick, comprehensive, and precise identification of putative MITE sequences in plant genomes; (2) filtration of false-positives from presumed MITE candidates; and (3) efficient categorization of analogous MITE sequences into separate MITE families. MITE discovery on a genome-wide scale can assist in understanding their origins, mode of transposition, and regulatory functions in plant genome organization and gene structure, expression, and regulation (Oki et al., 2008, Jiang et al., 2004; Yaakov et al., 2013). Several computational approaches for detecting MITEs in plant DNA sequences have been developed on the basis of clearly defined structures of MITEs and sequence similarity among distinct MITE homologs.

P-MITE (Chen et al., 2014) is a database of plant miniature inverted transposable elements (MITEs) that comprise 3527 MITE families derived from 41 plant species and is designed to have the information of all these species. The P-MITE database download links were broken (http://pmite.hzau.edu.cn/download/), but it may help to investigate the origin and replication of MITEs, its derived short RNAs, and its functions in plant gene and genome evolution.

IRF (Warburton et al., 2004) is a multiplatform software designed to identify inverted repeat elements. It has five required parameters and 19 optional parameters. Some of the IRF parameters are quite technical for inexperienced users, such as matching/mismatching scores, indel penalty, and match/indel probabilities. The output of IRF software is a .dat file, with coordinates and other information, therefore, users need to take one more step to obtain the FASTA file from the IRF output.

MITE-Hunter is a Perl program developed by Han and Wessler (2010) and its installation requires formatdb, blastall, mdust, and muscle as prerequisites. The software blastall and formatdb are outdated, and have been replaced by the software package BLAST+. MITE-Hunter contains 17 parameter options, some of which are redundant. One of the best features for MITE-Hunter is that this software can take checkpoints. The S parameter allows the user to start the program from a specific step. Therefore, if there is anything wrong in the running process, users do not need to begin with starting. MITE-Hunter creates hundreds of intermediate files, some of which are rather huge, in the same folder. Users may have difficulty obtaining the final output and organizing files in the target folder as a result of this.

TIRvish is a program module in the Genome Tools (Gremme et al., 2013) package. The binary version of this package makes it free of installation. The result of TIRvish is a GFF file with coordinates for the complete candidate element, the TIR regions of the element, and the TSDs of the element. TIRvish was the easiest and fastest TIR/MITE program we tested. However, the element boundaries defined by TIRvish are frequently shifted ± 1–2 bp, which can have major impacts on downstream analyses.

detectMITE is an open-source MATLAB program created by Ye et al., 2016 for *de novo* detection of MITEs. Its installation requires a third-party software CD-HIT (Fu et al., 2012) and the result/output is a FASTA file which includes the coordinates, TIR length, and TSD length.

GRF (Shi and Liang, 2019) is an open-source C++ software on Github. GRF can find multiple types of repeats including terminal inverted repeats (TIRs) and MITEs. GRF requires the CD-HIT software (Fu et al., 2012) for clustering of repeat candidates. Customers can select the length of TIR/TSD as well as the candidates' minimum and maximum lengths.

miteFinderII (Hu et al., 2018) is an open-source C++ software on Github. The miteFinderII does not have as many parameter options as others. miteFinderII has limited instructions in the README file and refers most questions to the publication.

MITE-Tracker (Crescente et al., 2018) is a Python3 package, which needs search (Rognes et al., 2016) for the clustering process. MITE-tracker parameters include the minimum and the maximum length of MITE elements, as well as the minimum and the maximum length of TSD. Various false-positive sequences with the terminal inverted repeat structure were found in the raw MITE-Tracker candidates that may not be suitable for TE annotation. The number of clustered MITE candidates is comparable to other programs.

MUSTv2 (Ge et al., 2017) is a Perl program that requires BLAST, BLAT (Kent, 2002) and several Perl packages such as Bioperl (Stajich et al., 2002). The output result file of MUSTv2 is a .txt file with MITE information, such as ID, length, strand, cluster, TSD, coordinates, TIR, and scores. Users need to further process the output .txt file to get the FASTA file of candidate MITEs.

TIR-Learner (Su et al., 2019) is a Python3 program that uses machine learning algorithms to facilitate identification and classification of TIR elements. TIR-Learner was originally designed for maize TIR element annotations using IRF (Warburton et al., 2004) as the TIR structure search engine.

4.9 HELITRON

Helitrons are only DNA transposons that do not duplicate target sites during transposition because of their asymmetric structure. The majority of Canonical Helitrons begin with a 5′ T (C/T) and conclude with the nucleotides CTRR (most commonly CTAG, though variations have been seen), but they lack terminal inverted repeats. A hairpin at 11 bp from the 3′ end of a palindromic sequence of 16–20 nt is also prevalent. In general, they bind to an AT host dinucleotide and integrates. Although, the nonautonomous Helitrons have most of the similar structural features including termini as autonomous Helitrons, but they do are not able to synthesize the same comprehensive range of proteins as autonomous Helitrons (Fig. 4.3). The rolling-circle (RC) replication initiator (Rep) and DNA helicase (Hel) domains, which are encoded a protein of 1000–3000 amino acids (Rep/Hel) expressed by all autonomous Helitron elements, are the key enzymatic markers of Helitrons. Helitrons in plants genome are also adjacent to an open reading frame that is similar to single-stranded DNA-binding proteins known as Replicon Protein A (RPA). RPA in Helitrons is approximately 150–500 AA long and is transcribed by multiple exons.

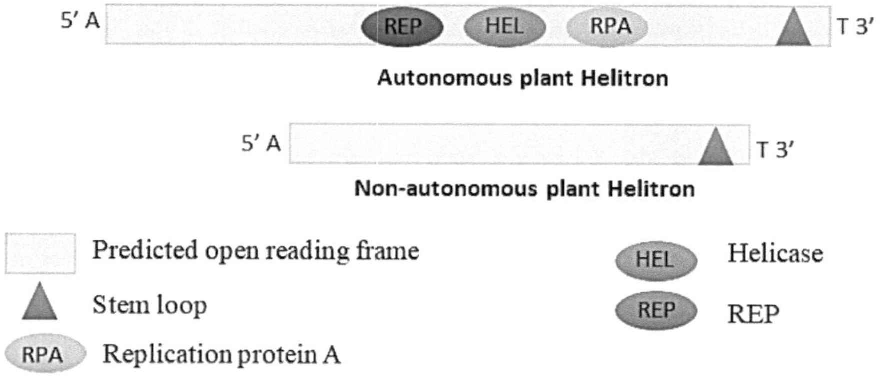

FIGURE 4.3 Structure of plant helitron

Helitrons are eukaryotic transposons that have a distinctive rolling-circle shape and the capacity to acquire gene sequences, making them evolutionarily significance. These are mostly found in angiosperms, where they regularly accumulate and occasionally express one or more gene segments. Unlike other DNA transposons, these are not finish in inverted repetitions or cause target site duplications, making them extremely difficult to detect. They are thought to make up less than 2% of sequenced genomes.

- ***Helitron's basic characteristics include***
  - At the 5' and 3' termini, there are conserved TC and CTAG sequences, respectively.
  - Palindromes ("hairpin loops" of 16 to 20 bps) Upstream of the 3' terminus, 10–15 bp.
  - The 5' and 3' termini of flanking A and T host genome nucleotides, respectively.

Helitron Finder—developed by Du C. et al., in 2008 is the first Helitron computational searching tool which helps in analyzing the Helitron present in maize. The FASTA-formatted DNA sequences work as input for this which helps in identifying the hairpin loop patterns. It is mostly based on conserved sequences at the termini of most Helitrons (5′-TC and CTAG-3′) and a conserved 16- to 20-bp palindromic structure located 10–15 bp upstream of the 3′ terminus.

HelSearch (Yang and Bennetzen, 2009) is a Perl program that relies on blast, all of which were extremely slow. It is basically identical to HelitronFinder in recognizing the 3′ end of Helitrons. Both HelitronFinder and HelSearch will search for the hairpin structure as well as the CTRR 3′ terminus except for some minor variations. The distinction is that HelSearch users must manually search for the 5′ end of Helitrons, whereas HelitronFinder can automatically find the 5′ end.

HelitronScanner (Xiong et al., 2014) is a Java program that utilizes the local combinational variable (LCV) algorithm to identify sequence patterns that are associated with Helitron transposons. The algorithm runs on both the positive and negative strands of the input genome and generates candidates with prediction scores that can be used to assess the prediction's credibility. HelitronScanner uses conserved nucleotides at potentially varying sites to minimize the divergence of Helitron termini between species. HelitronScanner has so far detected 64,654 Helitrons from a diverse set of plant genomes in a fully automated manner.

4.10 CONCLUSION

Identification of TEs in a plant genome is mostly dependent on the detection of repeat families existing in that target plant genome, followed by annotation that is either homology-based or structure-based, with tools employed to seek all detectable members of those families. The majority of the existing TEs of plant genomes are well described in terms of their proteins and the particular termini they encode. Despite the fact that many sequences are well recognized and much repeated, the precise extent of categorization is unknown. There are several TE databases available that characterize a wide range of TEs in plants. The comprehensive TE annotation generated by combination of the results from two or more separate modules would give more precise results. Multiple criteria were carried out to detect which copy among the overlapping repeats should be deleted. In addition, population-based and genome-wide analyses have the possibility to shed light on the function of TEs in plant adaptation, speciation, and evolution. Understanding the development of TEs in plant genomes requires a complete knowledge of TE activity and behavior functional and comparative genomic investigations focused on TE dynamics and interactions in the plant genome, assisted by machine learning and high-throughput bioinformatics techniques, in light of new findings and tools provided.

KEYWORDS

- **plant genomes**
- **transposable elements**
- **annotation tools**
- **bioinformatics**

REFERENCES

Achaz, G.; Boyer, F.; Rocha, E. P. C.; Viari, A.; Coissac, E. Repseek, a Tool to Retrieve Approximate Repeats from Large DNA Sequences. *Bioinformatics* **2007,** *23* (1), 119–121.

Altschul, S. F.; Gish, W.; Miller, W., Myers, E. W.; Lipman, D. J. Basic Local Alignment Search Tool. *J. Mol. Biol.* **1990,** *215*, 403–410.

Andrieu, O.; Fiston, A. S.; Anxolabehere, D.; Quesneville, H. Detection of Transposable Elements by Their Compositional Bias. *BMC Bioinform.* **2004,** *5* (1), 94.

Arkhipova, R. Using bioinformatic and Phylogenetic Approaches to Classify Transposable Elements and Understand Their Complex Evolutionary Histories. *Mob. DNA* **2017,** *8*, 19.

Bao, W.; Kojima, K. K.; Kohany, O. Repbase Update, a Database of Repetitive Elements in Eukaryotic Genomes. *Mob. DNA* **2015,** *6*, 11.

Bao, Z; Eddy, S. R. Automated *de novo* Identification of Repeat Sequence Families in Sequenced Genomes. *Genome Res.* **2002,** *12* (8), 1269–1276.

Bateman, A.; Birney, E.; Cerruti, L. et al., The Pfam Protein Families Database. *Nucl. Acids Res.* **2002,** *30*, 276–280.

Baucom, R. S.; Estill, J. C.; Chaparro, C.; Upshaw, N.; Jogi, A.; Deragon, J. M.; Westerman, R. P.; Sanmiguel, P. J.; Bennetzen, J. L. Exceptional Diversity, Non-Random Distribution, and Rapid Evolution of Retroelements in the B73 Maize Genome. *PLoS Genet.* **2009,** e1000732.

Berezikov, E.; Bucheton, A.; Busseau, I. A Search for Reverse Transcriptase-Coding Sequences Reveals New Non-LTR Retrotransposons in the Genome of *Drosophila melanogaster*. *Genome Biol.* **2000,** *1*, research0012.1.

Bergman, C. M.; Quesneville, H. Discovering and Detecting Transposable Elements in Genome Sequences. *Brief. Bioinform.* **2007,** *8*, 382–392.

Beule, T.; Agbessi, M. D.; Dussert, S.; Jaligot, E.; Guyot, R. Genome-Wide Analysis of LTR-Retrotransposons in Oil Palm. *BMC Genom.* **2015,** *16*, 795.

Bologna, N. G.; Voinnet, O. The Diversity, Biogenesis, and Activities of Endogenous Silencing Small RNAs in Arabidopsis. *Annu. Rev. Plant Biol.* **2014,** *65*, 473–503.

Bourque, G. et al., Ten Things You Should Know About Transposable Elements. *Genome Biol.* **2018,** *19*, 199.

Caspi, A.; Pachter, L. Identification of Transposable Elements Using Multiple Alignments of Related Genomes. *Genome Res.* **2006,** *16*, 260–270.

Chang, C. H.; Larracuente, A. M. Heterochromatin-Enriched Assemblies Reveal the Sequence and Organization of the Drosophila melanogaster Y Chromosome. *Genet.* **2019,** *211*, 333–348.

Chen, J.; Hu, Q.; Zhang, Y.; Lu, C.; Kuang, H. P-MITE: A Database for Plant Miniature Inverted-Repeat Transposable Elements. Nucl. Acids Res. 2014, 42, D1176–D1181.

Copeland, C. S.; Mann, V. H.; Morales, M. E.; Kalinna, B. H.; Brindley, P. J. The Sinbad Retrotransposon from the Genome of the Human Blood Fluke, Schistosomamansoni, and the Distribution of Related Pao-Like Elements. *BMC Evol. Biol.* **2005,** *5* (1), 20.

Crescente, J. M.; Zavallo, D; Helguera, M.; Vanzetti L. S. MITE Tracker: An Accurate Approach to Identify Miniature Inverted-Repeat Transposable Elements in Large Genomes. BMC Bioinform. 2018, 19, 348.

de Sá Pablo, H. C. G.; Guimarães, L. C.; Graças, D. A. das; de Oliveira, V. A. A.; Barh, D; Azevedo, V. et al., Chapter 11—Next-Generation Sequencing and Data Analysis: Strategies, Tools, Pipelines and Protocols. In Omics Technologies and Bio-Engineering; Academic Press, 2018; pp 191–207.

Domingues, D. S.; Cruz, G. M.; Metcalfe, C. J.; Nogueira, F. T.; Vicentini, R.; AlvesCde, S.; Van Sluys, M. A. Analysis of Plant LTR-Retrotransposons at the Fine-Scale Family Level Reveals Individual Molecular Patterns. *BMC Genom.* **2012,** *13*, 137.

Doolittle, W.F; Sapienza, C. Selfish Genes, the Phenotype Paradigm and Genome Evolution. *Nature* **1980,** *284*, 601–603.

Du, C.; Caronna, J.; He, L.; Dooner, H. K. Computational Prediction and Molecular Confirmation of Helitron Transposons in the Maize Genome. BMC Genom. 2008, 9, 51.

Du, J.; Tian, Z.; Hans, C. S.; Laten, H. M.; Cannon, S. B.; Jackson, S. A.; Shoemaker, R. C.; Ma, J. Evolutionary Conservation, Diversity and Specificity of LTR-Retrotransposons in Flowering Plants: Insights from Genome-Wide Analysis and Multi-Specific Comparison. *Plant J.* **2010,** *63*, 584–598.

Durbin, R.; Eddy, S. R.; Krogh, A. et al., *Biological Sequence Analysis: Probabilistic Models of Proteins and Nucleic Acids*; Cambridge University Press: Cambridge, 1999; p 368.

Eddy, S. R. Accelerated Profile HMM Searches. *PLoS Comput. Biol.* **2011,** *7*, e1002195.

Edgar, R. C.; Myers, E. W. PILER: Identification and Classification of Genomic Repeats. *Bioinformatics* **2005,** *21* (suppl 1), 152–158.

Ellinghaus, D.; Kurtz, S.; Willhoeft, U. LTRharvest, an Efficient and Flexible Software for de novo Detection of LTR Retrotransposons. BMC Bioinform. 2008, 9, 18.

Fattash, I. et al., Miniature Inverted-Repeat Transposable Elements: Discovery, Distribution, and Activity. *Genome* **2013,** *56*, 475–486.

Feschotte, C.; Keswani, U.; Ranganathan, N.; Guibotsy, M. L.; Levine, D. Exploring Repetitive DNA Landscapes Using REPCLASS, a Tool That Automates the Classification of Transposable Elements in Eukaryotic Genomes. *Genome Biol. Evol.* **2009,** *1*, 205–220.

Feschotte, C.; Pritham, E. J.; DNA Transposons and the Evolution of Eukaryotic Genomes. *Annu. Rev. Genet.* **2007,** *41*, 331–368.

Flutre, T.; Duprat, E.; Feuillet, C.; Quesneville, H. Considering Transposable Element Diversification in *de novo* Annotation Approaches. *PLoS ONE* **2011,** *6* (1), 16526.

Flynn, J. M.; Hubley, R.; Goubert, C.; Rosen, J.; Clark, A. G.; Feschotte, C.; Smit, A. F. Repeat Modeler2 for Automated Genomic Discovery of Transposable Element Families. *Proc. Natl. Acad. Sci.* **2020,** *117* (17), 9451–9457.

Fu, L.; Niu, B.; Zhu, Z.; Wu, S.; Li, W. CD-HIT: Accelerated for Clustering the Next-Generation Sequencing Data. *Bioinformatics*. 2012, 28, 3150–3152.

Galindo-Gonzalez, L; Mhiri, C; Deyholos, M.K; Grandbastien, M. A. LTR-Retrotransposons in Plants: Engines of Evolution. *Gene* **2017,** *626*, 14–25.

Gao, D.; Jimenez-Lopez, J. C.; Iwata, A.; Gill, N.; Jackson, S. A. Functional and Structural Divergence of an Unusual LTR Retrotransposon Family in Plants. *PLoS ONE* **2012,** *7*, e48595.

Gao, D.; Li, Y.; Kim, K. D.; Abernathy, B.; Jackson, S. A. Landscape and Evolutionary Dynamics of Terminal Repeat Retrotransposons in Miniature in Plant Genomes. *Genome Biol.* **2016,** *17*, 7.

Ge, R.; Mai, G.; Zhang, R.; Wu, X.; Wu, Q.; Zhou, F. MUSTv2: An Improved De Novo Detection Program for Recently Active Miniature Inverted Repeat Transposable Elements (MITEs). J. Integr. Bioinform. 2017, 14, 20170029.

Girgis, H. Z. Red: An Intelligent, Rapid, Accurate Tool for Detecting Repeats *de-novo* on the Genomic Scale. *BMC Bioinform.* **2015,** *16* (1), 1–19.

Grabundzija, A. B.; Hickman, F. Dyda. Helraiser Intermediates Provide Insight Into the Mechanism of Eukaryotic Replicative Transposition. *Nat. Commun.* **2018,** *9*, 1278.

Grandbastien, M. A. LTR Retrotransposons, Handy Hitchhikers of Plant Regulation and Stress Response. *Biochim. Biophys. Acta.* **2015,** *1849*, 403–416.

Gremme, G.; Steinbiss, S.; Kurtz, S. Genome Tools: A Comprehensive Software Library for Efficient Processing of Structured Genome Annotations. IEEE/ACM Trans. Comput. Biol. Bioinform. 2013, 10, 645–656.

Han, Y.; Wessler, S. R. MITE-Hunter: A Program for Discovering Miniature Inverted-Repeat Transposable Elements from Genomic Sequences. Nucl. Acids Res. 2010, 38, e199.

Hoede, C.; Arnoux, S.; Moisset, M.; Chaumier, T.; Inizan, O.; Jamilloux, V. PASTEC: An Automatic Transposable Element Classification Tool. *PLoS One* **2014,** *9*, e91929.

Hu, J.; Zheng,Y.; Shang, X. MiteFinderII: A Novel Tool to Identify Miniature Inverted-Repeat Transposable Elements Hidden in Eukaryotic Genomes. BMC Med Genom 2018, 11, 101.

Huang, R. L.; Burns, K. H.; Boeke, J. D. Active Transposition in Genomes. *Annu. Rev. Genet.* **2012,** *46*, 651–675.

Hua-Van, A.; Le Rouzic, C.; Maisonhaute, P. Capy, Abundance, Distribution and Dynamics of Retrotransposable Elements and Transposons: Similarities and Differences. *Cytogenet. Genome Res.* **2005,** *110*, 426–440.

Hubley, R.; Finn, R. D.; Clements, J.; Eddy, S. R.; Jones, T. A.; Bao, W.; Smit, A. F.; Wheeler, T. J. The Dfam Database of Repetitive DNA Families. *Nucl. Acids Res.* **2016,** *44* (D1), D81–D89.

Jiang, N.; Feschotte, C.; Zhang, X.; Wessler, S. R. Using Rice to Understand the Origin and Amplification of Miniature Inverted Repeat Transposable Elements (MITEs). *Curr. Opin. Plant Biol.* **2004,** 7, 115–119.

Jurka, J. Repbase Update: A Database and an Electronic Journal of Repetitive Elements. *Trends Genet.* **2000,** *16* (9), 418–420.

Jurka, J.; Kapitonov, V. V.; Kohany, O.; Jurka, M. V. Repetitive Sequences in Complex Genomes: Structure and Evolution. *Annu. Rev. Genomics Hum. Genet.* **2007,** *8*, 241–259.

Jurka, J.; Klonowski, P.; Dagman, V.; Pelton, P. Censor–a Program for Identification and Elimination of Repetitive Elements from DNA Sequences. *Comput. Chem.* **1996,** *20* (1), 119–121.

Kalyanaraman, A.; Aluru, S. Efficient Algorithms and Software for Detection of Full-Length LTR Retrotransposons. *J. Bioinform. Comput. Biol.* **2006,** *4* (2), 197–216.

Kapitonov, V. V.; Jurka, J. Rolling-Circle Transposons in Eukaryotes. *Proc. Natl. Acad. Sci. USA* **2001,** *98*, 8714–8719.

Kent, W. J. BLAT-The BLAST-Like Alignment Tool. Genome Res. 2002, 12, 656–664.

Koepfli, K.-P.; Paten, B.; O'Brien, S. J. Genome 10K Community of Scientists, The Genome 10K Project: A Way Forward. *Annu. Rev. Anim. Biosci.* **2015,** *3*, 57–111.

Krogh, A. An Introduction to Hidden Markov Models for Biological Sequences. In *Computational Methods in Molecular Biology*; Searls, D., Kasif, S, Eds.; Elsevier, 1998; pp 45–63.

Kumar, A.; Bennetzen, J. L. Plant Retrotransposons. *Annu. Rev. Genet.* **1999,** *33*, 479–532.

Kurtz, S.; Narechania, A.; Stein, J.; Ware, D. A New Method to Compute K-mer Frequencies and Its Application to Annotate Large Repetitive Plant Genomes. *BMC Genom.* **2008,** *9* (1), 517.

Lander, E. S. et al., International Human Genome Sequencing Consortium, Initial Sequencing and Analysis of the Human Genome. *Nature* **2001,** *409*, 860–921.

Lee, H.; Lee, M.; Mohammed I. W.; Rho, M.; Fox, G. C.; Oh, S. MGEScan: A Galaxy-Based System for Identifying Retrotransposons in Genomes. *Bioinformatics* 2016, 32, 2502–2504.

Lerat, E.; Rizzon, C.; Biémont, C. Sequence Divergence within Transposable Element Families in the Drosophila Melanogaster Genome. *Genome Res.* **2003,** *13*, 1889–1896.

Lewin, H. A. et al., Earth BioGenome Project: Sequencing Life for the Future of Life. *Proc. Natl. Acad. Sci. USA* **2018,** *115*, 4325–4333.

Li, R.; Ye, J.; Li, S.; Wang, J.; Han, Y.; Ye, C. et al., ReAS: Recovery of Ancestral Sequences for Transposable Elements from the Unassembled Reads of a Whole Genome Shotgun. *PLoS Comput. Biol.* **2005,** *1* (4), 43.

Llorens, C.; Futami, R.; Covelli, L.; Dominguez-Escriba, L.; Viu, J. M.; Tamarit, D.; Aguilar-Rodriguez, J.; Vicente-Ripolles, M.; Fuster, G.; Bernet, G. P. et al., The Gypsy Database (GyDB) of Mobile Genetic Elements: Release 2.0. *Nucl. Acids Res.* **2011,** *39*, D70–D74.

Lu, C., Chen, J., Zhang, Y., Hu, Q., Su, W., Kuang, H. Miniature Inverted–Repeat Transposable Elements (MITEs) Have Been Accumulated through Amplification Bursts and Play Important Roles in Gene Expression and Species Diversity in *Oryza sativa*.. *Mol. Biol. Evol.* **2012,** *29*, 1005–1017.

Mao, H.; Wang, H. SINE_scan: An Efficient Tool to Discover Short Interspersed Nuclear Elements (SINEs) in Large-Scale Genomic Datasets. *Bioinformatics* 2017, 33, 743–745.

Mao, L.; Wood, T. C.; Yu, Y. et al., Rice Transposable Elements: A Survey of 73,000 Sequence-Tagged-Connectors. *Genome Res.* **2000,** *10*, 982–990.

McCarthy, E. M.; McDonald, J. F. LTR_STRUC: A Novel Search and Identification Program for LTR Retrotransposons. *Bioinformatics* **2003,** *19* (3), 362–367.

McClintock, B. The Origin and Behavior of Mutable Loci in Maize. *Proc. Natl. Acad. Sci. USA* **1950,** *36* (6), 344–355.

McClintock, B. The Significance of Responses of the Genome to Challenge. *Sciences* **1984,** *226*, 792–801.

Miller, J. R.; Koren, S; Sutton, G. Assembly Algorithms for Next-Generation Sequencing Data. *Genomics* **2010,** *95*, 315327.

Ming, R.; Wai, C. M.; Guyot, R. Pineapple Genome: A Reference for Monocots and CAM Photosynthesis. *Trends Genet.* **2016,** *32*, 690–696.

Momose, M.; Abe, Y.; Ozeki, Y. Miniature Inverted-Repeat Transposable Elements of Stowaway Are Active in Potato. *Genetics* **2010,** *186*, 59–66.

Morgante, M.; Policriti, A.; Vitacolonna, N. et al., Structured Motifs Search. *J. Comput. Biol.* **2005,** *12*, 1065–1082.

Neumann, P.; Novak, P.; Hostakova, N.; Macas, J. Systematic Survey of Plant LTR-Retrotransposons Elucidates Phylogenetic Relationships of Their Polyprotein Domains and Provides a Reference for Element Classification. *Mob. DNA-UK* **2019,** *10.*

Oki, N. et al., A Genome-Wide View of Miniature Inverted-Repeat Transposable Elements (MITEs) in Rice. *GENES Genet. Syst.* **2008,** *83*, 321–329.

Orgel, L. E.; Crick, F. H. Selfish DNA: The Ultimate Parasite. *Nature* **1980,** *284*, 604–607.

Ou, S.; Jiang, N. LTR_FINDER_parallel: Parallelization of LTR_FINDER Enabling Rapid Identification of Long Terminal Repeat Retrotransposons. *Mob. DNA* **2019,** *10* (1), 1–3.

Ou, S.; Jiang, N. LTR_retriever: A Highly Accurate and Sensitive Program for Identification of Long Terminal Repeat Retrotransposons. Plant Physiol. 2018, 176, 1410–1422.

Price, A. L; Jones, N. C; Pevzner, P. A. *De novo* Identification of Repeat Families in Large Genomes. *Bioinformatics* **2005,** *21* (suppl_1), i351–i358.

Quesneville, H.; Bergman, C. M.; Andrieu, O.; Autard, D.; Nouaud, D.; Ashburner, M. et al., Combined Evidence Annotation of Transposable Elements in Genome Sequences. *PLoS Comput. Biol.* **2005,** *1* (2), 22.

Quesneville, H.; Nouaud, D.; Anxolabehere, D. Detection of New Transposable Element Families in Drosophila melanogaster and Anopheles gambiae Genomes. *J. Mol. Evol.* **2003,** *57*, S50–S59.

Rahman, A. Y. A.; Usharraj, A. O.; Misra, B. B.; Thottathil, G. P.; Jayasekaran, K.; Feng, Y.; Hou, S.; Ong, S. Y.; Ng, F. L.; Lee, L. S. *et al.,* Draft Genome Sequence of the Rubber Tree Hevea brasiliensis. *BMC Genom.* **2013,** *14*, 75.

Reeves, G. A; Talavera, D; Thornton, J. M. Genome and Proteome Annotation: Organization, Interpretation and Integration. *J. R. Soc. Interface* **2009,** *6*, 129–147.

Rey, O; Danchin, E; Mirouze, M; Loot, C; Blanchet, S. Adaptation to Global Change: A Transposable Element-Epigenetics Perspective. *Trends. Ecol. Evol.* **2016,** *31*, 514–526.

Rho, M.; Choi, J. H.; Kim, S.; et al., *De novo* Identification of LTR Retrotransposons in Eukaryotic Genomes. *BMC Genom.* **2007,** *8*, 90.

Rho, M.; Schaack, S.; Gao, X.; Kim, S.; Lynch, M.; Tang, H. LTR Retroelements in the Genome of Daphnia pulex. BMC Genom. 2010, 11, 425.

Rho, M.; Tang, H. MGEScan-non-LTR: Computational Identification and Classification of Autonomous Non-LTR Retrotransposons in Eukaryotic Genomes. Nucl. Acids Res. 2008, 37, e143.

Rognes, T.; Flouri, T.; Nichols, B.; Quince, C.; Mahé, F. VSEARCH: A Versatile Open Source Tool for Metagenomics. Peer J. 2016, 4, e2584.

Sampath, P.; Murukarthick, J.; Izzah, N. K.; Lee, J.; Choi, H. I.; et al., Genome-Wide Comparative Analysis of 20 Miniature Inverted-Repeat Transposable Element Families in *Brassica rapa* and *B. oleracea*. *PLoS ONE* **2014,** *29* (4), e94499.

Shi, J.; Liang, C. Generic Repeat Finder: A High-Sensitivity Tool for Genome-Wide *de novo* Repeat Detection. *Plant Physiol.* **2019,** *180* (4), 1803–1815.

Smit, A. F. A.; Hubley, R.; Green, P. RepeatMasker Genomic Datasets, 2019. http://www.repeatmasker.org/genomicDatasets/RMGenomicDatasets.html

Smit, A. F. A.; Hubley, R.; Green, P. RepeatModeler Open-1.0., 2008–2015; Institute for Systems Biology: Seattle. http:\\www.repeatmasker.org(accessed May 1, 2015), 2018.

Smit, F. Interspersed Repeats and Other Mementos of Transposable Elements in Mammalian Genomes. *Curr. Opin. Genet. Dev.* **1999,** *9*, 657–663.

Sreeskandarajan, S.; Flowers, M. M.; Karro, J. E.; Liang, C. A MATLAB Based Tool for Accurate Detection of Perfect Overlapping and Nested Inverted Repeats in DNA Sequences. *Bioinformatics* **2014,** *30*, 887–888.

Stajich, J. E.; Block, D.; Boulez, K.; Brenner, S. E.; Chervitz, S. A.; Dagdigian, C. The Bioperl Toolkit: Perl Modules for the Life Sciences. Genome Res. 2002, 12, 1611–1618.

Stein, L. Genome Annotation: from Sequence to Biology. *Nat. Rev. Genet.* **2001,** *2*, 493503.

Stitzer, M. C.; Anderson, S. N.; Springer, N. M.; Ross-Ibarra, J. The Genomic Ecosystem of Transposable Elements in Maize. *BioRxiv*. **2019.** DOI: 10.1101/559922.

Su, W.; Gu, X.; Peterson, T. TIR-Learner, a New Ensemble Method for TIR Transposable Element Annotation, Provides Evidence for Abundant New Transposable Elements in the Maize Genome. Mol. Plant 2019, 12, 447–460.

Szak, S. T.; Pickeral, O. K.; Makalowski, W. et al., Molecular Archeology of L1 Insertions in the Human Genome. *Genome Biol* **2002,** *3*, research0052.1.

Thomas, J.; Pritham, E. J. Helitrons, the Eukaryotic Rolling-Circle Transposable Elements. *Microbiol. Spectr*. **2015,** 3.

Tu, Z. Eight Novel Families of Miniature Inverted Repeat Transposable Elements in the African Malaria Mosquito, Anopheles gambiae. *Proc. Natl. Acad. Sci. USA* **2001,** *98* (4), 1699–1704.

Vassetzky, N. S.; Kramerov, D. A. SINEBase: A Database and Tool for SINE Analysis. Nucl. Acids Res. 2013, 41, D83–D89.

Vicient, C.M; Casacuberta, J. M. Impact of Transposable Elements on Polyploid Plant Genomes. *Ann. Bot.* **2017,** *120*, 195–207.

Warburton, P. E.; Giordano, J.; Cheung, F.; Gelfand, Y.; Benson, G. Inverted Repeat Structure of the Human Genome: The X-Chromosome Contains a Preponderance of Large, Highly Homologous Inverted Repeats That Contain Testes Genes. Genome Res. 2004, 14, 1861–1869.

Wei, L; Cao, X. The Effect of Transposable Elements on Phenotypic Variation: Insights from Plants to Humans. *Sci. China Life Sci.* **2016,** *59*, 24–37.

Wenke, T.; Döbel, T.; Sörensen, T. R.; Junghans, H.; Weisshaar, B.; Schmidt, T. Targeted Identification of Short Interspersed Nuclear Element Families Shows Their Widespread Existence and Extreme Heterogeneity in Plant Genomes. Plant Cell 2011, 23, 3117–3128.

Wessler, S. R.; Bureau, T. E.; White, S. E. LTR-retrotransposons and MITEs: Important Players in the Evolution of Plant Genomes. *Curr. Opin. Genet. Dev.* **1995,** *5*, 814–821.

Wheeler, T. J.; Clements, J.; Eddy, S. R.; Hubley, R.; Jones, T. A.; Jurka, J. Dfam: A Database of Repetitive DNA Based on Profile Hidden Markov Models. *Nucl. Acids Res.* **2013,** *41*, D70–D82.

Wicker, T.; Sabot, F.; Hua-Van, A.; Bennetzen, J. L.; Capy, P.; Chalhoub, B.; Flavell, A.; Leroy, P.; Morgante, M.; Panaud, O. et al., A Unified Classification System for Eukaryotic Transposable Elements. *Nat. Rev. Genet.* **2007,** *8*, 973–982.

Wright, S. I.; Agrawal, N.; Bureau, T. E. Effects of Recombination Rate and Gene Density on Transposable Element Distributions in *Arabidopsis thaliana. Genome Res.* **2003,** *13*, 1897–1903.

Xiong, W.; He, L.; Lai, J.; Dooner, H. K.; Du, C. HelitronScanner Uncovers a Large Overlooked Cache of Helitron Transposons in Many Plant Genomes. Proc. Natl. Acad. Sci. USA 2014, 111, 10263–10268.

Xu, Z.; Liu, J.; Ni, W. GrTEdb: The First Web-Based Database of Transposable Elements in Cotton (Gossypium raimondii). *Database (Oxford)*, **2017,** *1*, bax013. DOI: 10.1093/database/bax013

Xu, Z.; Wang, H. LTR_FINDER: An Efficient Tool for the Prediction of Full-Length LTR Retrotransposons. *Nucl. Acids Res.* **2007,** W265–W268.

Yaakov, B.; Ben-David, S.; Kashkush, K. Genome-Wide Analysis of Stowaway-Like MITEs in Wheat Reveals High Sequence Conservation, Gene Association, and Genomic Diversification. *Plant Physiol.* **2013,** *161*, 486–496.

Yandell, M.; Ence, D.; A beginner's Guide to Eukaryotic Genome Annotation. *Nat. Rev. Genet.* **2012,** *13*, 329–342.

Yang, L.; Bennetzen, J. L. Structure-Based Discovery and Description of Plant and Animal Helitrons. Proc Natl. Acad. Sci. USA 2009, 106, 12832–12837.

Ye, C.; Ji, G.; Li, L.; Liang, C. detectIR: A Novel Program for Detecting Perfect and Imperfect Inverted Repeats Using Complex Numbers and Vector Calculation. *PLoS One* **2014,** *9*, e113349.

Ye, C.; Ji, G.; Liang, C. detectMITE: A Novel Approach to Detect Miniature Inverted Repeat Transposable Elements in Genomes. Sci. Rep. 2016, 6, 19688.

CHAPTER 5

Retro Transposon Capture Sequencing (RC-Seq)

PRAVEEN KUMAR[1,2*], MAINU HAZARIKA[1], GAGAN KUMAR[3], and LONISHREE DUTTA[4]

[1]*Agricultural Research Station, Mandore, Agriculture University, Jodhpur 342304, Rajasthan, India*

[2]*Mayurakshi College of Agriculture, Mayurakshi Group of Institute, Jodhpur 342001, Rajasthan*

[3]*Krishi Vigyan Kendra, Narkatiaganj, Dr. Rajendra Prasad Central Agricultural University, Pusa, Samastipur 848125, Bihar, India*

[4]*Assam Agricultural University, Assam 785013, India*

**Corresponding author. E-mail: bkpraveen13@gmail.com*

ABSTRACT

Vastly repetitive DNA or RNA sequences of genome and mobile elements abound in eukaryotic genomes. With the discovery of transposable elements or jumping gene by B. Mc-Clintock in the 1950s, a new era started. Retrotransposons are a type of mobile genetic elements; this mobile element is first transcribed into an RNA molecule, then after its through to the mechanism of reverse transcription it will convert into the DNA. In eukaryotic genomes such as corn, retrotransposons amplify themselves quickly to become abundant through reverse transcription. Retrotransposons are categorized into LTRs and non-LTRs; this is divided on the basis of the existence of long terminal repeats (LTRs). Retrotransposon capture sequencing (RC-Seq) is a high-throughput

Plant Transposable Elements: Biology and Biotechnology, Deepu Pandita, Anu Pandita, (Eds.)

sequencing tool for decoding the sequence of gene/s or finding out the relative position or distance between the genes, analyzing, comparing retrotransposon insertions sequences by using the bioinformatics tools. When genomic DNA is fractionated, retrotransposon-binding sites on DNA combine with transposon-binding sites on a microarray. Novel retrotransposition events can be discovered by the information provided by deep sequencing that would be carried out by comparing the query sequence with the already present reference genomic sequence. RC-seq detects the known polymorphic insertions present in an individual, which also identify rare or private germline insertions. Nonreproductive tissue insertions of mobile DNA could be investigated by using RC-seq. It is a useful tool for comparing the similarities or differences between healthy and diseased cells or tissue; this could be done by RC-Seq. Transposable elements are commonly used in structural and functional genomics, as well as in developmental biology. Polymorphism in transposable elements and insertions of somatic mobile genetic elements would affect an individual's phenotype depending on their genomic positions and functional implications. RC-seq is mainly used to understand the role of various diseases, evolution of species, etc.

5.1 INTRODUCTION

Barbara McClintock discovered the jumping DNA elements in the maize genome in the year 1944 (McClintock, 1950). These DNA element phenomena are known by various names such as mutation-causing elements, unstable genes, mutation-causing segments, and "position-effect." Most of research and genetic instability on mobile element have been done in *Drosophila melanogaster.*

These mobile elements are essential contributors to crop species' genetic diversity and variation. Transposable elements (TEs) are the most repetitive sequences and found mostly in eukaryotic genomes (Muszewska et al., 2011). They have the ability to transfer across the genomes, cause mutations, and, of course, multiply their copies; resultant copy number variation is also arisen (Arango-López et al., 2017). They are generally categorized based on the coding regions that are involved in the replication of the element (Chaparro et al., 2015). Retrotransposons of the class I type (Fig. 5.1) move through RNA molecules whereas those of the class II type move through DNA molecules (Wicker et al., 2007). Because of the movement of transposable elements in the genome of plant, they account maximum part of plant genome. Long Terminal Repeat retrotransposons (LTR-RT), non-LTR retrotransposons (non-LTR-RT), Penelope-like elements (PLEs), and Dictyostelium repetitive sequence (DIRS) are the four orders of retrotransposons based on structural properties and the part's life cycle. According to current understanding, all these mobile elements occupy the maximum portion of eukaryotic genome and are having an important contribution in the regulation and control of the gene/s (Dashti and Masoudi-Nejad, 2010). Long Terminal Repeat retrotransposons are the important mobile elements in the plant genome and these increase their copy number by the production of RNA molecule through the copy-and-paste mechanism (Schulman, 2013; Li et al., 2018). For example, they make up 40–75% of the genomes in maize and sugarcane (Negi et al., 2016). LTR-RTs are also recognized for possessing a wide variety of structures, functions, and positions within genomes.

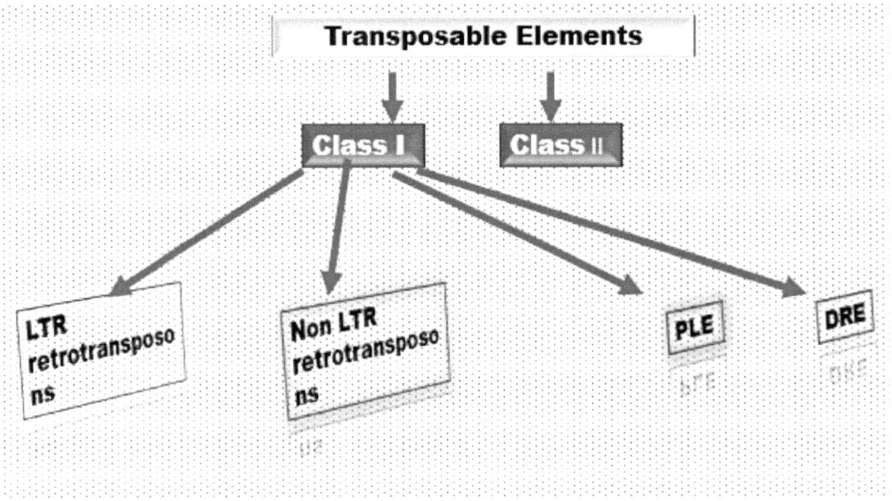

FIGURE 5.1 Classification of transposable elements.

5.2 RETROTRANSPOSONS STRUCTURE

The first publication of RC-seq was released in 2011 by Gao et al. (2012). According to Wicker's classification four types of retrotransposons are Long Terminal Repeat retrotransposons (LTR-RT), non-LTR retrotransposons (non-LTR-RT), Penelope-like elements (PLEs), and Dictyostelium repetitive sequence (DIRS), but some are exception such as long interspersed elements and short interspersed elements that are under the non-LTR retrotransposons (Casacuberta et al., 2013). There are great significant differences in the retrotransposon's structure and their composition, regulatory sequences, and number of enzyme domains (Kejnovsky et al., 2015).

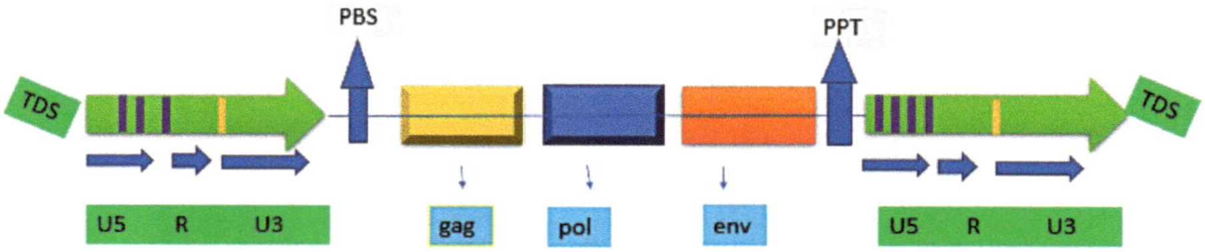

FIGURE 5.2 LTR retrotransposon structure.

5.2.1 LTR RETROTRANSPOSON

According to Grandbastien (2015), Zhang and Gao (2017) reported that the structural structure of LTR-RTs is identical to that of retroviruses, with the exception of the env gene. Size variation for the LTR-RTs ranges from 4 kb to more than 31–23 kb (Mascagni, 2017; Cossu, 2012). Long terminal repeats (LTRs) that are noncoding homologous DNA sequence are present and it is a key feature. One or more ORF (open reading frame) is present and this is used by the host for the transcription and it is coded to gag and pol (polyprotein), separated by stop codon. Major structural protein, packaging, or retrotransposons is done by the gag genes. Piednoël et al. (2013), Usai et al. (2017), and Paz et al. (2017) reported that various enzymes are helpful in the replication cycle, for example, aspartic proteinase (AP), reverse transcriptase (RT), RNase H, and integrase (INT) enzymes and all these enzymes are encoded by the Pol gene.

5.2.2 NON-LTR RETROTRANSPOSON

The LTRs are absent from non-LTR retrotransposons, which are transcribed from an internal promoter.

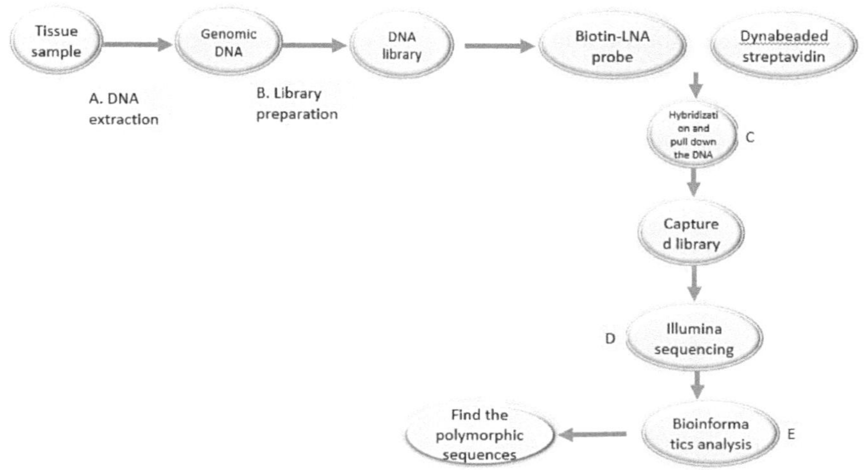

FIGURE 5.3 Steps of RC-seq procedure. (*A*) Extraction of genomic DNA. (*B*) Library preparation. (*C*) Hybridization and extraction of the hybridized DNA fragments, which are contained in the boundaries of transposable elements, for that they used the biotin-LNA probe and Dynabeads Streptavidin (*D*). After the pull down of the captured, sequencing will take place using an Illumina platform. (*E*) Fresh insertion is the subject of bioinformatics investigation.

5.3 PROCEDURE

Retrotransposon activity can be measured using the RC-seq procedure, which can help determine where it is involved in a tissue or cell. To find the expression retrotransposon in the tissue that will be used as a therapy, extract the genomic DNA and compare it to the DNA of the control tissue for detecting the unannoted polymorphic insertion in the treatment tissue. To begin with, we must sonicate the genomic DNA of treatment tissue, and then use this sheared tissue to prepare an illumine sequencing library (Fig. 5.3) After that, we would hybridize with a biotin labeled locked nucleic acid probe that targets the 5′-3′ retrotransposon consensus sequences, then streptavidin associate with the probe and pull it down with the junction between the retrotransposon and flanking genomic DNA. On a high-throughput platform, the post-hybridization library is amplified and sequenced, producing a set of sequencing reads enriched for retrotransposon genome junctions. Using a bioinformatics software, the amplified data is analyzed with the help of the previously defined annotated sequence data. In silico, we can generate insertion data sequence information from individual data to classify previously annotated data. Instead of sequence or polymorphic insertions knowledge, we will screen the organism for any previously unidentified polymorphic insertions. The remaining retrotransposon of genome junction's sequences identify putative somatic insertions specific to the tissue under study. RC-seq is a high-throughput method for mapping genes for a phenotype as well as for analyzing any annotated or unannotated retrotransposon insertions found in an organism (Baillie et al., 2011).

For the retrotransposon analysis, we first extracted the genomic DNA, then segmented it into small DNA fractions, and hybridized the probe with the DNA library containing the retrotransposon binding sites on a micro-array, followed by deep sequencing on an illumine sequencer, and finally it will provide precise information after the alignments, It is possible that new retrotransposition sequence knowledge will be discovered in genomic DNA (Upton et al., 2015).

5.4 METHOD OF RC-SEQ

We will study that unannotated polymorphic insertion sequence which is not identified earlier in both donner and recipient parent, and it is very important to compare the sequence of control tissue from the same donner. Figure 5.3 depicts the RC-Seq flow diagram and the subsequent steps that explain the RC-Seq approach are as follows.

Step 5.4.1: Genomic DNA extraction

Step 5.4.2: DNA shearing—Here sonication is carried out separately for each sample and whole amount of DNA is not used for sonication but some amount of DNA is remained for the PCR validation reactions. The sonicated sample can be preserved at −20/−80°C for storage for the longer duration and for the short duration it is done at 4°C.

Step 5.4.3: Library preparation—With the help of a thermocycler we can prepare the library.

Step 5.4.4: Agarose gel size selection-separation of fragmented, DNA bands are purified and you can do scoring of DNA bands on the next day and for that you have to store it at 4°C.

Step 5.4.5: Hybridization—Pool the same volume of treatment DNA (DNA in this case) and monitor treatment DNA libraries in the same tube in a 1:1 molecular mass ratio, using both 5′ end and 3′ end captures, to produce 1 g of amplified DNA. You can use less DNA (up to 100 ng), but the ratios for the different libraries should stay the same. Add the capture, the pooled DNA libraries, and the Universal Blocking Oligo to a new tube.

Step 5.4.6: Capture recovery and amplification—Pass each 5′ and 3′ capture hybridization reaction to each tube with Dynabeads streptavidin while keeping the tubes in the thermocycler block. Start the thermocyclers for amplification after closing the thermocycler lid and incubating for a few seconds.

Step 5.4.7: Sequencing—Illumina Instruments Sample Sheet with Illumina Experiment Manager programme for the Sample Sheet, use "FASTQ Only."

Step 5.4.8: Bioinformatics Analysis—End-to-end Two fastq files will have been developed by Illumina sequencing; these are files produced for every read. Before you begin analyzing these details, it is suggested that you develop a file name nomenclature that is consistent across projects, including the date of sequencing, the project identifier, and a sample identifier.

Step 5.4.9: PCR validation—This step is mostly hooked on the experience of scientist. It requires both ends of insertion fragment and new copy of mobile element. For confirmation, a simple PCR amplification within each insertion's missing junction could be conducted.

For the replication of the junction sequence, it is necessary to "recreate" the arrangement of the absent junction, which can be accomplished by integrating the annotated DNA sequences following the genomic segment of the established junction with the conserved genomic sequence of the L1 end unlike the one found in the junction. To do so, primers must be designed to anneal within the mobile elements sequence as well as the genomic area required to contain the missing junction.

5.5 MERITS

- It will help to detect the new retrotransposition events.
- We can analyze a particular transposon-binding site.
- A high-throughput approach.
- The rate of PCR validation is calculated to be 98.5%.

5.6 DEMERITS

- According to Xing et al. (2013) PCR experiments with different primers are needed for different types of mobile element insertions (MEIs).

- Hybridization errors might be leads in the sequencing of unnecessary DNA fragments.
- Due to PCR biases, GC-rich models may be underestimated.
- Sequence uncertainty and identification of a transposition event may be caused by similar transposition-binding sites.

5.7 CONCLUSION

Retrotransposons do not increase the genetic variation and species diversity although they increase the size of genome. In a short evolutionary time, it increases the copy number variation, mutation and alters the genomic configuration, so we could say it is a gold mining house for variation. Because of high-throughput sequencing in less time we can identify the polymorphic sequences, and also study the comparative genomics analysis, transcriptomic analysis, function genomics, epigenomics, metabolomics, etc. In future retrotransposons capture sequencing will be used to study the plant improvement, linkage mapping, phylogeny, gene transfer, and to identify the function of gene, genetic diversity, etc.

5.8 DEFINITIONS

- The method of defining the positions of genes and all of the coding regions in a genome, as well as deciding what those genes do, is known as DNA annotation or genome annotation.
- Using a gene transfer method, cloned genes are inserted into an organism's egg or sperm cell to perpetuate a desirable trait in its descendants, such as pest tolerance in a crop plant.

KEYWORDS

- **RC-Seq**
- **retrotransposons**
- **illumina sequencing**
- **RNA**
- **polymorphic sequence**

REFERENCES

Arango-López, J.; Orozco-Arias, S.; Salazar, J. A.; Guyot, R. Application of Data Mining Algorithms to Classify Biological Data: The *Coffea canephora* Genome Case. In *Colombian Conference on Computing*; Springer: Cham, Switzerland, 2017; pp 156–170.

Casacuberta, E.; González, J. The Impact of Transposable Elements in Environmental Adaptation. *Mol. Ecol.* **2013**, *22*, 1503–1517.

Chaparro, C.; Gayraud, T.; de Souza, R. F.; Domingues, D. S.; Akaffou, S.; Laforga Vanzela, A. L.; de Kochko, A.; Rigoreau, M.; Crouzillat, D.; Hamon, S. et al. Terminal-Repeat Retrotransposons with GAG Domain in Plant Genomes: A New Testimony on the Complex World of Transposable Elements. *Genome Biol. Evol.* **2015**, *7*, 493–504.

Cossu, R. M.; Buti, M.; Giordani, T.; Natali, L.; Cavallini, A. A Computational Study of the Dynamics of LTR Retrotransposons in the *Populus Trichocarpa* Genome. *TREE Genet. Genomes* **2012**, *8*, 61–75.

Dashti, T. H.; Masoudi-Nejad, A. Mining Biological Repetitive Sequences Using Support Vector Machines and Fuzzy SVM. *Iran. J. Chem. Chem. Eng.* **2010**, *29*, 1–17.

Gao, D.; Chen, J.; Chen, M.; Meyers, B. C.; Jackson, S. A Highly Conserved, Small LTR Retrotransposon That Preferentially Targets Genes in Grass Genomes. *PLoS One* **2012**, *7*, e32010.

Grandbastien, M.-A. LTR Retrotransposons, Handy Hitchhikers of Plant Regulation and Stress Response. *Biochim. Biophys. Acta* **2015**, *1849*, 403–416.

Kejnovsky, E.; Tokan, V.; Lexa, M. Transposable Elements and G-Quadruplexes. *Chromosome Res.* **2015**, *23*, 615–623.

Klawitter S.; Fuchs N. V.; Upton K. R. et al. Reprogramming Triggers Endogenous L1 and Alu Retrotransposition in Human Induced Pluripotent Stem Cells. *Nat Commun.* **2016**, *7*, 10286.

Li, Q.; Zhang, Y.; Zhang, Z.; Li, X.; Yao, D.; Wang, Y.; Ouyang, X.; Li, Y.; Song, W.; Xiao, Y. A D-Genome-Originated Ty1/Copia-Type Retrotransposon Family Expanded Significantly in Tetraploid Cottons. *Mol. Genet. Genom.* **2018**, *293*, 33–43.

Mascagni, F.; Giordani, T.; Ceccarelli, M.; Cavallini, A.; Natali, L. Genome-Wide Analysis of LTR-Retrotransposon Diversity and Its Impact on the Evolution of the Genus *Helianthus* (L.). *BMC Genom.* **2017**, *18*, 634.

McClintock, B. The Origin and Behavior of Mutable Loci in Maize. *Proc. Natl. Acad. Sci.* **1950,** *36*, 344–355.

Muszewska, A.; Hoffman-Sommer, M.; Grynberg, M. LTR Retrotransposons in Fungi. *PLoS ONE* **2011**, *6*, 12.

Negi, P.; Rai, A. N.; Suprasanna, P. Moving Through the Stressed Genome: Emerging Regulatory Roles for Transposons in Plant Stress Response. *Front. Plant Sci.* **2016**, *7*.

Paz, R. C.; Kozaczek, M. E.; Rosli, H. G.; Andino, N. P.; Sanchez-Puerta, M. V. Diversity, Distribution and Dynamics of Full-Length Copia and Gypsy LTR Retroelements in Solanum *Lycopersicum. Genetica* **2017**, *145*, 417–430.

Piednoël, M.; Carrete-Vega, G.; Renner, S. S. Characterization of the LTR Retrotransposon Repertoire of a Plant Clade of Six Diploid and One Tetraploid Species. *Plant J.* **2013**, *75*, 699–709.

Schulman, A. H. Retrotransposon Replication in Plants. *Curr. Opin. Virol.* **2013**, *3*, 604–614.

Shukla, R.; Upton, K. R.; Munoz-Lopez, M. et al. Endogenous Retrotransposition Activates Oncogenic Pathways in Hepatocellular Carcinoma. *Cell* **2013,** *153*, 101–111.

Solyom, S.; Ewing, A. D.; Rahrmann, E. P. et al. Extensive Somatic L1 Retrotransposition in Colorectal Tumors. *Genome Res.* **2012,** *22*, 2328–2338.

Upton, K. R.; Gerhardt, D. J.; Jesuadian, J. S. et al. Ubiquitous L1 Mosaicism in Hippocampal Neurons. *Cell* **2015,** *161*, 228–239.

Usai, G.; Mascagni, F.; Natali, L.; Giordani, T.; Cavallini, A. Comparative Genome-Wide Analysis of Repetitive DNA in the *Genus Populus* L. *Tree Genet. Genomes* **2017**, *13*, 96.

Wicker, T.; Sabot, F.; Hua-Van, A.; Bennetzen, J. L.; Capy, P.; Chalhoub, B.; Flavell, A.; Leroy, P.; Morgante, M.; Panaud, O. et al. A Unified Classification System for Eukaryotic Transposable Elements. *Nat. Rev. Genet.* **2007**, *8*, 973–982.

Xing, J.; Witherspoon, D. J.; Jorde, L. B. Mobile Element Biology: New Possibilities with High-Throughput Sequencing. *Trends Genet.* **2013,** *29*, 280–289.

Zhang, Q.-J.; Gao, L.-Z. Rapid and Recent Evolution of LTR Retrotransposons Drives Rice Genome Evolution During the Speciation of AA-Genome Oryza Species. *G3 Genes Genomes Genet.* **2017**, *7*, 1875–1885.

Part III

Transposon Biology

CHAPTER 6

Transposable Elements and Plant Development

BRAULIO EDGAR HERRERA-CABRERA[1], LUIS GERMÁN LÓPEZ-VALDEZ[2], RAFAEL SALGADO-GARCIGLIA[3], JORGE MONTIEL-MONTOYA[4], GONZALO GUILLERMO LUCHO-CONSTANTINO[5], FABIOLA ZARAGOZA-MARTÍNEZ[6], JOSÉ LORENZO-LAUREANO[7], and HEBERT JAIR BARRALES-CUREÑO[3*]

[1]*Colegio de Postgraduados, Campus Puebla. Boulevard Forjadores No 205, Santiago Momoxpan, Puebla. C.P. 72760, México*

[2]*Laboratorio de Productos Naturales, Área de Química. AP74 Oficina de correos Chapingo, Universidad Autónoma Chapingo, Km 38.5 Carretera México-Texcoco, 56230, Texcoco, Estado de México, México*

[3]*Instituto de Investigaciones Químico Biológicas, Universidad Michoacana de San Nicolás de Hidalgo, Edificio B-3, Ciudad Universitaria, 58030, Morelia, Michoacán, México*

[4]*Centro Interdisciplinario de Investigación para el Desarrollo Integral Regional, Instituto Politécnico Nacional Unidad Sinaloa, Boulevard Juan de Dios Bátiz 250, Col. San Joachín, Guasave, Sinaloa, México*

[5]*Universidad Tecnológica de Gutiérrez Zamora, Veracruz. Prolongación Dr. Miguel Patiño S/N, Centro, 93556, Veracruz, México*

[6]*Centro de Investigación y Estudios Avanzados del IPN. Av. Instituto Politécnico Nacional 2508, San Pedro Zacatenco, Gustavo A. Madero, 07360, Ciudad de México, México*

[7]*Universidad Intercultural del Estado de Puebla, Calle Principal a Lipuntahuaca S/N, 73475, Puebla, México*

[*]*Corresponding author.*
E-mail: hebert.jair@uiep.edu.mx; hebert.barrales@umich.mx

Plant Transposable Elements: Biology and Biotechnology, Deepu Pandita, Anu Pandita, (Eds.)

ABSTRACT

Transposons are DNA sequences that can change position to different chromosomal locations in the plant genome. They can frequently copy themselves according to this process, so they are also known as mobile genetic elements. With the advances made in large-scale DNA sequencing, transposons constitute a significant proportion of the genetic material of many eukaryotes. In this respect, they comprise about 45% of the human genome itself and, in the case of grasses, a proportion within the range of 50–80% of genetic material. Transposable elements (TEs) often produce mutations, as well as chromosomal rearrangements, with the potential to modify the cell identity itself. They can also regulate other genes and modify the size of the genome, among other functions. Transposons can be used as a convenient way of generating other lines from existing ones, which is achieved by remobilizing the inserted element toward a different location from that occupied by the original insert. When transposons change from their original position, a state of deletion or loss of bases occurs. If the TE is inserted into a gene (a condition that would result in the disruption of the gene itself), the function of the gene is repositioned. In another scenario, high additions of nucleotides occur that produce a mutant phenotype derived from greater dysfunctionality of the gene. This condition arises from the insertion of the mobile genetic element when its position is modified (disrupting the given gene). Superior plants are organisms that are distinguished by an elevated level of flexibility in terms of their growth and development, as well as their capacity to adapt quickly and continuously to environmental changes, especially during their postembryonic development. TEs can generate new phenotypes through exaptation. From an evolutionary point of view, exaptation is the subversion of a series of naturally selected adaptive characteristics to achieve a new function. Sequences of TEs can be directly exapted by specific phenotypic functions in plants, and there are some examples that, although rare, play a critical role in plant growth and development. Therefore, this chapter discusses the main TEs, regulatory genes, and transcription factors involved in plant and flower development, as well as reproductive fitness, and metabolism.

6.1 INTRODUCTION

Transposable genetic elements (TGEs) were discovered in maize plants by Barbara McClintock in 1951. TGEs are DNA fragments that can be inserted into new chromosomal sites, generally making copies multiples of themselves in the plant bioprocess (Feschotte et al., 2002). The first TGE was associated with the site of chromosome breakage in maize and was, therefore, named *Dissociation* (*Ds*). This element is only able to move or cause chromosome breakage events in the presence of another genetic locus, called *Activator* (*Ac*), which can also promote its own transposition (Feschotte et al., 2002). Transposons are DNA fragments that can jump from one region of DNA to another in vivo. A good example is the Ac element, which creates an 8 base pair duplication at the integration site and can precisely cleave itself from one part of the genome and integrate into another, taking into account a distal position. The discovery of this phenomenon prompted the possibility of using transposons as gene transfer vectors. TGEs typically have poor target site selectivity and, therefore, can insert into variable sites within DNA (Alberts et al., 2002). During transposition, a specific enzyme known as a transposase, which is typically encoded by the TGE, acts on specific DNA sequences at each end of the TGE, first detaching it from the insertion site and subsequently inserting the TGE into a new site. Sequence similarity between the ends of the TGE and its insertion site is not required (Alberts et al., 2002). Insertion of transposable elements (TEs) is another tool that has been used to generate mutations. Although *Arabidopsis thaliana* has endogenous transposons (Miura et al., 2004), systems using heterologous transposons from maize, Ac/Ds (Activator/Dissociation) (Bancroft et al., 1992) and En/I (Enhancer/Inhibitor), also known as Spm-dSpm (Suppressor-mutator) (Aarts et al., 1995), prove to be more useful. TEs are a unique—and the most abundant—type of repetitive DNA. Owing to their diversity, ubiquity, and mobilization capacity, TEs have actively participated in the evolution of genomes, by shaping them and providing them with significant plasticity. Nonallelic relationships occur when more than one gene contributes to the expression of a phenotype. Most of the characteristics that constitute the phenotype of an organism are the result of interactions among several different genes. These interactions can occur at different levels, thus distinguishing the phenomena of epistasis and pleiotropy from TGEs. Barbara McClintock defended the idea that transposons function an active role in the evolution of plant genomes. She described transposons as highly controlling elements, capable of modifying gene expression patterns during development, by inserting themselves near any locus, as well as by remodeling genomes in response to stress, thus favoring the appearance of new variants with greater biological efficiency, that could adapt to or overcome these unfavorable conditions more easily

(McClintock, 1987, 1984). In contrast to this functional view of TEs, other authors have developed the selfish DNA hypothesis, which describes TEs as parasitic sequences that have accumulated throughout evolution, owing to their ability to replicate autonomously, without benefiting the host genome (Doolittle and Sapienza, 1980; Orgel and Crick, 1980). Recently, however, evidence has shown that, although not all TEs exert beneficial functions for the genomes in which they are hosted—since transposition is a fundamentally mutagenic phenomenon—several are essential to ensure proper cellular functioning. Therefore, they play a principal role in gene and plant genome evolution. Therefore, this chapter discusses the main TEs, regulatory genes, and transcription factors implicated in plant and flower development, as well as reproductive fitness, and metabolism.

6.2 FLORAL VARIEGATION AND ITS RELATIONSHIP WITH TGEs

Variegation is a phenotypic phenomenon generated by the unstable expression of anthocyanins, which results in the generation of cyanic spots, sectors, or bangs on cornflower or pale backgrounds. Variegation in flower colors is the subject of studies in various plant species, including "Dragon's Mouth" (*Antirrhinum majus*), whose variegation was first described by Darwin and other scientists in the ninth century; however, only in 1980 was it established that the movement of TGEs was associated with this phenomenon (Kidwell, 2005). Variegation is a characteristic analyzed in natural populations and, in some cases, is originated by TGEs (Liu et al., 2001; Lönning and Saudler, 2002).

Some types of variegation are caused by viral infections and mericlinal or periclinal chimerism, but the most interesting form is caused by the movement of TGEs (Itoh et al., 2001). Mutation can cause the repression of gene expression associated with the anthocyanin biosynthetic pathways, disrupting pathway activation and resulting in the production of cyanin flowers. For example, in the carnation variety Kaly, the flowers are white due to the accumulation of a flavanone glycoside caused by a deficiency in the flavone 3-hydroxylase gene transcripts (Yoshida et al., 2004).

In yellow carnation flowers with fringe and white sectors, the *CHI* (Chalcone isomerase) and *DFR* (Dihydroflavonol reductase) genes are disrupted by a TGE called *dTdic1*, resulting in the generation of a variegated flower phenotype in these plants (Itoh et al., 2002). The insertion and excision of TGEs within genes involved in the anthocyanin synthetic pathway can produce phenotypes, such as mosaics or variegations, whose pattern depends on the frequency and timing of TGE excision (Liu et al., 2001).

Variegations that involve genes associated with the synthetic pathways of floral pigments include those found in the "Dragon's mouth" (*A. majus*) plants, in which insertions of the TGEs *Tam1*, *Tam2*, and *Tam3*, within the *CHS* (Chalcone synthetase) and *DFR* (Dyhidroflavonol reductase) genes, generate different color sectors within the flower petals, due to the separation of the element (Coen and Carpenter, 1986).

Another model involving a regulatory gene is the variegated petunia line (W138), which contains a TGE in the *An1* gene, which encodes a transcription factor that regulates anthocyanin biosynthesis (Quattrocchio et al., 1993).

In *Ipomoea purpurea* (morning glory), TGE insertions have been identified that originated phenotypic changes, including those responsible for many flower color variants (Yoshishige et al., 1994). A TGE from the *Ac/Ds* group (*Tip100*) was inserted into an intron of the gene encoding the CHS enzyme, and the flowers produced by this mutant are white with pigmented sectors in the corolla (Durbin et al., 2001).

Another example can be found in *Ipomoea nil* (Japanese morning glory), in which the TGE *Tpn1,* which belongs to the *En/Spm* group, was isolated from plants with variegated flowers. In these plants, the TGE was found to interrupt the *DFR* gene, which is involved in the anthocyanin biosynthetic pathway. Similarly, in "Don Diego de Día" (*I. purpurea*) plants, the elements *Tip100* and *Tip201,* which belong to the *Ac/DsyTpn1* family, and the elements *Tpn2*, *Tpn3*, and *Tpn4* of the *En/Spm* family, which affect anthocyanin biosynthesis, have been identified (Yoshishige et al., 1994; Shigeru et al., 1999).

The insertion of TGEs generally results in the generation of new alleles that are associated with modifications in gene expression or timing, which can result in phenotypic effects (Durbin et al., 2001).

The analysis of various mutant phenotypes induced by the insertion or excision of class II TGEs in plant genes has revealed a variety of methods through which these elements can regulate gene expression (Feschotte et al., 2002), including insertions associated with previous transposition events (typically a few extra base pairs) that are retained in the target when the element is transposed, nonautonomous TGEs that function as introns, and TGEs that are inserted into promoters or other regulatory sequences, which can alter tissue-specific expression patterns (Feschotte et al., 2002).

6.3 FLOWER STAINING AND TGEs

Discovery of the presence of TGEs in anthocyanin regulation or biosynthesis genes has aroused great interest in their potential biotechnological use in the generation of new floral genotypes (Liu et al., 2001; To and Wang, 2006). The effect of TGEs on flower coloration has been studied in several species, including *Nicotiana tabacum*, *Ipomea* sp., and *Petunia* sp. (Liu et al., 2001; Lida et al., 2004; Van Houwelingen et al., 1998). TGEs are used in biotechnological breeding techniques such as insertional mutagenesis and transposon tagging (Babiychuk et al., 1997; Chuck et al., 1993). Recently, the dTdic1 element has been used as a plant breeding tool through tagging (Patent Abstracts of Japan, 2007). TGEs are classified in two ways, according to the degree of self-sufficiency (autonomous and nonautonomous) and the transposition mechanism (Class I or Class II) (Galun, 2003). Class II elements, or DNA elements, are subdivided into families that include the hAT superfamily, which includes the dTdic1 element (Ac/Ds superfamily, group II) (Itoh et al., 2002), the Activator (Ac) in maize, and Tam3 in Snapdragon (Itoh et al., 2002; Galun, 2003; Huang et al., 2009).

In a recent study of the carnation cultivar "Rhapsody," variegation of petal color characterized by randomly patterned red bangs on a white background was observed, which is believed to have been caused by a reversal of DFR (dihydro-flavonol reductase) gene activity following excision of EGTdTdic1 (Itoh et al., 2002). Similarly, in yellow carnation flowers with fringe and white sectors, the CHI (chalcone isomerase) and DFR genes also showed dTdic1 activity associated with variegation patterns in the petals (Yoshida et al., 2004; Itoh et al., 2002). The main pigments involved in floral coloration are carotenoids (tetraterpenoids, red, orange, and yellow pigments), bioflavonoids (yellow polyphenolic pigment), and betalains (yellow and red pigment of indoles), although other pigments such as phenylphenalenones, quinochalcones, and chlorophylls, are also involved in petal pigmentation (Davies, 2009).

Anthocyanins type secondary metabolites are the basis for almost all blue, purple, scarlet, orange, red, and pink coloration in flowers petals (Davies, 2009), and in carnations they are responsible for red colors. In carnations, chalcones are associated with yellow flower coloration. In carnation flowers with yellow petals, chalcone 2′-O-glucoside has been identified as the main flavonoid, a pigment also known as isosalipurposide (narigenin-chalcon-2′-glucoside) (Ogata et al., 2004; Yoshida et al., 2004; Itoh et al., 2002). For this chalcone to accumulate, plants must be incomplete in chalcone isomerase enzyme activity (Gatt et al., 1998).

6.4 INTERSPECIFIC HYBRIDIZATION AND GENOMIC SHOCK

One of the most interesting consequences of hybridization is the activation of "genomic shock," which results in the mobilization of TGE elements, such as transposons and retrotransposons. The massive activation of TGEs was observed in hybrids between *Helianthus* species (Ungerer et al., 2006). The extensive activation of TGEs accompanied by epigenetic instabilities has been described during intergeneric crosses in rice (Wang et al., 2009). The mechanisms that activate TGEs are not known; however, hybridization may result in the removal or remodeling of epigenetic markers in promoter regions, causing the massive mobilization of TGEs. Michalak (2009) states that the activation of TGEs is closely linked to epigenetic and small RNA regulation. Most of the repetitive elements identified in moderately repetitive eukaryotic DNA consist of TGEs.

In general, transposon activity is unfavorable, and both DNA methylation and small RNAs may have developed as mechanisms to control and repress these elements. Hybridization induces a series of genetic and epigenetic changes to the genome, some of which are mediated by TEs (Chase et al., 2010). These changes result in the generation of epigenetic variations, which are metastable (eventually reversible) and are potentially influenced by the environment; this type of heritable epigenetic variation is understood to have relevant implications for the evolution of natural populations.

Phenotypic novelties, such as (a) changes in the flowering time, (b) alterations in floral structures (flower symmetry, flower color, or floral aberrations), or (c) decreases in pollen fertility, can result in the reproductive isolation of hybrids relative to the parent species; if the hybrid plant possesses sufficient fertility, it could establish itself as a new species. However, how natural selection acts on these types of epialleles, which can potentially respond to environmental changes to reverse specific phenotypes, remains unclear.

6.5 IN VITRO SOMACLONAL VARIATION AND TGE

Somaclonal variation results from random genetic or epigenetic events that occur during the in vitro tissue culture technique, modifying the phenotype of regenerated plants and/or tissues. Although such variation in phenotypic traits can, in some cases, produce undesirable outcomes, in others, it constitutes a practical method for generating variability in the gene pool of most plant species.

The detection of TGEs in maize in vitro tissue cultures techniques suggests a possible relationship between somaclonal variation and these TGEs (Lee and Phillips, 1988). TGEs, particularly retrotransposons, are ubiquitous within plant genomes; they can represent 50% of a plant's genomic sequence, or up to 90% in species with very high DNA content but a similar number of genes (Kumar and Bennetzen, 1999; Heslop-Harrison et al., 1997).

In situ hybridization information on metaphase chromosomes and prophase nuclei have manifested that the sequence of retrotransposons such as Ty1-copy 19 is scattered throughout the euchromatin, either randomly or nonrandomly, depending on the plant species and the type of TGEs under study (Heslop-Harrison et al., 1997). However, several of these TGEs are present in very low numbers or are absent in certain regions (e.g., ribosomal DNA localizations interstitial, terminal heterochromatic regions, and centromeres) (Kumar and Bennetzen, 1999).

Similar to TGEs, retrotransposons generate mutations by inserting themselves into or in the vicinity of genes. These mutations are relatively stable as they are transposed for the replication, retaining the sequence at the insertion site. They can cause gene inactivation, alter gene expression patterns, or modify the structure of the proteins they encode (Kumar and Bennetzen, 1999). Genomes that carry TEs should be expected to be more unstable in in vitro culture compared to those with none, or those with fewer elements (Karp, 1995).

There is sufficient evidence to indicate that a significant percentage of mutations in plants manipulated in vitro occurs due to the induced participation of retrotransposons during the plant cell and tissue culture technique (Kumar and Bennetzen, 1999). However, Kubis et al. (2003) used PCR to clone fragments of retroelements of the gypsy and LINE groups, as well as of the En/Spm transposons, from the oil palm genome and found no difference in the genomic organization of the different classes of TGEs between palm trees grown from seeds (parental) and those regenerated by tissue culture that showed abnormal morphology and floral abortion.

6.6 TRANSGENIC PLANTS AND TGEs

Transposition refers to the modification of position of certain base pairs in the DNA sequence, or, from a genetic point of view, to the translocation of a chromosome region to another position within the same chromosome. Transposons can produce copies of themselves in any part of the genome or other genomes (if they are present in the same cell).

The movement of a transposon results in "transposition," a process used in genetics and genetic engineering to move genes, especially in bacteria and, to some extent, in plants. Many transposons carry genes with utility, as well as "selfish" DNA fragments that propagate themselves throughout the genome.

The mechanism by which transposons move is reminiscent of the way retroviruses reproduce, in that the transposon is transcribed into RNA that is retrocopied into the genome as DNA. Because of this similarity, transposons and retroviruses are often grouped under the heading "retrotransposons."

In plants, gene transfer methods can be classified into direct and indirect systems. Direct transformation systems include direct DNA uptake by protoplasts, electroporation, microinjections, and by using microprojectiles (i.e., bioballistic technology) (Klein et al., 1987 and 1988; From et al., 1985). Indirect systems utilize vectors (e.g., *Agrobacterium tumefaciens*), viruses, bacteria, fungi, and mobile genetic elements or transposons (Klee et al., 1987).

Studies on the mobilization of the Tnt1 retrotransposon from *N. tabacum* or Tos17 from *Oryza sativa* indicate that plant retrotransposons also appear to insert into transcriptionally active areas (Miyao et al., 2003; Grandbastien et al., 2005). Moreover, plant TEs can exhibit insertion specificity, although this process is relatively less rigorous than that observed in yeast. For example, class II transposons appear to insert (and be retained) in genetically active, unmethylated (presumably euchromatic) regions (Bennetzen, 2000).

Maize (*Zea Mays*) DNA TGEs, such as the Ac/Ds, Mu1, or Spm elements, typically transpose into transcriptionally active regions. Although elements that have few copies in a genome can be inserted into transcriptionally active regions to facilitate their transcription and transposition, they eventually accumulate in heterochromatic regions.

There appears to be a clear association between retrotransposons and the inactive and methylated chromatin (presumably heterochromatic DNA), which occupies centromeres, pericentromeric regions, telomeres, or knobs of heterochromatin located in the chromosome arms of many plants (Bennetzen, 2000). This is exemplified in the *O. sativa* or *A. thaliana* genomes, where retrotransposons mainly occupy the centromeric regions (*A. thaliana* Genome Initiative, 2000; Goff et al., 2002). Typically, transposons tend to insert next to or within the sequence of other TGEs, in regions rich in such elements (San Miguel et al. 1996; Hua-Van et al., 2005). Retroelements are presumed to transpose into heterochromatic or gene-poor regions to minimize the effects of their replicative transposition.

6.7 STRESS AND TGEs

It has been shown that biotic and abiotic stresses influence activation of TEs in plants, particularly in wheat (Hirochika, 1995; Mansour, 2007). This is of particular relevance given that wheat chromosome 6A is located close to the telomeric zone, a region characterized by being very rich in transposons (Devos et al. 2005). Additionally, it is known that plants can perceive stress during the vegetative state and memorize it through epigenetic mechanisms. Thus, environmentally induced epigenetic changes can be passed on to progeny (Mirouze and Paszkowski, 2011).

The formation of alternative epigenetic states not only induces the origin of new epialleles, but also promotes the mobility of DNA transposons and retrotransposons, which are abundant in plant genomes (Mirouze and Paszkowski, 2011).

6.8 TGEs IN ARABIDOPSIS

The length of the genetic map of *A. thaliana* is about 600 cM, each of which is equivalent to about 210 kb. Prior to sequencing, it had been estimated that the nuclear genome of *A. thaliana* should contain 10% highly repetitive sequences (4000–6000 copies) and another 10% moderately repetitive ones (10–1000 copies, mostly tandem repeats of ribosomal RNA genes) (Meyerowitz, 1994). The remaining 80% would include virtually all protein-producing genes. Based on these indirect estimates, it has been assumed that the *A. thaliana* genome is almost devoid of repetitive elements and that those that exist are generally very distant from each other (Pruitt and Meyerowitz, 1986). Whole genome sequencing confirms the scarcity of repetitive DNA in this species, since only 10% (Bevan et al., 2001) of its sequence is related to transposons and only 1% of its sequence corresponds to retrotransposons of the copy-like family (Terol et al., 2001). Most of these repetitive elements, mainly microsatellites, retroelements, and transposons, are very poorly represented in the euchromatin and are accumulated in the pericentromeric areas (*A. thaliana* Genome Initiative, 2000). In the human genome, on the other hand, these types of elements account for 45% of the sequence IHGSC (International Human Genome Sequencing Consortium, 2001). In *A. thaliana*, the predominant histone methylation at lysine 9 consisted of monomethylated (H3K9me1) and demethylated (H3K9me2) forms (located in regions of repeated sequences) and transposons (regions associated with heterochromatin) (Lippman et al., 2004; Johnson, 2004). It was also shown that H3K9 methylation and its interaction with DNA methylation are essential for the maintenance of transcriptional silencing and genome stability (Vaillant et al., 2007). Likewise, trimethylation of lysine 27 (H3K27me3) has been shown to repress embryogenesis and meristem formation in the calluses of various plants (Ikeuchi et al., 2013). It has furthermore been described how Polycomb-type proteins, through their methyltransferase activity, maintain the repression of certain genes in embryo development and in the endosperm of *A. thaliana* (Köhler et al., 2003). Figure 6.1 shows the main TGEs and their roles in development in *A. thaliana.*

6.9 TGEs IN A. MAJUS

Some lines have been produced in a laboratory setting from *A. majus* cultivars, and there also exists a great collection of mutants, most of which are on an isogenic genetic background of *A. majus*: Sippe 50. In particular, there is a collection of *A. majus* mutants that have been selected for their high transposon activity. In some cases, these selected lines have been utilized to isolate genes via transposon tagging (Hudson et al., 2008; Noda et al., 1994; Bradley et al., 1993; Stubbe, 1966).

MITEs (miniature inverted-repeat TEs) are present in low copy numbers in all *A. majus* species and have been termed IDLE due to relatively low activity. MITE-type transposons have often been found in areas of the genome with high gene content, in contrast to the many transposon families that are major elements of heterochromatin. This phenomenon was also observed in *A. majus*, in which the insertions of IDLEs were interspersed in coding areas but absent from centromeres or telomeres (Schwarz-Sommer et al., 2010; Cartolano et al., 2007).

An interesting line of study that has been developed in peaches uses *A. tumefaciens* strains known as "shooty mutants." These are unarmed strains with a mutation consisting of a disruption with the tn5 transposon (in a gene involved in auxin synthesis) but an intact ipt gene (involved in cytokinin synthesis). Infection with a "shooty mutant" strain causes the target plant to develop a series of tumors from which transgenic shoots differentiate. This type of transformation produces plants with increased endogenous cytokinin synthesis, resulting in a more branched phenotype with shorter internodes.

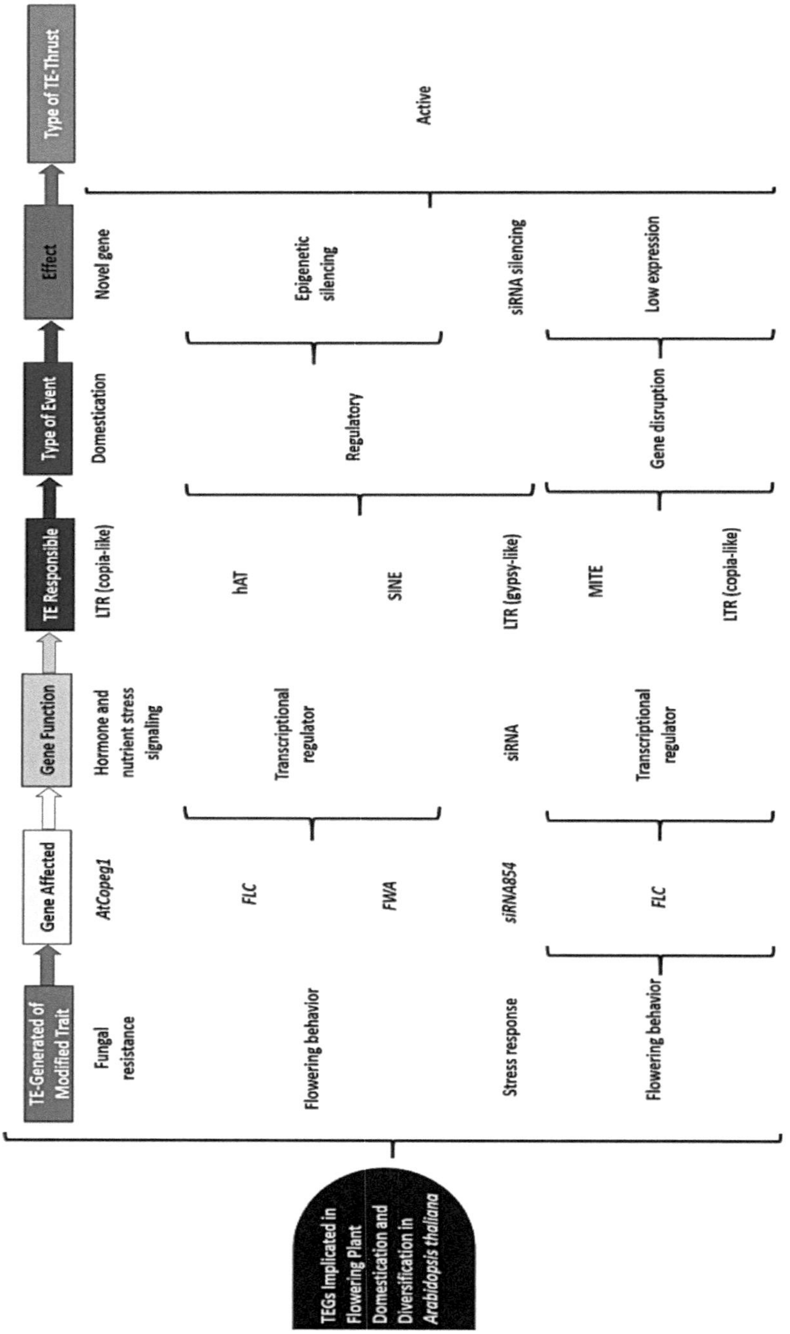

FIGURE 6.1 Transposable genetic elements implicated in development plant domestication and diversification in *A. thaliana*.

6.10 TGEs IN WHEAT

Hexaploid wheat is one of the largest plant genomes, containing a total of 15,966 Mb (megabases), which is 8 times larger than that of *Zea mays* and 40 times larger than that of *O. sativa* (Arumuganathan and Earle, 1991). Given the great complexity of this genome, the complete sequencing of wheat is not yet public. It is estimated that each wheat genome contains between 40,000 and 50,000 genes, although its complexity lies in the fact that more than 80% of its DNA consists of repeated sequences, mainly TGEs and retrotransposons (Devos et al., 2005; Wanjugi et al., 2009; Safár et al., 2010).

Deletion line mapping reveals that the genes are not reparted homogeneously in the chromosomes, but are located mainly in the telomeric areas (Qi et al., 2004). Despite having chromosomes in triplicate, their genes are specific to each chromosome and are not repeated in three copies. For example, genes for resistance to pests such as fungi (rusts, mildews) and insects (Russian aphid, green aphid, wheat fly) are assumed to be present in only one of the homologous groups (Gill et al., 2004).

Consequently, the assumption that each gene is represented in at least three copies loses accuracy due to permanent recombination between chromosomes and the presence/absence of retrotransposons, processes that together are responsible for the decrease in synteny between genomes and the high complexity of wheat gene mapping based on cloning techniques (Qi et al., 2004; Akhunov et al., 2003). Figure 6.2 shows the main TGEs and their roles in development of *Triticum turgidum* and *Triticum aestivum*.

6.11 TGEs IN MAIZE

In maize (*Zea mays*), Mutator and Spm transposon insertions modify the expression of the a1 gene, which encodes dehydroflavonol reductase, an enzyme involved in the formation of the pigments phlobaphene red and anthocyanin purple. Additionally, it is indistinctly regulated by the transcription factors C1 and P of the MYB family, which affect the coloration of *Zea mays* (Chuck, 2008). Figure 6.3 shows the main TGEs and their roles in development in *Zea mays*.

6.12 HORIZONTAL TRANSFER AND TGEs

The new gene pool is no longer that rigid set that can only transform mutations and recombinations and, to a small extent, gene flow. Mobile or

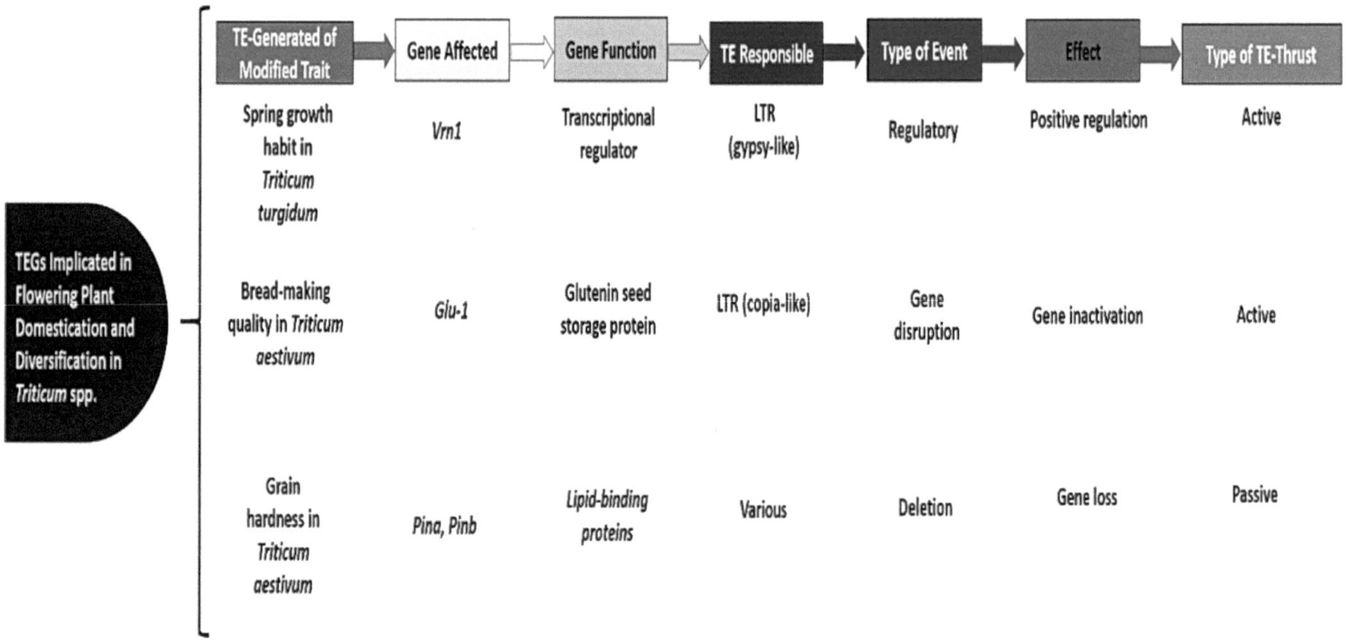

FIGURE 6.2 Transposable genetic elements implicated in development plant domestication and diversification in *Triticum turgidum* and *Triticum aestivum*.

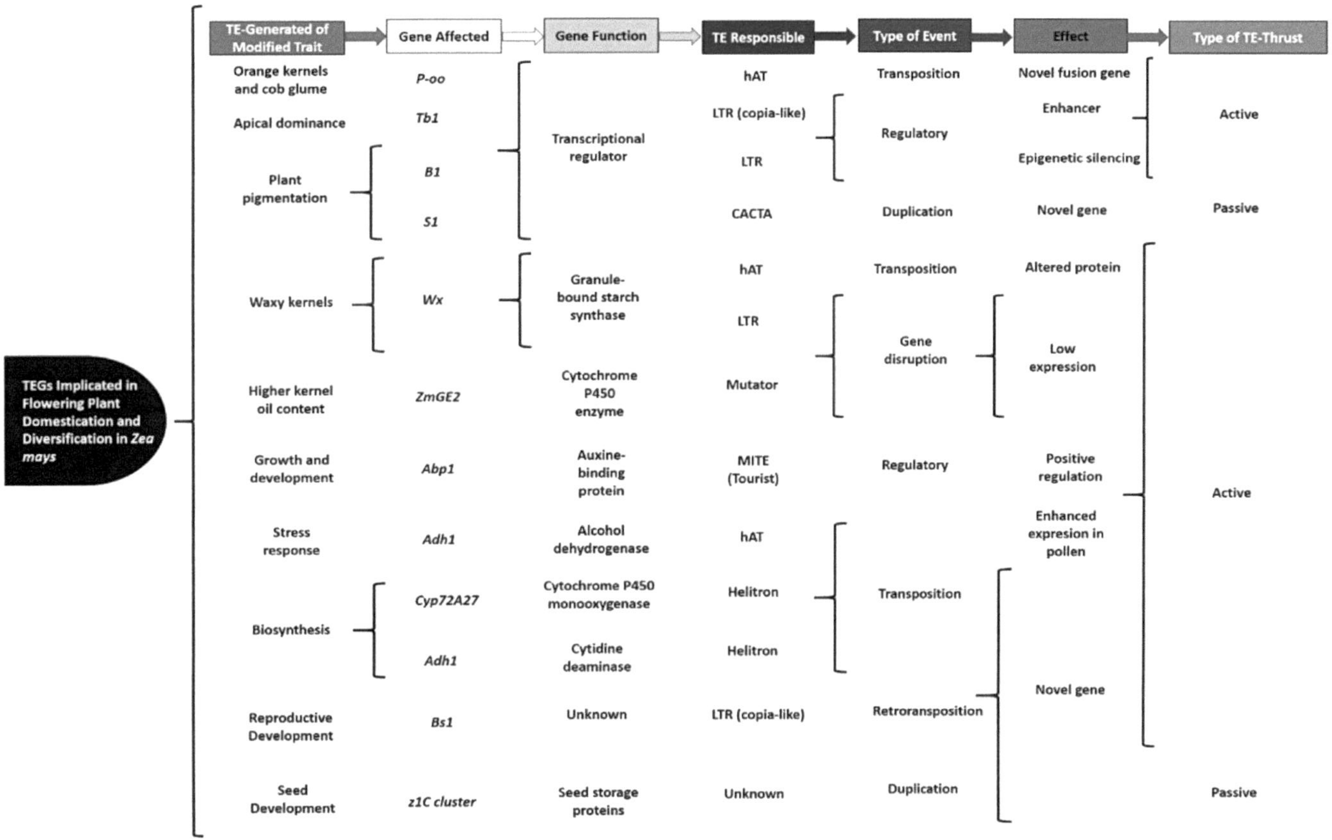

FIGURE 6.3 Transposable genetic elements implicated in development plant domestication and diversification in *Zea mays*.

TGEs constitute a large fraction of the DNA of plant species, sometimes comprising the majority (Gaut et al., 2001). Although most mutations that generate these transpositions are neutral or deleterious, they can provide an important avenue for increasing genetic variation among members of a population and even produce competitive advantages. For example, the small mobile elements known colloquially as Tourist and Stowaway are associated with hundreds of genes in angiosperms and may have played a role in the plant evolution of dicots and monocots (Rudall and Buzgo, 2002).

Horizontal intracellular gene transfer from the chloroplasts and mitochondria of cells to the nuclei has become increasingly common throughout the evolutionary history of angiosperms, particularly over the last few million years (Palmer et al., 2000).

Given their abundance in plants, one might assume that transposons and other mobile elements would create cluttered plant genomes. However, more than a decade of work involving Solanaceae family, grasses, *Brassica napus*, and *A. thaliana* shows a surprisingly conservative gene order (Fedoroff, 2001). Research concerning horizontal transfer in plants suggests that angiosperm evolution has very often gone the way of genomic growth (Palmer and Delwiche, 1998, Palmer et al., 2000). For example, the accumulation of retrotransposon packages at the intergenic level is a principal factor in explaining the size differences between the maize genome and the genomes of closely related species (Singh and Krimbas, 2000).

6.13 EVOLUTION AND *ZEA MAYS*

Transposons are widely distributed in both prokaryotes and eukaryotes, and are observed to have significantly modified the course of evolution by incorporating genetic variations or by generating chromosomal instability (Kazazian, 2004).

When TEs move, they often show little sequence selectivity at the insertion site. As a result, transposons can insert themselves into genes, completely disrupting gene function, or into regulatory sequences, where they can lead to modifications in the way a gene is expressed (Watson et al., 2014).

Although in nature most transposons are silenced, many of them have evolutionary roles of great importance through altering gene function by insertion, inducing rearrangements in the chromosome, or inserting novel genes or gene regulatory sequences (Feschotte and Pritham, 2007). Some notable examples are the transfer of genes that confer antimicrobial resistance in bacteria (Bennett, 2009), phenotypic alterations in fish (Koga et al., 2006), and mutations in plants (Wicker et al., 2016).

6.14 CONCLUSION

Transposons (or TEs) are DNA sequences that can change position to new chromosomal locations in the genome. They can frequently copy themselves according to this process, so they are also known as mobile genetic elements. With the advances made in large-scale DNA sequencing, transposons constitute a significant proportion of the genetic material of many eukaryotes. In this respect, they comprise about 45% of the human genome itself and, in the case of grasses, a proportion within the range of 50–80% of genetic material. TEs often produce mutations, as well as chromosomal rearrangements, with the potential to modify the cell identity itself. They can also regulate other genes and modify the size of the genome, among other functions. Transposons can be used as a convenient way of generating other lines from existing ones, which is achieved by remobilizing the inserted element toward a different location from that occupied by the original insert. When transposons change from their original position, a state of deletion or loss of bases occurs. If the TE is inserted into a gene (a condition that would result in the disruption of the gene itself) the function of the gene is repositioned. In another scenario, high additions of nucleotides occur that produce a mutant phenotype derived from greater dysfunctionality of the gene. This condition arises from the insertion of the mobile genetic element when its position is modified (disrupting the given gene). Superior plants are organisms that are characterized by a high level of flexibility in terms of their growth and development, as well as their capacity to adapt quickly and continuously to environmental changes, especially during their postembryonic development. TEs can generate new phenotypes through exaptation. The exaptation is the subversion of a series of naturally selected adaptive characteristics to achieve a new function. Sequences of TEs can be directly exapted by specific phenotypic functions in plants, and there are some examples that, although rare, play a critical role in plant growth and development.

KEYWORDS

- **exaptation**
- **metabolism**
- **plant development**
- **reproductive fitness**
- **signaling**
- **transposons**

REFERENCES

Aarts, M.; Corzaan, P.; Stiekema, W.; Pereira, A. A Two-Element Enhancer-Inhibitor Transposon System in *Arabidopsis thaliana*. *Mol. Gen. Genet.* **1995,** *247*, 555–564.

Akhunov, E. D.; Akhunova, A. R.; Linkiewics, A. M.; Dubcovsky, J.; Hummel, D.; Lazo, G. et al. Synteny Perturbation Between Wheat Homeologous Chromosomes Caused by Locus Duplications and Deletions Correlate with Recombinantion Rates Along Chromosome Arms. *Proc. Natl. Acad. Sci.* **2003,** *100*, 10836–10841.

Alberts, B.; Johnson, A.; Lewis, J.; Raff, M.; Roberts, K.; Walter, P. *Molecular Biology of the Cell*, 4th ed.; Garland Science, Taylor & Francis group, 2002.

Arabidopsis Genome Initiative (AGI). Analysis of the Genome Sequence of the Flowering Plant *Arabidopsis thaliana*. *Nature* **2000,** *408*, 796–815.

Arumuganathan, K.; Earle, E. D. Nuclear DNA Content of Some Important Plant Species. *Plant Mol. Biol. Rep.***1991,** *9*, 208–218.

Babiychuk, E.; Fuangthong, M.; Van Montagu, M.; Inzé, D.; Kushnir, S. Efficient Gene Tagging in *Arabidopsis thaliana* Using a Gene Trap Approach. *Proc. Natl. Acad. Sci. USA* **1997,** *94*, 12722–12727.

Bancroft, I.; Bhatt, A.; Sjodin, C.; Scofield, S.; Jones, J.; Dean, C. Development of an Efficient Two-Element Transposon Tagging System in *Arabidopsis thaliana*. *Mol. Gen. Gen.* **1992,** *233*, 449–461.

Bennett, P. M. Plasmid Encoded Antibiotic Resistance: Acquisition and Transfer of Antibiotic Resistance Genes in Bacteria. *Br. J. Pharmacol.* **2009,** *153*, 347–357.

Bennetzen, J. L. Transposable Element Contributions to Plant Gene and Genome Evolution. *Plant Mol. Biol.* **2000,** *42*, 251–269.

Bevan, M.; Mayer, K.; White, O.; Eisen, J. A.; Preuss, D.; Bureau, T.; Salzberg, S. L.; Mewes, H. Sequence and Analysis of the Arabidopsis Genome. *Curr. Opin. Plant Biol.* **2001,** *4*, 105–110.

Bradley, D.; Carpenter, R.; Sommer, H.; Hartley, N.; Coen, E. Complementary Floral Homeotic Phenotypes Result from Opposite Orientations of a Transposon at the Plena Locus of Antirrhinum. *Cell* **1993,** *72*, 85–95.

Cartolano, M.; Castillo, R.; Efremova, N.; Kuckenberg, M.; Zethof, J.; Gerats, T.; Schwarz-Sommer, Z.; Vandenbussche, M. A Conserved MicroRNA Module Exerts Homeotic Control Over *Petunia hybrida* and *Antirrhinum majus* Floral Organ Identity. *Nat. Genet.* **2007,** *39*, 901–905.

Chase, M.; Paun, O.; Fay, M. Hybrid Matic Histone H3 Methylation Patterns on the Arabidopsis Geneization and Speciation in Angiosperms: A Role for Pollinator Shifts. *BMC Biol.* **2010,** *8*, 45.

Chuck, G.; Robbins, T.; Nijjar, C.; Ralston, E.; Courtney-Gutterson, N.; Donner, H. K. Tagging and Cloning of a Petunia Flower Color Gene with the Maize Transposable Element Activator. *Plant Cell* **1993,** *5*, 371–378.

Coen, E. S.; Carpenter, R. Transposable Elements in *Antirrhinum majus*: Generators of Genetic Diversity. *Trends Genet.* **1986,** *2*, 292–296.

Davies, K. *Modifying Anthocyanin Production in Flowers en Anthocyanins Biosynthesis, Functions and Applications.* In *An-thocyanins*; Gould, K., Davies, K., Winefield, C., Eds.; Springer Science Business Media: New Zealand, 2009; p. 83.

Devos, K. M.; Ma, J.; Pontaroli, A. C.; Pratt, L. H.; Bennetzen, J. L. Analysis and Mapping of Randomly Chosen Bacterial Artificial Chromosome Clones for Hexaploid Bread Wheat. *Proc. Natl. Acad. Sci. USA* **2005,** *102*, 19243–19248.

Doolittle, W. F.; Sapienza, C. Selfish Genes, the Phenotype Paradigm and Genome Evolution. *Nature* **1980,** *284*, 601–603.

Durbin, M.; Denton, A.; Clegg, M. Dynamics of Mobile Element Activity in Chalcone Synthase Loci in the Common Morning Glory (*Ipomoea purpurea*). *Proc. Natl. Acad. Sci. USA* **2001,** *98*, 5084–5089.

Fedoroff, N. Transposons and Genome Evolution in Plants. In *Variation and Evolution in Plants and Microorganisms. Toward a New Synthesis 50 Years* After Stebbins; Ayala, F. J., Fitch, W. M., Clegg, M. T., Eds.; National Academic Press: Washington, DC, 2001, p 186.

Feschotte, C.; Jiang, N.; Wesler, S. Plant Transposable Elements: Where Genetics Meets Genomics. *Nature* **2002,** *3*, 329–341.

Feschotte, C.; Pritham, E. J. DNA Transposons and the Evolution of Eukaryotic Genomes. *Annu. Rev. Genet.* **2007,** *41*, 331–368.

Fromm, M.; Taylor, L. P.; Walbot, V. Expression of Genes Transforred Into Monocot Anddicot Plant Cells by Electroporation. *Proc. Natl. Acad. Sci.* **1985,** *82*, 5824–5828.

Galun, E. *Class II Transposable Elements in Eucariotes en Transposable Elements, a Guide to the Perplexed and the Novice*; The Weisman Institute of Science, Kluwer Academic Publishers: The Netherlands, 2003; p 183.

Gatt, M.; Hammett, K.; Markham, K.; Murria, B. Yellow Pinks: Interespecific Hybridization Between Dianthus Plumarius and Related Species with Yellow Flowers Scientia. *Acta Hortic.* 1998, *77*, 207–218.

Gaut, B. S.; Le Thierry, M.; Peek, A. S.; Sawkins, M. C. Maize as a Model for the Evolution of Plant Nuclear Genomes. In *Variation and Evolution in Plants and Microorganisms. Toward a New Synthesis 50 Years After Stebbins*; Ayala, F. J., Fitch, W. M., Clegg, M. T., Eds.; National Academic Press: Washington, DC, 2001, p 210.

Gill, B. S.; Appels, R.; Botha A. M.; Oberholster, C. R.; Buell, J. L.; Bennetzen, B.; Chalhoub, F. et al. A Workshop Report on Wheat Genome Sequencing: International Genome Research on Wheat Consortium. *Genetics* **2004,** *168*, 1087–1096.

Goff, S. A.; Ricke, D.; Lan, T. H.; Presting, G.; Wang, R.; Dunn, M. et al. A Draft Sequence of the Rice Genome (*Oryza sativa* L. ssp. japonica). *Science* **2002,** *296*, 92–100.

Grandbastien, M. A.; Audeon, C.; Bonnivard, E.; Casacuberta, J. M.; Chalhoub, B.; Costa, A. P.; Le, Q. H. et al. Stress Activation and Genomic Impact of Tnt1 Retrotransposons in Solanaceae. *Cytogenet. Genome. Res.* **2005,** *110*, 229–241.

Heslop, J. S.; Brandes, A.; Taketa, S.; Schmidt, T.; Vershinin, A. V.; Alkhimova, E. G.: Kamm, A.; Doudrick, R .L.; Schwarzacher, T.; Katsiotis, A.; Kubis, S.; Kumar, A.; Pearce, S. R.; Flavell, A. J.; Harrison, G. E. The Chromosomal Distributions of Ty1-Copia Group Retrotransposable Elements in Higher Plants and Their Implications for Genome Evolution. *Genetic* **1997,** *100*, 197–204.

Hirochika, H. Activation of Plant Retrotransposons by Stress. In *Modification of Gene Expression and Non-Mendelian Inheritance*; NIAR: Japan, 1995; pp 15–21.

Huang, J.; Zhang, K.; Shen, Y.; Huang, Z.; Li, M.; Tang, D.; Gu, M.; Cheng, Z. Identification of a High Frequency Transposon Induced by Tissue Culture, nDaiZ, a Member of the hAT Family in Rice. *Genomics* **2009,** *93*, 274–281.

Hua-Van A.; Le Rouzic, A.; Maisonhaute, C.; Capy. P. Abundance, Distribution and Dynamics of Retrotransposable Elements and Transposons: Similarities and Differences. *Cytogenet. Genome Res.* **2005,** *110*, 426–440.

Hudson, A.; Critchley, J.; Erasmus, Y. The Genus Antirrhinum (Snapdragon): A Flowering Plant Model for Evolution and Development. *Cold Spring Harb. Protoc.* **2008,** *11*, 1–12.

I. I. B. Genome Sequencing and Analysis of the Model Grass Brachypodium distachyon. *Nature* **2010,** *463*, 763–768.

Ikeuchi, M.; Sugimoto, K.; Iwase, A. Plant Callus: Mechanisms of Induction and Repression. *Plant Cell* **2013,** *25*, 3159–3173.

Itoh, Y.; Higeta, D.; Suzuki, A.; Yoshida, H.; Ozeki, Y. Excision of Transposable Elements from the Chalcone Isomerase and Di-Hydroflavonol 4-Reductase Genes May Contribute to the Variegation of the Yellow-Flowered Carnation (*Dianthus caryophyllus*). *Plant Cell Physiol.* **2002,** *43*, 578–585.

Johnson, L.; Mollah, S.; García, B. A.; Muratore, T. L.; Shabanowitz, J.; Hunt, D. F.; Jacobsen, S. E. Mass Spectrometry Analysis of Arabidopsis Histone H3 Reveals Distinct Combinations of Post-Translational Modifications. *Nucl. Acids Res.* **2004,** *32*, 6511–6518.

Karp, A. Somaclonal Variation as a Tool for Crop Improvement. *Euphytica* **1995,** *85*, 295–302.

Kazazian, H. H. Mobile Elements: Drivers of Genome Evolution. *Science*, **2004,** *303*, 1626–1632.

Kidwell, M. Transposable Elements en the Evolution of Genome Size in Eukaryotes. *Genetic.* **2005,** *115*, 49–63.

Klee, H.; Horsch, R.; Rogers, S. Agrobacterium Mediated Plant Transformation and Its Further Applications to Plant Biology. *Ann. Rev. Plant Physiol.* **1987,** *38*, 487–496.

Klein, T. M.; Gradziel, T.; Fromm, M. E.; Sanford, J. C. Factors Influencing Gene Delivery Into Zea Mays Cell by High-Velocity Microprojectiles. *Biotech* **1988,** *6*, 559–663.

Klein, T. M.; Wolf, E. D.; Wu, R.; Sandorf, J. C. High Velocity Microprojectiles for Delivering Nucleic Acids Into Living Cells. *Nature* **1987,** *327*, 70–73.

Koga, A.; Iida, A.; Hori, H.; Shimada, A.; Shima, A. Vertebrate DNA Transposon as a Natural Mutator: The Medaka Fish Tol2 Element Contributes to Genetic Variation Without Recognizable Traces. *Mol. Biol. Evol.* **2006,** *23*, 1414–1419.

Köhler, C.; Hennig, L.; Spillane, C.; Pien, S.; Gruissem, W.; Grossniklaus, U. The Polycomb-Group Protein MEDEA Regulates Seed Development by Controlling Expression of the MADS-Box Gene PHERES1. *Genes Dev.* **2003,** *17*, 1540–1553.

Kubis, S. E.; Castilho, A. M.; Vershinin, A. V.; Heslop, P. S. Retroelements, Transposons and Methylation Status in the Genome of Oil Palm (*Elaeis guineensis*) and the Relationship to Somaclonal Variation. *Plant Mol. Biol.* **2003,** *52*, 69–79.

Kumar, A.; Bennetzen, J. L. Plant Retrotransposons. *Annu. Rev. Genet.* **1999,** *33*,479–532.

Lee, M.; Phillips, R . The Chromosomal Basis of Somaclonal Variation. *Annu. Rev. Plant Physiol.* **1988,** *39*, 413–437.

Lida, S.; Morita, Y.; Choi, J. D.; Park, K. I.; Hoshino, A. Genetics and Epigenetics in Flower Pigmentation Associated with Transposable Elements in Morning Glories. *Adv. Biophys.* **2004,** *38*, 141–159.

Lippman, Z.; Gendrel, A. V.; Black, M.; Vaughn, M. W.; Dedhia, N.; McCombie, W. R.; Lavine, K.; Mittal, V.; May, B.; Kasschau, K. D.; Carrington, J. C.; Doerge, R. W.; Colot, V.; Martienssen, R. Role of Transposable Elements in Heterochromatin And Epigenetic Control. *Nature* **2004,** *430*, 471–476.

Liu, D.; Galli, M.; Crawford, N. Engineering Variegated Floral Patterns in Tobacco Plants Using Arabidopsis Transposable Element Tag 1. *Plant Cell Physiol.* **2001,** *42*, 419–423.

Lonnig, W. E.; Saedler. H. Chromosome Rearrangements and Transposable Elements. *Annu. Rev. Genet.* **2005,** *36*, 389–410.

Mansour, A. Epigenetic Activation of Genomic Retrotransposones. *J. Mol. Cell Biol.* **2007,** 6: 99–107.

McClintock, B. *The Discovery and Characterization of Transposable Elements: The Collected Papers of Barbara McClintock*; Garland Publishing Co.: New York, 1987.

McClintock, B. The Significance of Responses of the Genome to Challenge. *Science* **1984,** *226*, 792–801.

Meyerowitz, E. M. Structure and Organization of the *Arabidopsis thaliana* Nuclear Genome. En *Arabidopsis*, Meyerowitz, E. M., Somerville, C. R., Eds.; Cold Spring Harbor Laboratory Press, 1994, pp 21–36.

Michalak, P. Epigenetic, Transposon and Small RNA Determinants of Hybrid Dysfunctions. *Heredity*. **2009,** *102*, 45–50.

Mirouze, M.; Paszkowski. J. Epigenetic Contribution to Stress Adaptation in Plants. *Curr. Opin. Plant Biol.* **2011,** *14*, 267–274.

Miura, A.; Kato, M.; Watanabe, K.; Kawabe, A.; Kotani, H.; Kakutani, T. Genomic Localization of Endogenous Mobile CACTA Family Transposons in Natural Variants of *Arabidopsis thaliana. Mol. Genet. Genom.* **2004,** *270*, 524–532.

Miyao, A.; Tanaka, K.; Murata, K.; Sawaki, H.; Takeda, S.; Abe, K.; Shinozuka, Y.; Onosato, K.; Hirochika. H. Target Site Specificity of the Tos17 Retrotransposon Shows a Preference for Insertion Within Genes and Against Insertion in Retrotransposon-Rich Regions of the Genome. *Plant Cell* **2003,** *15*, 1771–1780.

Noda, K.; Glover, B. J.; Linstead, P.; Martin, C. Flower Colour Intensity Depends on Specialized Cell Shape Controlled by a Myb-Related Transcription Factor. *Nature* **1994,** *369*, 661–674.

Ogata, J.; Itoh, Y.; Ishida, M.; Yoshida, H.; Yoshihiro, O. Cloning and Heterologous Expresion of cDNAs Encoding Flavonoid Glucosyltransferases from *Dianthus caryophyllus*. *Plant Biotech J.* **2004,** *21*, 367–375.

Orgel L. E.; Crick. F. H. Selfish DNA: The Ultimate Parasite. *Nature* **1980,** *284*, 604–607.

Palmer, J. D.; Adams, K. L.; Cho, Y.; Parkinson, C. L.; Qiu, Y. L.; Song, K. Dynamic Evolution of Plant Mitochondrial Genomes: Mobile Genes and Introns, and Highly Variable Mutation Rates. *Proc. Natl. Acad. Sci. USA* **2000,** *97*, 6960–6966.

Palmer, J. D.; Delwiche, C. F. The Origin and Evolution of Plastids and Their Genomes. In *Molecular Systematics of Plants II. DNA Sequences*; Soltis, D. E., Soltis, P. S., Doyle, J. J., Eds.; Kluwer Academic Publishers: Norwell, 1998; p 409.

Pruitt, R. E.; Meyerowitz, E. M. Characterization of the Genome of *Arabidopsis thaliana*. *J. Mol. Biol.* **1986,** *187*, 169–183.

Qi, L. L.; Echalier, B.; Chao, S.; Lazo, G. R.; Butler, G. E.; Anderson, O. D. et al. A Chromosome Bin Map of 16,000 Expressed Sequence Tag Loci and Distribution of Genes Among the Three Genomes of Polyploid Wheat. *Genetics* **2004,** *168*, 701–712.

Quattrocchio, F.; Wing, J.; Leppen, H.; Mol, J.; Koes R. Regulatory Genes Controlling Antocyanin Pigmentation Are Functionally Conserved Among Plant Species and Have Distinct Sets of Target Genes. *Plant Cell* **1993,** *5*, 1497–1512.

Rudall, P. J.; Buzgo, M. Evolutionary History of the Monocot leaf. In *Developmental Genetics and Plant Evolution*; Cronk, Q. C. B., Bateman, R. M., Hawkins, J. A., Eds.; Taylor & Francis; London, 2002; p 458.

Safár, J.; Simková, H.; Kubaláková, M.; Cíhalíková, J.; Suchánková, P.; Bartos, J.; Dolezel, J. Development of Chromosome-Specific BAC Resources for Genomics of Bread Wheat. *Cytogenet. Genome Res.* **2010,** *129*, 211–223.

San Miguel P., Tikhonov, A.; Jin, Y. K.; Motchoulskaia, N.; Zakharov, D.; Melake-Berhan, A.; Springer, P. S.; Edwards, K. J.; Lee, M.; Avramova, Z.; Bennetzen. J. L. Nested retrotransposons in the Intergenic Regions of the Maize Genome. *Science* **1996,** *274*, 765–768.

Schwarz-Sommer, Z.; Gübitz, T.; Weiss, J.; Gómez-di-Marco, P.; Delgado-Benarroch, L.; Hudson, A.; Egea-Cortines, M. A Molecular Recombination Map of *Antirrhinum majus*. *BMC Plant Biol.* **2010,** *10*, 268–275.

Shigeru, I.; Atsushi, H.; Yasuyo, J.; Yoshiki, H.; Yoshishige I. Floricultural Traits and Transposable Elements in the Japanese and Common Morning Gloriesa. *Ann. NY Acad. Sci.* **1999,** *870*, 265–274.

Singh, R. S.; Krimbas, C. B. *Evolutionary Genetics: From Molecules to Morphology*; Cambridge University Press: Cambridge, 2000.

Stubbe, H. Genetics and Cytology of Antirrhinum L. *sect. Antirrhinum.* **1966,** *421*, 1–12.

Terol, J.; Castillo, M. C.; Bargues, M.; Pérez-Alonso, M.; de Frutos, R. Structural and Evolutionary Analysis of the Copia-Like Elements in the *Arabidopsis thaliana* Genome. *Mol. Biol. Evol.* **2001,** *18*, 882–892.

To, K. Y.; Wang, C. K. *Molecular Breeding of Flower Colour*. In *Floriculture, Ornamental and Plant Biotechnology*, Vol. 1; Global Science Books, 2006.

Ungerer, M. C.; Strakosh, S. C.; Zhen, Y. Genome Expansion in Three Hybrids Sunflower Species Is Associated with Retrotransposon Proliferation. *Curr. Biol.* **2006,** *16*, 876–873.

Vaillant, C.; Palmeira, L.; Chevereau, G.; Audit, B.; d'Aubenton-Carafa, Y.; Thermes, C.; Arneodo, A. A Novel Strategy of Transcription Regulation by Intragenic Nucleosome Ordering. *Genome Res.* **2010,** *20*, 59–67.

Van Houwellingen, A.; Souer, E.; Spelt, K.; Kloods, D.; Mol, J.; Koes, R. Analysis of Flower Pigmentation Mutants Generated by Random Transposition Mutagenesis in *Petunia hybrida*. *Plant J.* **1998,** *13*, 39–50.

Wang, H.; Chai, Y.; Chu, X.; Zhao, Y.; Wu, Y.; Zhao, J.; Ngezahayo, F.; Xu, C. G.; Liu, B. Molecular Characterization of a Rice Mutator-Phenotype Derived from an Incompatible Cross-Pollination Reveals Trans-Generational Mobilization of Multiple Transposable Elements and Extensive Epigenetic Instability. *BMC Plant Biol.* **2009,** *9*, 63.

Wanjugi, H., Coleman-Derr, D.; Huo, N.; Kianian, S. F.; Luo, M. C.; Wu, J.; Anderson, O.; Gu, Y. Q. Rapid Development of PCR-Based Genome-Specific Repetitive DNA Junction Markers in Wheat. *Genome* **2009,** *52*, 576–587.

Watson, J.; Gann, A.; Baker, T.; Levine, M.; Bell, S.; Losick, R.; Harrison, S. *Molecular Biology of the Gene*, 7ta ed.; Cold Spring Harbor Laboratory Press, 2014.

Yoshida, H.; Itoh, Y.; Ozeki, Y.; Iwashina, T.; Yamaguchi, M. Variation in Chalcononarigenin 2′-O-glucoside Contents in the Petals of Carnations (*Dianthus caryophyllus*) Bearing Yellow Flowers. *Sci. Hortic.* **2004,** *99*, 175–186.

Yoshishige, I.; Yasuyo, H.; Tetsuya, S.; Kichiji, K.; Shigeru, I. Isolation of a Supresor-Mutator/Enhancer-Like Transposable Element, Tpn1, from Japanese Morning Glory Bearing Variegated Flowers. *Plant Cell* **1994,** *6*, 375–383.

CHAPTER 7

Transposons as Natural Genetic Engineers of Genome Mutation, Evolution, and Speciation

HIMANSHU PANDEY[1], DIVYA CHAUHAN[2], NGUYEN TRAN HAI BANG[3], RAJESH KUMAR SINGHAL[4*], and UDIT NANDAN MISHRA[5]

[1]YSP University of Horticulture and Forestry, Solan, Himachal Pradesh, 173230, India

[2]ICAR-National Bureau of Plant Genetic Resource, New Delhi, 110012, India

[3]Cuu Long Delta Rice Research Institute, Can Tho City, Vietnam

[4]ICAR-Indian Grassland and Fodder Research Institute, Ministry of Agriculture and Farmer Welfare, JHANSI, Uttar Pradesh, 284003, India

[5]Faculty of Agriculture, Sri Sri University, Sri Sri Vihar, GodiSahi, Cuttack, Odisha, 754006, India

[*]Corresponding author. E-mail: rajasinghal151@gmail.com

ABSTRACT

Repetitive DNA sequences that form a large portion of higher eukaryotic genome and can change positions within the genome are known as transposons. The movement of these genetically mobile elements occurs through cut and paste mechanism and is guided by enzymes. Transposable elements are considered as DNA transfer vehicles as they are capable of introducing a new sequence of DNA into the genome of an organism. Moreover, their utilization in the development of transgenic lines and in process of insertional

Plant Transposable Elements: Biology and Biotechnology, Deepu Pandita, Anu Pandita, (Eds.)

mutagenesis has been observed in certain plants and animals as they can move independently of the host system and hence act as an important tool to develop mutants. In addition, the study of the evolution process in particular species requires the knowledge at both phenotypic and molecular level so to properly understand genetic diversity and gene flow at species level. Transposable elements are capable of affecting genome evolution by promoting recombination and transferring genes to new locations. Coevolution of transposons together with plant genome had led to overcome unfavorable environmental conditions or process of hybridization and polyploidy. This chapter deals with the role of transposons in evolution and species diversity.

7.1 INTRODUCTION

Transposable elements (TEs) have the ability to move and cause mutation within the host genome. Such mutations which can arise at unusually high frequencies are insertion, excision, and chromosomal rearrangements. Hence, the expression of genes present within or adjacent to the TE can be altered. In particular biological systems, the mobility of TEs has been well elucidated in activator and dissociation (Ac/Ds) elements of maize (Shepherd et al., 1988) and P-M dysgenic hybrid in *Drosophila melanogaster* (Finnegan, 1992). Further, spontaneous mutations caused by TE insertions have been reported in Drosophila while many are observed in the humans (Kazazian et al., 1988), yeast (Lambert et al., 1988), and mice (Gridley et al., 1987) also. Previous studies on TEs display that the rigor of the mutational effects is influenced by the structure of the inserted sequence and location of the TE inserted within the gene. The insertion of TE near or at the gene of interest can possibly increase the intensity of the mutant phenotype (Engels, 1989). Therefore, extreme phenotypes or nulls can be obtained in the coding region during insertional mutagenesis (Chia et al., 1986).

7.2 SIGNATURE TAGGED TRANSPOSON MUTAGENESIS

It is the negative selection technique given by David Holden (1995), used to determine genes, that are essential under a given condition. In this technique, all mutants are created by random transposon insertion and subjected to further tested for their survival in laboratory media. Therefore, mutants formed by this technique contain a molecular "tag" sequence that uniquely defines it. Further, tags used in the experimental studies can be searched through computational analysis. Later, this inserted tagged TE is used to find and locate the gene it has disrupted. As a result of this insertion, protein structure and function will be adversely affected. Transposon tagging is a specialized application of mutagenesis where a gene can be inactivated by the insertion of the transposon into the gene sequence itself. Transposon tagging was primarily reported by Bingham et al. (1981) in *Drosophila melongastor*. In this study, a gene from the cDNA of white-eyed mutant was used by using a probe homologous to Copia element. Later, this technique was applied to plants (Walbot, 1992) and mice (Friedrich and Soriano, 1991). Mutagenesis by using this strategy involves three genetic procedures: (1) first step includes the incorporation of the transposon into a specific DNA, (2) screening of altered phenotype having the desired trait that depicts that alteration in DNA sequence which was caused due to a specific TE insertion, (3) and hence identifying and cloning of specific sequence (Ivics and Izsvak, 2010).

This recent approach holds advantages over traditional mutagenesis. Primarily, mutations induced by using the processes of irradiation or chemicals are difficult to generate, whereas mutations induced by tagging of TE are generally utilized for gene cloning purposes. Secondly, it is possible to clone a gene without having its prior information regarding gene of interest. This specific advantage is essential for genes controlling developmental processes at different stages (Friedli and Trono, 2015). Tagging of TE has shown to be a very effective technology yet, it has some complications. TEs can induce mutations in multiple ways, but when they are present in a greater amount in genome, it often leads to unstable mutations. Thus, it is difficult to deduce which portion of the DNA segment in the genome of an individual is mutagenized with transposons (Dupuy et al., 2009). Some examples of different transposon systems causing mutagenesis are discussed below.

7.2.1 SLEEPING BEAUTY MUTAGENESIS

This system uses nonviral vectors for incorporating a gene cassette into a vertebrate genome. It follows a cut and paste mechanism, to achieve sustained expression of the transgene, transposing of gene cassette into the organism's genome (Aronovich et al., 2011) is required. Further modifications can be made by altering the transposon sequences and the transposase enzymes used for the study. Various determining elements for the success of sleeping beauty mutagenesis are shown in Figure 7.1. This system contains a cassette of genes to be expressed along with its own transposase enzyme.

Here, the transposon itself is flanked by the sequences known as inverted/direct repeats (IR/DR), repeated twice in a straight fashion, whereas the internal region forms the sequence to be transposed and can also comprise the transposase gene. The transposition process involves the following steps: (1) primary substrates binding to the IRs, (2) resulting in the formation of a synaptic complex (SCF) followed by (3) excision from the donor DNA thereby generating 3′ overhangs. (4) The excised elements are then insert into TA dinucleotide as target site for reintegration (Sakai et al., 1995).

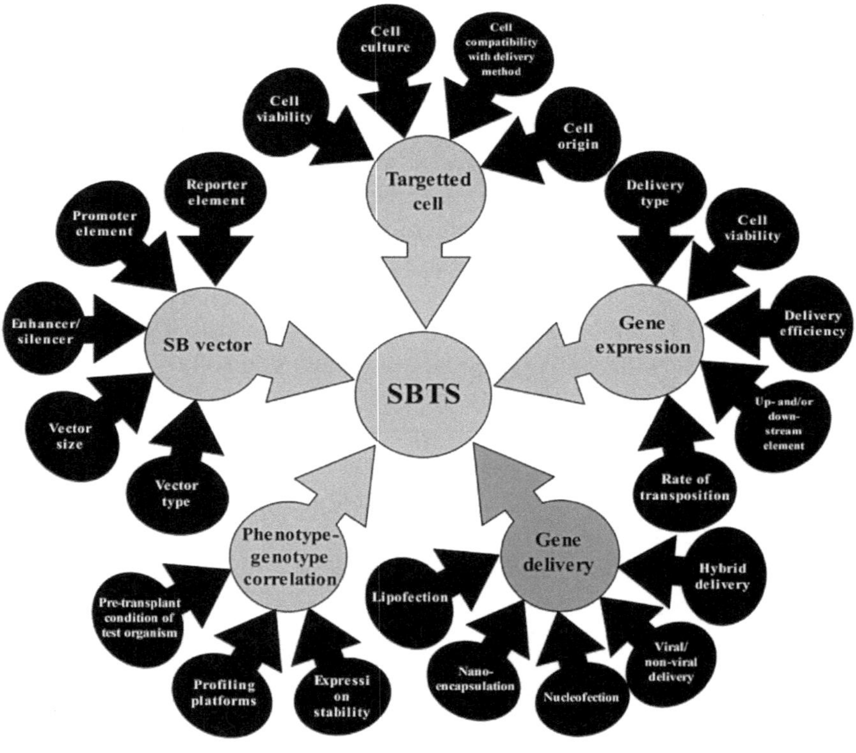

FIGURE 7.1 Elements of sleeping beauty transposon system (SBTS).

Some of the applications of the SB system include routine cell culture techniques for creating gene knockdowns and transgenic cell lines (Miskey et al., 2005). Also, the plasmid-based system can be integrated with conventional nonviral delivery techniques. Although, a sustained expression of transgenes holds a great challenge for big-scale biotechnological applications. For such an instance, it represents a highly suitable transgene expression vector having the ability to encourage an effective genomic integration in different mammalian cell types. Transposon systems, including SB and piggyBac are appropriate candidates for producing clonal cell lines for huge-scale (industrial) production of recombinant proteins in Chinese Hamster Ovary (CHO) cells (Balasubramanian et al., 2016). Further, the expression of the gene cassette can be altered by combining the SB cassette with other molecular engineering tools (Grabundzija et al., 2013). In addition, pharmacokinetic and pharmacodynamics characterization of different synthetic analogs of vitamin D_3 was done using an SB system developed from a protein-based sensor system (Staunstrup et al., 2011).

7.2.2 TRANSPOSON TN5

This system is comprised of a bacterial composite transposon where the genes are bordered by two similar insertion repeats, namely, IS50R and IS50L belonging to the right and left flanking regions of the transposon system accordingly (Reznikoff, 1993). The IS50R is responsible for the coding of two different types of proteins, namely, inh and Tnp. Inh protein odes for an inhibitor of transposase and lacks the 55 NH_2-terminus of amino acids, whereas Tnp encodes for an enzyme transposase. This system begins with Tnp binding to the outside and inside ends of OE and IE (IS50) sequence. Once, this sequence is cut out of the chromosome the two terminals are combined through the process of oligomerization of the DNA sequence. Therefore, after insertion of 9-base pair 5′ends into the targeted sequence, along with the transposon, its incorporated genes are introduced into the desired location and perform duplication on different sides of the transposon. Expression of the genes will be achieved by placing the transposon under the control of a host promoter. Besides genes of interest, incorporated genes usually include, a marker gene to find the transformants, a eukaryotic promoter, and terminator sequence, and 3′ UTR region to differentiate the genes from a polycistronic DNA region (Wilson et al., 2007). For example, the genes that are linked with virulence during Mycobacterium tuberculosis infection were recognized through the process mediated by transposonal mutagenesis-using gene knockout technique. A plasmid, *pCG113* having kanamycin resistance genes along with IS1096 insertion sequence was engineered to constitute a 80-base pair variable tag. Transformed cells were later selected for resistant cells from the colonies plated on kanamycin-containing medium. Further, colonies were screened for candidate virulence genes having mutations. The attenuated phenotypes with mutations were mapped by amplifying adjoining regions to the IS1096 sequences and a comparison was made with the published Mycobacterium tuberculosis genome (Camacho et al.,1999).

7.3 APPLICATIONS

Many plant and animal breeders and other geneticists dream to clone quantitative trait loci (QTLs). There is a possibility to clone QTLs in some species that include Drosophila, maize, and mice through transposon tagging in the regions where TE is available. Therefore, loci influencing a quantitative trait can be detected by measuring the alterations in the mean attribute value of an inbred line into which a transposon has been introduced (Soller and Beckmann, 1990). The location of the introduced sequences along with the QTL can be mapped accurately such that these transposons can easily tag the QTL. Also, the practicality of this strategy has been verified in Fruitfly (Mackay et al., 1992).

Tagging of the transposon is a crucial technique in the field of "reverse genetics," that is, by elucidating a gene's function even after it has been cloned. A deletion caused by transposons from a particular location may lead to an altered phenotype, which may reveal the function of the gene. Therefore, screening of natural or TE-mutagenized population can be done by a particular PCR method. This can be achieved by using two oligonucleotide primers, where one sequence should be synthesized from the cloned gene of interest and the other should be homologous to the terminal sequence of the transposon. For example, a gene expressed in the compound eye during P-elements insertions in Drosophila can be detected by using this technique (Ballinger and Benzer, 1989). In addition, TEs can be considered to reconstruct phylogenies by means of presence and absence analyses. Identification of transgenes can be performed through the PCR technique on the organism's genome using a transposon-specific primer and an open reading frame-specific primer (Hamer et al., 2001).

Sleeping beauty has the potential to generate induced pluripotent stem cells. This can be achieved by reprogramming somatic cells to produce patient/disease-specific pluripotent stem cells that will genetically modify, expand, and can differentiate into multiple cell types for gene therapy applications. A study by Okita et al. (2007) showed the generation of pluripotent stem cells (iPSCs) by using retroviral and lentiviral vector systems. However, efforts have been made to generate iPS cells by means of nonviral approaches such as transposon-based systems (Muenthaisong et al., 2012). The SB system is convenient and economical and hence has been used for the production of iPSCs in different organisms and models (Fatima et al., 2016; Talluri et al., 2015). Its reprograming efficiency is quite similar when compared to viral vectors (Grabundzija et al., 2013).

7.4 ROLES OF TRANSPOSONS IN GENOME EVOLUTION

Genome evolution is being driven by many processes that involve breakage and rejoining of different chromosomes, segmental and gene duplication events, shuffling of functional domains in the coding region followed by gene conversion. It is reported that non-LTR retrotransposon over a history of some 500–600 million years is known to carry a reverse transcriptase (RT) that is quite similar to the RT of the mobile group II introns that occur in mitochondrial and chloroplast genomes of fungi and plants (Havecker et al., 2004; Lambowitz and Zimmerly, 2011). Although, there is no significant information regarding the evolutionary function of transposons in plants but in the case of mammals, there have been significant evidences that prove these genetic mobile elements are the source of genome variation. There are a number of ways by which TEs can bring changes in genes and genomes.

7.4.1 ALTERED PATTERN OF INHERITANCE

Various chromosome rearrangements that include translocations, inversions, and deletions can be encouraged by events involving mobile elements. These events can also lead to recombination between Copia elements distributed around the genome. In *Saccharomyces cerevisiae,* the occurrence of yeast transposable element (Ty) was reported during inversion, deletions, duplications, and translocation events. Probably, these changes were observed during the recombination events between pre-existing elements. The chromosome organization is dependent on location and orientation of the transposons involved in the process (Mieczkowski et al., 2006). There are few cases where rearrangements can directly result from the activity of TEs. For example, transposition of P elements in Drosophila results from crossing males of a P strain containing complete elements with females of a strain that does not contain an M stain during P-M hybrid dysgenesis (Khurana et al., 2011).

A study conducted by Martienssen (1998) on several strains of *A. thailana* reported the molecular basis of epigenetic inheritance and its interaction with different regulatory networks such as DNA methylation, RNA interference, and histone modification. While converging on a particular group of transposons with mutations interfering in the above processes, they found that some mutations affected all the elements while other mutations affected only a subset. The results revealed the varied quality of responses, as some mutations were lost while others were inherited in the next generation.

7.4.2 DIFFERENTIAL GENE EXPRESSION

Chromosome rearrangements can result in differential expression of individual genes. It can lead to small alterations in the expression of a gene as its coding region is falls under the influence of the regulatory sequence of the other gene. For example, a Antp 73b mutation in *D. melanogaster* has resulted in transformation of antennae into leg structure. This mutation involves inversion that breaks antennapedia complex, thereby fusing the whole exon with the promoter of the other gene (Frischer et al., 1986). Studies in the past have reported the role of TE-mediated epigenetic effects on gene expression. A study by author Slotkin et al. (2009) reported that differences in the expression of epigenetic modifiers involved in downstream targets are responsible for decrease in the expression of genes in *A. thaliana* that correlates with activation and methylation of TEs. In addition, some observations made by McClintock depicted that transposons have the ability to influence nearby gene expression in maize following a heritable fashion, and hence called as controlling elements (McClintock, 1951, 1956). Further, studies in Drosophila of Hoppel that belongs to the member of the P-element indicate that these mechanisms are not only limited to plants but are also present in animals. These studies revealed that DNA transposons are mobile characters for local heterochromatin formation as a byproduct of their repetitive nature (Haynes et al., 2006; Slotkin and Martienssen, 2007). Some other sequences known as boundary or insulator elements are also derived from repeats and are associated with siRNAs, such transposons participate in transcriptional activity. These events are inheritable and stable in nature but may lead to dynamic changes in response to genetic stress and environment cues leading to interspecific hybridization or polyploidization (Kashkush et al., 2003; Noor and Chang, 2006). It can trigger amplification and movement of TEs causing epigenetic reshuffling and structural changes in the genome. It also offers a chance to favor natural selection to establish new chromosomal domains and regulatory circuits (McClintock, 1984).

7.4.3 AMPLIFICATION OF DNA SEQUENCES

TEs have the capability to amplify DNA sequences within a genome during replicative transposition events, or throughout duplication events that took place during recombination between elements. In this process, the recombination takes place between copies of transposons that reside in the similar order at distinct locations on a chromosome and cleave the intervening DNA when it appears as an interchromatid. During interchromatid recombination, intervening DNA gets deleted from one strand and duplicates on other. For example, in eukaryotes short-interspersed sequences and processed pseudogenes are usually amplified by transposition events. TE is generally bordered by site-specific duplications that contain A-rich nucleotide strands at the 3′terminal sequence and thereby revealing insertion at a new specific location after reverse transcription of RNA. mRNAs act as an intermediates for pseudogenes since they do not contain introns.

7.4.4 VARIATIONS IN GENE SEQUENCE

TEs that perform excision often result in sequence alterations within the set of genes that can hold an evolutionarily advantage. Usually, when the number of nucleotides is inserted or deleted from the insertion site must be in multiples of three and the original insertion was in an exon that will lead to allele coding for the different gene product. Like other TEs, DNA transposons have properties that differ from those of retrotransposons that influence the possibility of their involvement in these mechanisms. The most probable outcome of transposon insertion can lead to disruption of the exon hindering the production of the viable gene product. Many transposons that exhibit cut-and-paste mechanism display a preference for insertion into or within the surrounding of genes and this property makes it powerful gene tagging tool that can be routinely used by geneticists (Muñoz-López and García-Pérez, 2010). Previous studies on P elements of Drosophila (Spradling et al., 1995), mutator elements in maize (Dietrich et al., 2002), and the Tc3 element in nematodes (Rizzon et al., 2003) has depicted a bias insertion into genic neighborhoods.

A breakthrough study on Miniature inverted-repeat transposable elements (MITE) explosion in rice has demonstrated a pattern of insertion, which was primarily because of targeting rather than the result of natural selection (Feschotte and Pritham, 2007). MITES are usually present in high copy-number in genomic regions and gene-rich environments of both plants and animals (Feng, 2003). Allelic diversity in the natural population can result from genic proximity of DNA transposon. Insertional mutations are caused by DNA transposons that result in spontaneous excisions and unstable reversible phenotypes (Wessler, 1988). DNA transposon excision sites influence the nature of variations by introducing random fillers, small deletions, and inversions. The excision/insertion property of DNA transposon plays a major role in modifying gene and altering promotor sequences that result in allelic diversity, hence its absence during these processes would prove difficult to illustrate (Lin and Waldman, 2001).

Speciation is a natural process that involves the occurrence of new species which is distinct from the existing species and is reproductively isolated (Coyne and Orr, 2004). The focused area of speciation is to understand the mechanism of reproductive isolation between species (Abbott et al., 2013; De Loof, 2015). Various methodologies have been applied in identifying barriers that are responsible for causing reproductive isolation between different species having a common origin (Coyne and Orr, 2004; Harrison, 2012), evolution forces (Seehausen et al., 2014), and the rate of speciation (Wang et al., 2015). Despite of all these problems, progress in the area of identifying gene loci that are related to reproductive isolation and evolutionary forces that drive it has been explored. It is still unclear what kind of mutations leads to reproductive isolation and the mechanism governing it.

There are two ways to detect reproductive isolation. In the first method, different crosses are made and genetic maps are prepared to find out loci related to reproductive isolation, such studies help in determining genetic changes that are useful in maintenance of species identity and molecular alterations that were formerly involved in process of speciation. This method is generally utilized when closely related species are studied at different stages of development (Widmer et al., 2009).

Another technique to study the reproductive isolation generally involves detection of changes or mutation occurring at molecular level and its association in relation to isolation for different genotypes. If changes at molecular level are responsible for creating obstruction to gene flow, then it may be considered as a possible reason for origin of new species or its persistence for collapse of gene flow. An example of this approach includes chromosomal inversions that are commonly linked with a decrease in recombination frequency and are associated with gene combinations that help in creation of reproductive isolation between different species (Noor et al., 2001; Hoffmann and Rieseberg, 2008). However, this method has not been applied to a greater extent to study the importance of mechanisms bringing out reproductive isolation at the molecular level. The present chapter provides an insight into TEs that are considered as important agents responsible for reproductive isolation along with their contribution in different crops to study the evolutionary relationships.

TEs are considered as mobile DNA elements (sequences) that are able to move within the genome by mechanism of copy and paste. TEs constitute approximately 80% of total nuclear DNA in plants, 3–20% in case of fungus, and 3–52% in metazoans (Bennetzen and Wang, 2014). TEs are classified based on the mechanism they utilize for transpose. In Class I type of transposons, RNA act as intermediate in order to replicate themselves within a nuclear genome, while in another type, that is Class II elements, there is no involvement of RNA intermediate and they move through cut-and-paste method, furthermore their replication and movement occur directly via by DNA to DNA. Different classes of TEs are given in. Evidently, the most important class of TEs differs between taxa (Kidwell 2002; Lockton and Gaut, 2010) and species, and their activity, position, and frequency in genome vary at different levels of population. TEs were first discovered in maize crop by Barbara McClintock in the year 1950 (McClintock, 1950), where they were responsible for causing somatic mutations influencing phenotypic and genotypic expression which greatly depend upon the time of transposition and its insertion site on chromosomes. The position of TE is responsible for altering gene expression, which may have deleterious effects on gene expression (Woodhouse et al., 2006; Feschotte et al., 2009). Transposons normally occur in all higher organisms (Voytas et al., 1992; Aziz et al., 2010) and it has a negative impact on plant and animal health furthermore disrupt important gene function, hence TEs are also referred to as selfish DNA. Generally, functional gene interruption is not considered as primary consequence of transposons as they migrate throughout the genome of an organism. TEs are responsible for changes occurring in the regulatory region of DNA, expansion in genome, and help in creation of chromosomal mutants through the inversions process. Moreover, transposons generate changes in the genome at a rapid rate (Kapusta et al., 2013; Flutre et al., 2011) under abiotic stress conditions—a hypothesis proposed by McClintock (McClintock, 1950). These alterations caused in the genome act as a source for the creation genotypic and phenotypic novelty upon which selection pressure can act (Hollister and Gaut, 2009). Due to its efficacy in creation of mutation within DNA sequences when required, some hypotheses suggest that transposons are present in entire genomes through the process of multilevel selection (Brunet and Doolittle, 2015).

Thus, TEs are an integral element of eukaryotic genomes that play a crucial role in rates of speciation and divergence. TEs have been reported as a major cause of RI between germplasm (Kidwell et al., 1977). However, the functional importance of TEs at a molecular level for generation of new species remains unclear and unexplored. The basic ideology behind the involvement of transposons in the creation of new species has shown significant importance (Hurst and Schilthuizen, 1998). RI which is caused by the process of intragenomic conflict seems to occur naturally in the environment but is supposed to be rare and not well studied. Preferential selection during meiosis, endosymbionts, and maternal inheritance is considered as an important source of RI (Coyne and Orr, 2004), and various different theories have been applied to examine the functional significance of these factors in the speciation process (Ginzburg et al., 1984). Yet, the purpose of TEs in process of initiation and maintenance of the speciation needs extensive studies. In this chapter, we highlight the functional importance of TEs as an important factor for the origin and maintenance of species, with a focus on other important events like mutation and evolution. The chapter also provides an insight on to how the mutations caused by TEs will cause mutations and invariably lead to evolution and speciation. Further, the role of TEs in the creation of reproductive barriers and genotypic variation has also been discussed.

7.5 SPECIATION: RAPID OR GRADUAL PROCESS?

The important question in evolutionary studies generally arises whether process of speciation is expeditious or occurs at a slow rate. Darwin, in his theory of evolution and speciation, describes that speciation occurs gradually at a very slow rate on a time scale due to mutations that are heritable and involves the organism that is geographically separated and eventually lead to entirely different organisms that are reproductively separated from the main species. The function of transposons in creating heritable changes that are better known as mutation leads genetic alterations in the genomes is well known. This chapter deals with genetic alteration in relation to speciation phenomena. Another method of speciation occurs rapidly which includes breeding procedures like polyploidization and hybridization in plants. Approximately 35% of phanerograms evolved by the process of polyploidy (Wood et al., 2009). Furthermore, research conducted on hybrids and polyploidy, which are comparable to developing species, has proven that TE act as precise detectors of genomic shock that are generated due to divergent genome blending. Unrevealing TE in hybrids genotypes or polyploids may be considered parallel to hybrid dysgenesis, contributing to a decrease in hybrid vigor, and prevention of gene-flow across closely linked but divergent species (Feschotte and Pritham, 2007). In similar cases, transposons switch functions from being the supporter of the speciation process and maintainer of species integrity. This chapter deals with importance of transposons in gradual and rapid methods of speciation in relation with mutation and evolution. Finally, here it is discussed how mutation due to transposons causes evolution and how the origin of new species occurs due to genomic shock and other factors.

7.5.1 PROCESS OF SPECIATION BY DNA CUT-AND-PASTE MECHANISM

DNA cut & paste sequences usually create mutations having a profound effect which include point mutations & indels in the genome, or they are responsible for causing chromosomal aberration which includes inversions & translocations. The discovery of McClintock on functional aspects of TEs including chromosomal breakage, and recombination (McClintock, 1946; 1948) is a significant illustration of how transposons can generate rapid genetic evolution that can cause genomic isolation and further responsible for the speciation process. McClintock's chromosomal breakage of double-Ds sequences emerged out to have a distinctive structure, constituting fa-Ds sequences ingrained within other Ds sequences (Jiang and Wessler, 2001). While these are not considered as typical sequences, such double-Ds elements are not found commonly, due to their ability to move to a nearby location on chromosomes (Dooner and Belachew, 1989). TEs generally cut out without generating any rearrangement in chromosome, producing a DNA double-strand break that is repaired by the process of nonhomologous-end-joining, leaving behind some footmark (Rinehart et al., 1997). This footmark can lead to slow changes in karyotype of chromosomes. Such minute mutation that accumulates over time can greatly affect chromosomal recombination process, which can eventually lead to reproductive polymorphism. Indeed, DNA point mutations eventually decrease the rate of meiotic homologous crossing over in the genomic variable site of chromosome (Emmanuel et al., 2006).

As it was reported in the case *Oryza australiensis* transposons proliferation in centromeric and noncentromeric portion of chromosomes causes a doubling of genome size during the process of speciation (Piegu et al., 2006). In cotton, TE proliferation varies among distinct species, showing mobilization of transposons at the time or after speciation help in generation of divergent genomes (Hawkins et al., 2006). Despite of wide morphological divergence of chromosomes during speciation process still gene order is preserved with the passage of time. Retro sequences proliferation occurred within a species as derived from intervarietal discrepancy of *Zea mays* (Wang and Dooner, 2006) and *Oryza sativa* (Huang et al., 2008).

7.6 TRANSPOSABLE ELEMENTS AND REPRODUCTIVE ISOLATION

Characteristics involved in the maintenance of species isolation can be grouped in accordance with its occurrence in the reproductive phases. Premating obstructions that are involved in reproductive isolation include behavioral attributes that decrease the mating ability between two individuals will mate and include habitat isolation and mating choice. Prezygotic factors are required in interactions between different gametes and further, involve pollen-egg cell incompatibility (self-incompatibility and male sterility). Lastly, postzygotic hindrance effect occurs after fertilization process and involves in reduced fitness of hybrids resulted from crossing (Coyne and Orr, 2004; Orr and Presgraves, 2000). The prezygotic and postzygotic studies in relation with reproductive isolation process have been extensively utilized at varying level and their connection with transposons was reported in some of the studies. The following book chapter deals with process and mechanism how TEs are responsible for affecting a trait that is significantly involved in reproductive isolation, therefore emphasizing the process of speciation due to mutation and evolution.

7.7 FUTURE DIRECTIONS

Transposon technology falls in a promising arena and its application is emerging in the field of gene therapy. Transposon insertion sequencing facilitates the knowledge of evolution and its understanding in different magnitudes of cancer development, its progression, response to therapy and genetic interactions, etc., in the future (Fig. 7.2). Transposon mutagenesis will be used to reflect deviations in the population by including alterations in mouse models. It will facilitate precise and suitable therapy to be delivered to the patients suffering from chronic disease like cancer and obesity. Moreover, transposon mutagenesis related studies can support to propose of changes in cancer therapy by discovering different resistance mechanisms to targeted therapies and novel ideas for improved drug design. In addition, transposon screens can be used to detect metastasis-specific drivers and tumor types in providing further treatment choices for patients with high-risk of disease. Later, future screens can be proposed to study the effects of aging on tumor development. Initial reports on the application of sleeping beauty transposon system for gene therapeutic purposes are now available and published in public domain. Two major areas for the initial point of research in the near future include improvement and optimization of SB for an efficient and safe gene transfer. The latter is developing a better understanding of molecular interactions between the transposon and the cell. Further, it also has a broad scope in plant systems like development of T-DNA insertion line as studied in the case of *Arabidopsis thaliana* (model organism) and many other model plants. In future, T-DNA insertion lines could be possible utilized to know the function of genes in already sequenced crops. Recently, it is used to study phylogeny of various angiosperm plants, which is important for a taxonomic point of view.

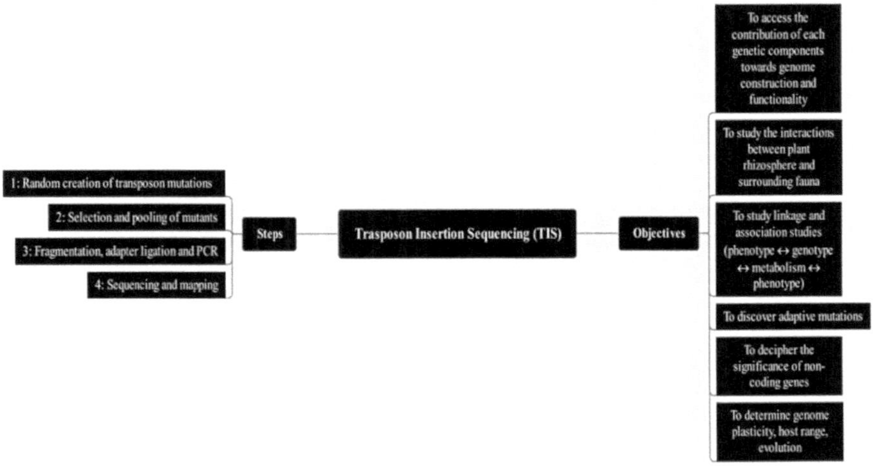

FIGURE 7.2 Steps and objectives of transposon insertion sequencing (TIS).

KEYWORDS

- **transposable elements**
- **coevolution**
- **mutagenesis**
- **polyploidy**

REFERENCES

Abbott, R.; Albach, D.; Ansell, S.; Arntzen, J. W.; Baird, S. J.; Bierne, N.; Boughman, J.; Brelsford, A.; Buerkle, C. A.; Buggs, R.; Butlin, R. K. Hybridization and Speciation. *J. Evol. Biol.* **2013,** *26* (2), 229–246.

Aronovich, E. L.; McIvor, R. S.; Hackett, P. B. The Sleeping Beauty Transposon System: A Non-Viral Vector for Gene Therapy. *Human Mol. Genet.* **2011,** *20* (R1), R14–R20.

Aziz, R. K.; Breitbart, M.; Edwards, R. A. Transposases Are the Most Abundant, Most Ubiquitous Genes in Nature. *Nucl. Acids Res.* **2010,** *38* (13), 4207–4217.

Balasubramanian, S.; Rajendra, Y.; Baldi, L.; Hacker, D. L.; Wurm, F. M. Comparison of Three Transposons for the Generation of Highly Productive Recombinant CHO Cell Pools and Cell Lines. *Biotechnol. Bioeng.* **2016,** *113* (6), 1234–1243.

Ballinger, D. G.; Benzer, S. Targeted Gene Mutations in Drosophila. *Proc. Natl. Acad. Sci.* **1989,** *86* (23), 9402–9406.

Bennetzen, J. L.; Wang, H. The Contributions of Transposable Elements to the Structure, Function, and Evolution of Plant Genomes. *Annu. Rev. Plant Biol.* **2014,** *65*, 505–530.

Bingham, P. M.; Levis, R.; Rubin, G. M. Cloning of DNA Sequences from the White Locus of D. Melanogaster by a Novel and General Method. *Cell* **1981,** *25* (3), 693–704.

Brunet, T. D.; Doolittle, W. F. Multilevel Selection Theory and the Evolutionary Functions of Transposable Elements. *Genome Biol. Evol.* **2015,** *7* (8), 2445–2457.

Camacho, L. R.; Ensergueix, D.; Perez, E.; Gicquel, B.; Guilhot, C. Identification of a Virulence Gene Cluster of Mycobacterium Tuberculosis by Signature-Tagged Transposon Mutagenesis. *Mol. Microbiol.* **1999,** *34* (2), 257–267.

Chia, W.; Howes, G.; Martin, M.; Meng, Y. B.; Moses, K.; Tsubota, S. Molecular Analysis of the Yellow Locus of Drosophila. *EMBO J.* **1986,** *5* (13), 3597–3605.

Coyne, J. A.; Orr, H. A. Speciation: A Catalogue and Critique of Species Concepts. *Phil. Biol.* **2004,** 272–92.

De Loof, A. From Darwin's on the Origin of Species by Means of Natural Selection to the Evolution of Life with Communication Activity as Its Very Essence and Driving Force (= Mega-Evolution). *Funct. Genom.* **2015,** *3*, 153–187.

Dietrich, C. R.; Cui, F.; Packila, M. L.; Li, J.; Ashlock, D. A.; Nikolau, B. J.; Schnable, P. S. Maize Mu Transposons Are Targeted to the 5′ Untranslated Region of the gl8 Gene and Sequences Flanking Mu Target-Site Duplications Exhibit Nonrandom Nucleotide Composition Throughout the Genome. *Genetics.* **2002,** *160* (2), 697–716.

Dooner, H. K.; Belachew, A. Transposition Pattern of the Maize Element Ac from the bz-m2 (Ac) allele. *Genetics*. **1989,** *122* (2), 447–457.
Dupuy, A. J., Rogers, L. M., Kim, J., Nannapaneni, K., Starr, T. K., Liu, P., Largaespada, D. A., Scheetz, T. E., Jenkins, N. A. and Copeland, N. G. A Modified Sleeping Beauty Transposon System That Can Be Used to Model a Wide Variety of Human Cancers in Mice. *Cancer Res*. **2009,** *69* (20), 8150–8156.
Emmanuel, E.; Yehuda, E.; Melamed-Bessudo, C.; Avivi-Ragolsky, N.; Levy, A. A. The Role of AtMSH2 in Homologous Recombination in *Arabidopsis thaliana. EMBO Rep*. **2006,** *7* (1), 100–105.
Engels, W. R. In *Mobile DNA*; Berg, D. E., Howe, M. M., Eds.; American Society for Microbiology: Washington, DC, 1989; pp 437–484.
Fatima, A.; Ivanyuk, D.; Herms, S.; Heilmann-Heimbach, S.; O'Shea, O.; Chapman, C.; Izsvák, Z.; Farr, M.; Hescheler, J.; Šarić, T. Generation of Human Induced Pluripotent Stem Cell Line from a Patient with a Long QT Syndrome Type 2. *Stem Cell Res*. **2016,** *16* (2), 304–307.
Feng, Y. Plant MITEs: Useful Tools for Plant Genetics and Genomics. *Genom. Proteom. Bioinform*. **2003,** *1* (2), 90–100.
Feschotte, C.; Keswani, U.; Ranganathan, N.; Guibotsy, M. L.; Levine, D. Exploring Repetitive DNA Landscapes Using REPCLASS, a Tool That Automates the Classification of Transposable Elements in Eukaryotic Genomes. *Genome Biol. Evol*. **2009,** *1*, 205–220.
Feschotte, C.; Pritham, E. J. DNA Transposons and the Evolution of Eukaryotic Genomes. *Annu. Rev. Genet*. **2007,** *1* (41), 331–368.
Finnegan, D. J. Transposable Elements. *Curr. Opin. Genet. Dev*. **1992,** *2* (6), 861–867.
Flutre, T.; Duprat, E.; Feuillet, C.; Quesneville, H. Considering Transposable Element Diversification in *de novo* Annotation Approaches. *PloS One*. **2011,** *6* (1), e16526.
Friedli, M. and Trono, D. The Developmental Control of Transposable Elements and the Evolution of Higher Species. *Annu. Rev. Cell Dev. Biol*. **2015,** *31*, 429–451.
Friedrich, G.; Soriano, P. Promoter Traps in Embryonic Stem Cells: A Genetic Screen to Identify and Mutate Developmental Genes in Mice. *Genes Dev*. **1991,** *5* (9), 1513–1523.
Frischer, L. E.; Hagen, F. S.; Garber, R. L. An Inversion That Disrupts the Antennapedia Gene Causes Abnormal Structure and Localization of RNAs. *Cell* **1986,** *47* (6), 1017–1023.
Garrels, W.; Talluri, T. R.; Ziegler, M.; Most, I.; Forcato, D. O.; Schmeer, M.; Schleef, M.; Ivics, Z.; Kues, W. A. Cytoplasmic Injection of Murine Zygotes with Sleeping Beauty Transposon Plasmids and Minicircles Results in the Efficient Generation of Germline Transgenic Mice. *Biotechnol. J*. **2016,** *11* (1), 178–184.
Ginzburg, L. R.; Bingham, P. M.; Yoo, S. On the Theory of Speciation Induced by Transposable Elements. *Genetics* **1984,** *107* (2), 331–341.
Grabundzija, I.; Wang, J.; Sebe, A.; Erdei, Z.; Kajdi, R.; Devaraj, A.; Steinemann, D.; Szuhai, K.; Stein, U.; Cantz, T.; Schambach, A. Sleeping Beauty Transposon-Based System for Cellular Reprogramming and Targeted Gene Insertion in Induced Pluripotent Stem Cells. *Nucl. Acids Res*. **2013,** *41* (3), 1829–1847.
Gridley, T.; Soriano, P.; Jaenisch, R. Insertional Mutagenesis in Mice. *Trends Genet*. **1987,** *3*, 162–166.
Guio, L.; Vieira, C.; González, J. Stress Affects the Epigenetic Marks Added by Natural Transposable Element Insertions in *Drosophila melanogaster*. *Sci. Rep*. **2018,** *8* (1), 1–10.
Hamer, L.; DeZwaan, T. M.; Montenegro-Chamorro, M. V.; Frank, S. A.; Hamer, J. E. Recent Advances in Large-Scale Transposon Mutagenesis. *Curr. Opin. Chem. Biol*. **2001,** *5* (1), 67–73.

Harrison, R. G. The Language of Speciation. *Evol.: Int. J. Organ. Evol.* **2012,** *66* (12), 3643–3657.

Havecker, E. R.; Gao, X.; Voytas, D. F. The Diversity of LTR Retrotransposons. *Genome Biol.* **2004,** *5* (6), 1–6.

Hawkins, B. A.; Diniz-Filho, J. A.; Jaramillo, C. A.; Soeller, S. A. Post-Eocene Climate Change, Niche Conservatism, and the Latitudinal Diversity Gradient of New World Birds. *J. Biogeogr.* **2006,** *33* (5), 770–780.

Haynes, K. A.; Caudy, A. A.; Collins, L.; Elgin, S. C. Element 1360 and RNAi Components Contribute to HP1-Dependent Silencing of a Pericentric Reporter. *Curr. Biol.* **2006,** *16* (22), 2222–2227.

Hoffmann, A. A.; Rieseberg, L. H. Revisiting the Impact of Inversions in Evolution: From Population Genetic Markers to Drivers of Adaptive Shifts and Speciation? *Annu. Rev. Ecol. Evol. System.* **2008,** *39*, 21–42.

Hollister, J. D.; Gaut, B. S. Epigenetic Silencing of Transposable Elements: A Trade-Off Between Reduced Transposition and Deleterious Effects on Neighboring Gene Expression. *Genome Res.* **2009,** *19* (8), 1419–1428.

Huang, X.; Lu, G.; Zhao, Q.; Liu, X.; Han, B. Genome-Wide Analysis of Transposon Insertion Polymorphisms Reveals Intraspecific Variation in Cultivated Rice. *Plant Physiol.* **2008,** *148* (1), 25–40.

Hurst, G. D.; Schilthuizen, M. Selfish genetic elements and speciation. *Heredity.* **1998,** *80* (1), 2–8.

Ivics, Z. and Izsvák, Z. The Expanding Universe of Transposon Technologies for Gene and Cell Engineering. *Mobile DNA* **2010,** *1* (1), 1–15.

Jiang, N.; Wessler, S. R. Insertion Preference of Maize and Rice Miniature Inverted Repeat Transposable Elements as Revealed by the Analysis of Nested Elements. *Plant Cell* **2001,** *13* (11), 2553–2564.

Kapusta, A.; Kronenberg, Z.; Lynch, V. J.; Zhuo, X.; Ramsay, L.; Bourque, G.; Yandell, M.; Feschotte, C. Transposable Elements Are Major Contributors to the Origin, Diversification, and Regulation of Vertebrate Long Noncoding RNAs. *PLoS Genet.* **2013,** *9* (4), e1003470.

Kashkush, K.; Feldman, M.; Levy, A. A. Transcriptional Activation of Retrotransposons Alters the Expression of Adjacent Genes in Wheat. *Nat. Genet.* **2003,** *33* (1), 102–106.

Kazazian, H. H.; Wong, C.; Youssoufian, H.; Scott, A. F.; Phillips, D. G.; Antonarakis, S. E. Haemophilia A Resulting from *de novo* Insertion of L1 Sequences Represents a Novel Mechanism for Mutation in Man. *Nature* **1988,** *332* (6160), 164–166.

Khurana, J. S.; Wang, J.; Xu, J.; Koppetsch, B. S.; Thomson, T. C.; Nowosielska, A.; Li, C.; Zamore, P. D.; Weng, Z.; Theurkauf, W. E. Adaptation to P Element Transposon Invasion in Drosophila Melanogaster. *Cell* **2011,** *147* (7), 1551–1563.

Kidwell, M. G. Transposable Elements and the Evolution of Genome Size in Eukaryotes. *Genetica* **2002,** *115* (1), 49–63.

Kidwell, M. G.; Kidwell, J. F.; Sved, J. A. Hybrid Dysgenesis in Drosophila Melanogaster: A Syndrome of Aberrant Traits Including Mutation, Sterility and Male Recombination. *Genetics* **1977,** *86* (4), 813–833.

Lambert, M. E.; McDonald, J. F.; Weistein, I. B. *Eukaryotic Transposable Elements As Mutagenic Agents*; Cold Spring Harbor Laboratory Press, Cold Spring Harbor: New York, 1988.

Lambowitz, A. M.; Zimmerly, S. Group II Introns: Mobile Ribozymes That Invade DNA. *Cold Spring Harbor Perspect. Biol.* **2011,** *3* (8), a003616.

Lee, Y. C.; Karpen, G. H. Pervasive Epigenetic Effects of Drosophila Euchromatic Transposable Elements Impact Their Evolution. *eLife* **2017,** *6*, e25762.

Lin, Y.; Waldman, A. S. Capture of DNA Sequences at Double-Strand Breaks in Mammalian Chromosomes. *Genetics* **2001,** *158* (4), 1665–1674.

Lockton, S.; Gaut, B. S. The Evolution of Transposable Elements in Natural Populations of Self-Fertilizing *Arabidopsis thaliana* and Its Outcrossing Relative *Arabidopsis lyrata. BMC Evol. Biol.* **2010,** *10* (1), 1–11.

Mackay, T.; Lyman, R. F.; Jackson, M. S. Effects of P Element Insertions on Quantitative Traits in Drosophila Melanogaster. *Genetic*s **1992,** *130* (2), 315–332.

Martienssen, R. A. Functional Genomics: Probing Plant Gene Function and Expression with Transposons. *Proc. Natl. Acad. Sci.* **1998,** *95* (5), 2021–2026.

McClintock, B. Carnegie Inst. *Wash. Year Book.* **1948,** *47*, 155–69.

McClintock, B. Chromosome Organization and Genic Expression. In *Cold Spring Harbor Symposia on Quantitative* Biology, Vol. 16; Cold Spring Harbor Laboratory Press, 1951; pp 13–47.

McClintock, B. Controlling Elements and the Gene. In *Cold Spring Harbor Symposia on Quantitative Biology*, Vol. 21; Cold Spring Harbor Laboratory Press, 1956; pp 197–216.

McClintock, B. The Origin and Behavior of Mutable Loci in Maize. *Proc. Natl.Acad. Sci.* **1950,** *36* (6), 344–355.

McClintock, B. The Significance of Responses of the Genome to Challenge, 1984.

McClintock, B. Maize Genetics, 1946.

Mieczkowski, P. A.; Lemoine, F. J.; Petes, T. D. Recombination Between Retrotransposons as a Source of Chromosome Rearrangements in the Yeast Saccharomyces Cerevisiae. *DNA Repair.***2006,** *5* (9–10), 1010–1020.

Miskey, C., Izsvák, Z., Kawakami, K. and Ivics, Z. DNA Transposons in Vertebrate Functional Genomics. *Cell. Mol. Life Sci.* **2005,** *62* (6), 629–641.

Muenthaisong, S.; Ujhelly, O.; Polgar, Z.; Varga, E.; Ivics, Z.; Pirity, M. K.; Dinnyes, A. Generation of Mouse Induced Pluripotent Stem Cells from Different Genetic Backgrounds Using Sleeping Beauty Transposon Mediated Gene Transfer. *Exp. Cell Res.* **2012,** *318* (19), 2482–2489.

Muñoz-López, M.; García-Pérez, J. L. DNA Transposons: Nature and Applications in Genomics. *Curr. Genom.* **2010,** *11* (2), 115–128.

Noor, M. A.; Chang, A. S. Evolutionary Genetics: Jumping Into a New Species. *Curr. Biol.* **2006,** *16* (20), R890–R892.

Noor, M. A.; Grams, K. L.; Bertucci, L. A.; Reiland, J. Chromosomal Inversions and the Reproductive Isolation of Species. *Proc. Natl. Acad. Sci.* **2001,** *98* (21), 12084–12088.

Okita, K.; Ichisaka, T.; Yamanaka, S. Generation of Germline-Competent Induced Pluripotent Stem Cells. *Nature* **2007,** *448* (7151), 313–317.

Orr, H. A.; Presgraves, D. C. Speciation by Postzygotic Isolation: Forces, Genes and Molecules. *BioEssays*. **2000,** *22* (12), 1085–1094.

Piegu, B.; Guyot, R.; Picault, N.; Roulin, A.; Saniyal, A.; Kim, H.; Collura, K.; Brar, D. S.; Jackson, S.; Wing, R. A.; Panaud, O. Doubling Genome Size Without Polyploidization: Dynamics of Retrotransposition-Driven Genomic Expansions in Oryza australiensis, a Wild Relative of Rice. *Genome Res.* **2006,** *16* (10), 1262–1269.

Reznikoff, W. S. The TN5 Transposon. *Annu. Rev. Microbiol.* **1993,** *47* (1), 945–963.

Rinehart, T. A.; Dean, C.; Weil, C. F. Comparative Analysis of Non-Random DNA Repair Following Ac Transposon Excision in Maize and Arabidopsis. *Plant J.* **1997,** *12* (6), 1419–1427.

Rizzon, C.; Martin, E.; Marais, G.; Duret, L.; Ségalat, L.; Biémont, C. Patterns of Selection Against Transposons Inferred from the Distribution of Tc1, Tc3 and Tc5 Insertions in the Mut-7 Line of the Nematode Caenorhabditis Elegans. *Genetics* **2003,** *165* (3), 1127–1135.
Sakai, J., Chalmers, R. M. and Kleckner, N. Identification and Characterization of a Pre-Cleavage Synaptic Complex That Is an Early Intermediate in Tn10 Transposition. *EMBO J.* **1995,** *14* (17), 4374–4383.
Seehausen, O.; Butlin, R. K.; Keller, I.; Wagner, C. E.; Boughman, J. W.; Hohenlohe, P. A.; Peichel, C. L.; Saetre, G. P.; Bank, C.; Brännström, Å.; Brelsford, A. Genomics and the Origin of Species. *Nat. Rev. Genet.* **2014,** *15* (3), 176–192.
Shepherd, N. S.; Sheridan, W. F.; Mattes, M. G.; Deno, G. The Use of Mutator for Gene-Tagging: Cross-Referencing Between Transposable Element Systems. In *Plant Transposable Elements*; Springer: Boston, MA, 1988; pp 137–147.
Slotkin, R. K.; Martienssen, R. Transposable Elements and the Epigenetic Regulation of the Genome. *Nat. Rev. Genet.* **2007,** *8* (4), 272–285.
Slotkin, R. K.; Vaughn, M.; Borges, F.; Tanurdžić, M.; Becker, J. D.; Feijó, J. A.; Martienssen, R. A. Epigenetic Reprogramming and Small RNA Silencing of Transposable Elements in Pollen. *Cell* **2009,** *136* (3), 461–472.
Soller, M.; Beckmann, J. S. Marker-Based Mapping of Quantitative Trait Loci Using Replicated Progenies. *Theor. Appl. Genet.* **1990,** *80* (2), 205–208.
Spradling, A. C.; Stern, D. M.; Kiss, I.; Roote, J.; Laverty, T.; Rubin, G. M. Gene Disruptions Using P Transposable Elements: An Integral Component of the Drosophila Genome Project. *Proceed. Natl. Acad. Sci.* **1995,** *92* (24), 10824–10830.
Staunstrup, N. H.; Sharma, N.; Bak, R. O.; Svensson, L.; Petersen, T. K.; Aarenstrup, L.; Kristiansen, K.; Bolund, L.; Mikkelsen, J. G. A Sleeping Beauty DNA Transposon-Based Genetic Sensor for Functional Screening of Vitamin D3 Analogues. *BMC Biotechnol.* **2011,** *11* (1), 1–3.
Talluri, T. R.; Kumar, D.; Glage, S.; Garrels, W.; Ivics, Z.; Debowski, K.; Behr, R.; Niemann, H.; Kues, W. A. Derivation and Characterization of Bovine Induced Pluripotent Stem Cells by Transposon-Mediated Reprogramming. *Cell. Reprogram* (Formerly "*Cloning and Stem Cells*) **2015,** *17* (2), 131–140.
Voytas, D. F.; Cummings, M. P.; Koniczny, A.; Ausubel, F. M.; Rodermel, S. R. Copia-Like Retrotransposons Are Ubiquitous Among Plants. *Proc. Natl. Acad. Sci.* **1992,** *89* (15), 7124–7128.
Walbot, V. Reactivation of Mutator Transposable Elements of Maize by Ultraviolet Light. *Mol. Gen. Genet.* **1992,** *234* (3), 353–360.
Wang, Q.; Dooner, H. K. Remarkable Variation in Maize Genome Structure Inferred from Haplotype Diversity at the bz Locus. *Proc. Natl. Acad. Sci.* **2006,** *103* (47), 17644–17649.
Wang, R. J.; White, M. A.; Payseur, B. A. The Pace of Hybrid Incompatibility Evolution in House Mice. *Genetics* **2015,** *201* (1), 229–242.
Wessler, S. R. Phenotypic Diversity Mediated by the Maize Transposable Elements Ac and Spm. *Science* **1988,** *242* (4877), 399–405.
Widmer, A.; Lexer, C.; Cozzolino, S. Evolution of Reproductive Isolation in Plants. *Heredity* **2009,** *102* (1), 31–38.
Wilson, M. H., Coates, C. J. and George Jr, A. L. PiggyBac Transposon-Mediated Gene Transfer in Human Cells. *Mol. Ther.* **2007,** *15* (1), 139–145.

Wood, T. E.; Takebayashi, N.; Barker, M. S.; Mayrose, I.; Greenspoon, P. B.; Rieseberg, L. H. The Frequency of Polyploid Speciation in Vascular Plants. *Proc. Natl. Acad. Sci.* **2009,** *106* (33), 13875–13879.

Woodhouse, M. R.; Freeling, M.; Lisch, D. Initiation, Establishment, and Maintenance of Heritable MuDR Transposon Silencing in Maize Are Mediated by Distinct Factors. *PLoS Biol.* **2006,** *4* (10), e339.

CHAPTER 8

Transposable Elements and Epigenetic Regulation

MUHAMMAD SHAN, AKASH FATIMA, ZARMEENA AMJAD, and ZULQURNAIN KHAN*

Institute of Plant Breeding and Biotechnology, MNS University of Agriculture Multan, Old Shujabad Road, Multan 60000, Pakistan

**Corresponding author. E-mail: zulqurnain.khan@mnsuam.edu.pk*

ABSTRACT

Genomes include many repeated sequences, a significant count of them are transposable elements (TEs). They have multiple activities: genome size restructuring, chromosome rearrangement, gene mutation induction, and gene activities altered by the insertion near or within the promoter, intronic, or enhanced regions. Mobilization of TEs is generally triggered when an organism is exposed to stress, including both biotic and abiotic stress, polyploid conditions, and interspecific hybridizations. TEs were considered genomic parasites, but this is also suggested that they may also play a valuable role in developing various biological processes. Consequently, a pattern of punctuated balance is generated by the evolutionary tug of war between host genomes and epigenetically mediated TEs. In eukaryotic cells, superlative epigenetic mechanisms have developed to silence the expression of TEs and their mobility. TEs are the backbone of epigenetic phenomena at the single gene level and across more significant chromosomal regions due to their ability to recruit silencing machinery. The combination of TE mobilization and the epigenetic landscape changes could lead to quick adaptation of phenotypes in response to global environmental changes. In this chapter,

Plant Transposable Elements: Biology and Biotechnology, Deepu Pandita, Anu Pandita, (Eds.)

epigenetic control of TEs as a source of phenotypically selectable variations will be discussed, along with their linkages to defense responses.

8.1 INTRODUCTION

DNA sequences that can modify their position within a genome can be transposed elements (TEs). Liang et al. (2021) and McClintock (1950) found transposable elements (TEs) through an analysis of maize (corn) genetic stability. According to him, a DNA segment jumped into the genes which code pigmented kernels, and thus, deactivated pigmented kernels were the cause of the light-colored kernels. These moving genes are known as transposons or TEs. These parasites in most organisms' genomes can boost their genome number, expanding, and evolving the genome. For example, TEs account for almost 50% of the human genome and > 80% of the maize genome (Bourque et al., 2018; Hermant and Torres-Padilla, 2021).

8.2 STRUCTURE OF TRANSPOSABLE ELEMENTS

Transposons are stretches of DNA with repeated segments of DNA at both ends. A transposon is made up of a central sequence with transpositions of genes and specific other genes. The short-repeated DNA segments are flanked on both sides. The repeated segments can be repeated directly or inverted, which helps in the identification of transposons. The number of nucleotides repeats is uneven. The insertion method at the target site is 5 or 7 and 9 nucleotides (Pray, 2008) (Fig. 8.1).

8.3 TYPES OF TRANSPOSONS

TEs were classified into the following types based on their transposition methods:
Class 1 retrotransposon.
Class 2 DNA transposon.

8.3.1 CLASS 1 RETROTRANSPOSON

Class I elements are mobilized as intermediate RNA, also known as retrotransposons. All retrotransposons, commonly referred to as copying

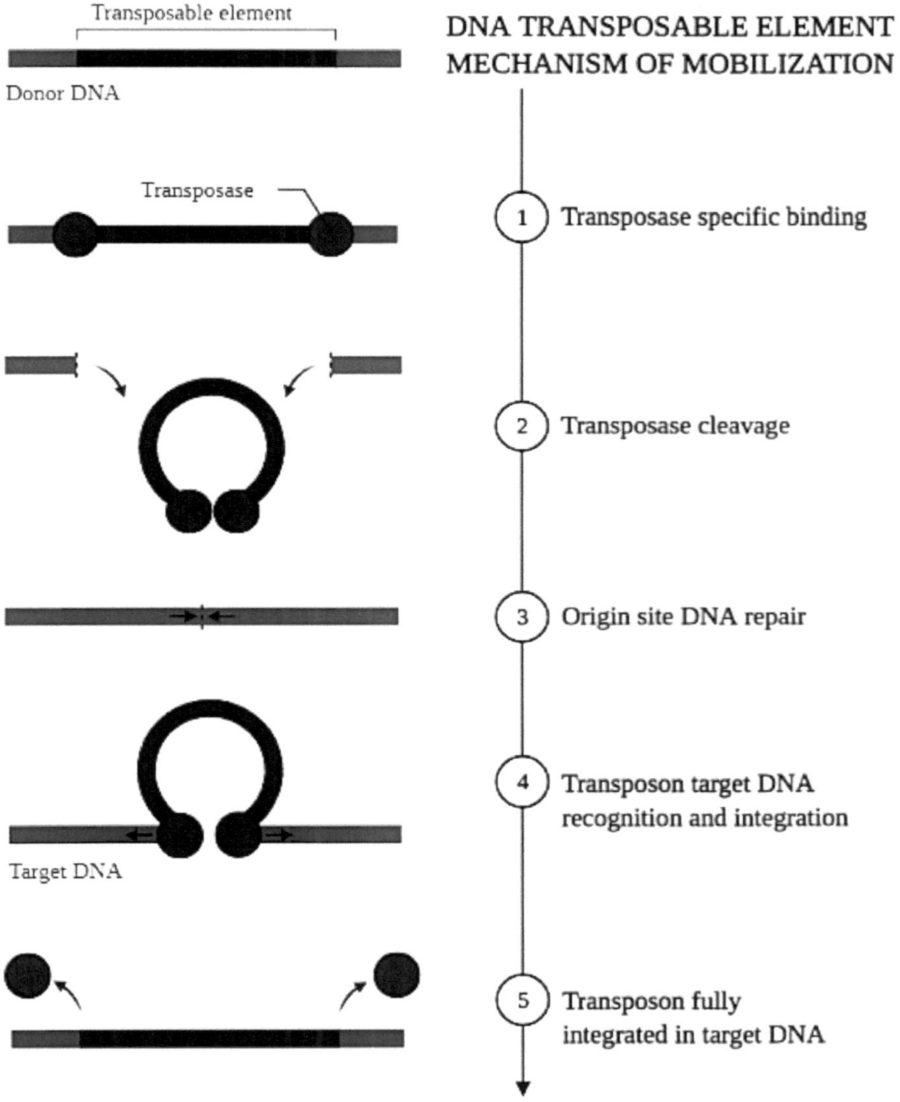

FIGURE 8.1 Mechanism of transposable elements mobilization. (1) Transposase enzyme specifically binds to donor DNA element. (2) Transposase enzyme helps in the cleavage of transposable elements, (3) origin site of DNA join end-to-end and repair itself, (4) DNA-transposase complex recognized the target DNA and integrated into target DNA. (5) Transposon fully integrated in the target DNA site.

and pasting elements, generate new copies by transcribing them in the genome. Two transposase subunits are used for the cut-and-paste transposition. Each subunit binds at the two ends of the transposon in the specific sequences. The subunits of the protein of transposase are then combined and lead to transposon excision. This excised "transposon–transposase complex" is then incorporated into the destination site. The transposon is thus cut from one place and then pasted on another by a transposase protein-mediated mechanism. Examples of transposon cut and paste are IS elements, P elements in maize, Drosophila hobo elements, etc. Two main retrotransposons are long terminal repeat (LTR) and non-LTR characterized by 100–300 bp of the repeated direct terminal presence or lack. LTR components are from 100 bp to 10 Kb in size and are similar to the retroviruses, including endogenous retroviruses (ERVs). LTR encodes proteins with the name of the elements. LTR-retrotransposons replicate and mobilize the Gag gene-encoded viral particle via tRNA-printed template switching (Platt et al., 2018).

The non-LTR elements consist of long and short interspersed elements LINEs and SINEs, respectively. The presence of multiple tails, generally poly-a, and the lack of LTRs can be recognized in Line and SINE. The LINEs are 4 to 7 kb and codify the enzyme machine required to move one to three proteins. A nuclear chaperone protein (ORF1) is one of the most common 2-ORF mammalian lines and reverse transcriptase. Recently in L1 elements, a third significantly shorter protein (ORF0) was described, but its role is not known (Denli et al., 2015). The SINE can be between 150 and 500 bp and lack the necessary machinery to mobilize itself, that is, nonself-sufficient. Most SINE mammals come from a 5′ head combination that comes from a pseudogenic ribosomal or tRNA with a 3′ head specific to the line. The SINE region, like the LINE, is employed for mobilization by the enzymatic machinery of LINEs. In contrast to LINEs, the de novo origin of SINEs in mammals is relative.

8.3.2 CLASS II TRANSPOSONS

The "DNA transposons" of Class II TEs have been identified as translating via intermediate DNA. They are transposed to an additional locus in the genome, which means the excision of the element. This pathway is often referred to as "conservative" because it is known as the original element itself as "cut-and-paste." Theory shows that this conservatory transposition is a less powerful tool to invade a genome but that any duplication occurs to increase the transposon population in the particular genome and its transposition during the replication of chromosomes. Class II TEs are not only cut-and-paste TEs (subclass 1); some are DNA TEs that use a replicate transposing mechanism to remove only one DNA strand (subclass 2). Class II is divided into four orders. The TIR sequence includes nine superfamilies, including the sequence of prokaryote insertions (IS630-Tc1 - mariner, hAT, Mutator, Merlin, Transib, P, PiggyBac, PIF-Harbinger, and CACTA) with all of the TIR areas on each end. The orders of Crypton, Helitron, and Maverick comprise only one superfamily (Jurka et al., 2011).

Class I elements are found in most eukaryotic lines and seldom prokaryotic, and prokaryotic and eukaryotic class II elements are readily available. It indicates that both progenitors were probably present in the common ancestor of all eukaryotes. Apart from ubiquities of TEs in vertebrates, differential amplification, genetic drift, and recombination in isolated populations can lead to dragging genomic TE landscapes between close-related taxa (Jurka et al., 2011; Staunstrup et al., 2012) (Fig. 8.2).

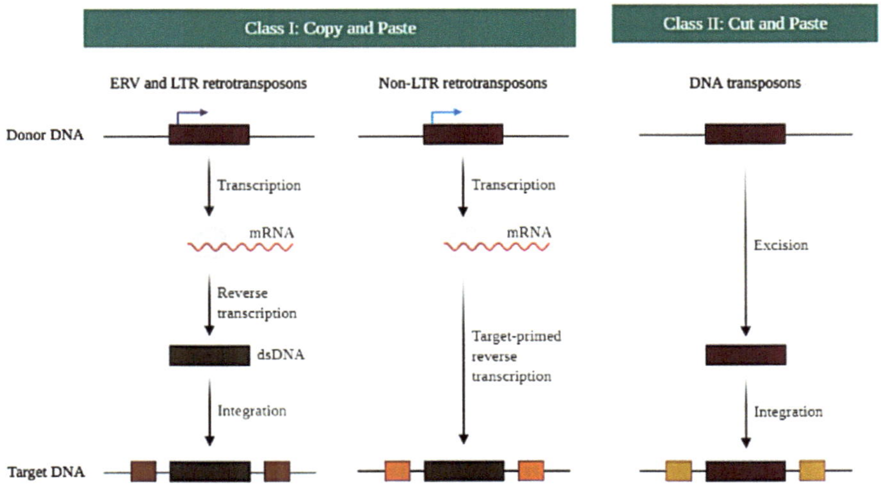

FIGURE 8.2 Transposable elements classes. Transposable elements are divided into two classes on basis of the replication mechanism. (a) Class 1 copy and paste and gives rise to two identical copies; one in the donor site and one in the target site. Class 1 further comprises of two classes LTR retrotransposon and non-LTR retrotransposons. Donor DNA is transcribed into mRNA and then through reverse transcription dsDNA formed which is then integrated into target DNA, (b) Class 2 cut and paste and translocates the TE element in the target site leaving a free TE donor site.

8.4 TEs AS GENOMIC PARASITES

Parasites are not only found in the kingdoms of plants and animals; they also belong to us. Our genome contains innumerable short portions of DNA that spread at the expense of your genome. These transposons are therefore also attributed to DNA parasites. When transposons are present, genetic data are transported from one genome location to another. This process occurs when germ cells develop, especially when the early embryo has strong divisions. At all times, the DNA sequence is redesigned with some severe implications. A transposon may be located in a gene segment containing information on metabolism-critical protein. The gene cannot be properly read.

An example of this is the APC gene. It encodes the protein to prevent cancer development. Colon cancers can occur if they are disturbed by the addition of a transposon. The proteins must be read, transferred to RNA, and translated into new proteins to the DNA's information. Transposons need to be replicated and inserted into the genome by different proteins, as well. The structure of the two essential parasite DNA sneakers, LINE-1 and Alu retrotransposon are two ubiquitous in the human genome (Oliver and Greene, 2009). The LINE-1 segment consists of about 6000 pairs of DNA baseline—about as long as a medium gene. The LINE-1 genome is comprised of around 500,000 copies and pieces. The LINE-1 element is only 17% of the genome. It is possible to produce more than half of our genome from TEs. The proportion of transposons may be even higher in other organisms. For instance, transposons can be traced in maize plants to up to 85% of the entire genome. For example, the transpositions can only be transmitted due to cancer if they do not kill the individual. "With every transposition, a person or his descendants are not necessarily fatal. The part of the genome in the LINE-1 copy is dependent on." In addition, many mutations have been disabled over time. Of the 500,000 LINE-1 sections, only approximately 100 are active and parasitic in the human genome. The rest is no longer working.

The ALU retrotransposons are hypocritical and distributed across the genome since they are over 1 million and represent about 10% of the genome and a step closer to parasites. They are parasites by hiding and replication of the LINE-1 equipment. It may replicate the LINE-1 itself. In contrast, for its purposes, Alu needs and uses LINE-1 proteins. Transposons, therefore, relate only to their spread and appear to contribute nothing to the organism's subservience. The body is just a way to an end. However, transposons can nonetheless profit following a closer examination. When transpose copies are inserted into the genome, DNA is converted to promote and flexibly preserve the genome. The body can adapt more quickly if environmental changes are significant due to climate change or natural disasters. Transposons have since been shown in plants, especially if plants are stressed, for example, during hot spelling. Many transposons are dangerous, but new characteristics are sometimes vital to the organism's survival. Evolution is long assumed to be based on exchanging individual genetic code letters, referred to as point mutations. Many of his colleagues (Oliver and Greene, 2009) think that transposons are a significant factor in reshaping the genome and creating new variations.

8.5 MULTIPLE ACTIVITIES OF TRANSPOSABLE ELEMENTS

Different activities related to gene function involve TEs. Transposons can modify genome structural and functional characteristics by changing their genome position. The mutation causes TEs through insertion, deletion, etc. Transposons contribute positively to evolution which has a significant effect on the change in the genetic organisms. They are also helpful in gene cloning as cloning vectors. As a vector for introducing transgenes to Drosophila, P elements are commonly used. During genome mapping, transposons may also be used as genetic markers. Transposon-mediated gene marking is performed to search for and isolate a given gene.

In addition to interrupting genes or disrupting their regulation, TEs can also lead to further rearrangements in the genome. There can be homologous recombination between two almost identical sequences. The two copies must have potential substrates for recombination when transposition makes other copying of a TE. Recombination results depend on the direction between the two TEs. Combining two transposable aspects with the same chromosome orientation can lead to deletion while the product is a reversal if they are in opposite directions.

8.6 GENOME SIZE RESTRUCTURING AND CHROMOSOME REARRANGEMENT

C-value is a quantitative measure to define the genome size. A whole chromosome supplement shows the DNA contents of the organism; hence 1C represents the amount of DNA in a monoploid genome that is not replicated. The parameter is specific to a specific species and is highly chosen during development (Lee and Kim, 2014; Canapa et al., 2015). The prokaryotic genome sizes are smaller than the eukaryotic ones, while the lower eukaryotic generations are smaller than the high eukaryotic genomes. The C values are usually well correlated with the complexities of cells and organism genomes;

the prokaryotic genome size is smaller than that of eukaryotic genomes, while the smaller the eukaryotic genome size is smaller the higher genome. However, large genome sizes are often among closely related species, and C values are inferior to organism complexity and suggest the paradox of C values. Large genomes can come from one or more amplification mechanisms from small genomes. However, the rate of DNA loss in small genomes is greater than that of TE amp for DNA gains, as demonstrated in cotton and its families (Canapa et al., 2015).

Possible change in the genome dimension can be considered the plausible mechanism for entire genome duplication (WGD). WGD (polyploidization) is not the only means to modify the gene content, increasing the genome size in a single generation. In the Eleocharis, for instance, the rapid increase in the number of copies in some families of translatable elements for such species is attributed to considerable variations in genome size. The linear relationship between genome size and genome content of TEs has been numerously reported. Class I retrotransposons appear to influence the semiconservative conversion of copy significantly and paste to genome size. The TE content for the low genome *A. thaliana* is roughly 15%, with only 4% retroposons.

On the other hand, large genome species have high TE content, a large proportion of Class 1 TE. For example, about 85% of *Zea mays* contain TE (1C = 2.3 pg), 8.6% of which are Class 1 and Class 2, respectively.

Transposition is a robust genome expansion mechanism countered over time by the deletion of DNA. The balance of both processes is a crucial driver for the development of the eukaryotic genome. In several studies on plant and animal genomic development, the effect and extent of this genomic shuffling and cycling content were demonstrated. As inserting and deleting TEs are often inaccurate, these processes can indirectly influence host sequences. These cases frequently lead to large volumes of host duplication and revision, including genes and regulative sequences. For example, a single DNA transposon group captured and rehabilitated ~1000 gene fragments in the rice genome (MULEs). The transposition rate of TEs, partially controlled by the host, was an essential driver of the genome (Biscotti et al., 2019). After losing their mobility and reordering as a byproduct of transposition, TE can endorse genome structural changes. In particular, recombination variations may result in large-scale deletions between very similar areas, spread through related TEs at distant genomic positions. Besides, TEs offer microchemistry areas that can be switched to another source of structural variants while repairing replication errors. These structural variations, not based on transposition, have substantially contributed to the evolution of the genome. Those processes can also make it more difficult to identify active conversion elements in population studies that determine functional aspects by detecting nonreference insertions.

The TEs are also contributing to specialized chromosome properties. An inspiring example is Drosophila, which has a LINE-similar form in retrotransposons and keeps the telomeres as a substitute for the telomerase enzyme lost during the development of Dipterists. This domestication was carried out, so a chromosome linearization caused the "end problem" reconstruction of what could have happened much earlier in Eucarian development. The reverse transcriptase part of telomerase was believed to originate from an old retrolental lineage. Structural roles also apply to the central sequences and the domesticated transposase genes (McCullers and Steiniger, 2017). First, Barbara McClintock showed that transposable maize elements might induce significant Chromosome reordering, including duplications, deletions, reversals, and translocations. In recent years, researchers have made considerable progress in clarifying how transposons can cause the genome to reorder. Rearrangements shall be made when the terminals of different elements are used for the TE system Ac/Ds. The resulting alternative translation reaction directly generates several rearrangements. Depending on the transposon's location and direction, insert the size and type of the rearrangements produced. A single locus containing a few alternative conversion-competent elements can create virtually unlimited rearrangements of the genome. With a fundamental understanding of the mechanisms involved, scientists begin to use both natural and in vitro generated chromosome structure-manipulating configurations of TEs (Zhang et al., 2012).

8.7 GENE MUTATION INDUCTION

Although some TE levels may foster adjustment, over TE may be a factor in the reciprocal decline. The mutual collapse results in positive retrograde in populations more susceptible to the aggregation by drift for further lethal mutations, causing reduced fitness and population size. The meltdown is not easy to see because it leads to a feed-forward extinction. The mammoth genome of Wrangel Island represents the best example of a mammalian mutational meltdown. This genome shows an increase in gene deletion, early stop codon, and minimizes heterozygotic activity compared to other large genomes in these groups. In addition, in the Mammoth on Wrangel Island, excess retrogenes are available. The retrogenes being a significant byproduct of LINE, their existence immediately before the extinction of the mammoth population at Wrangle Island 15 indicated a retrotransposon breakdown. The final explosion of TE activity could have resulted in a mutation of the mammoth population of Wrangel Island and the formation of retrogenes. Interestingly, this explosion in TE activity is anticipated under the epidemiological hypothesis (Rogers and Slatkin, 2017).

8.8 TEs CAN BE DAMAGING WITHOUT TRANSPOSITION

TEs are known most for their mobility, that is, their capacity to transpose into new sites. Although the breakdown and insertion of DNA associated with translation is an apparent cause of cell damage, it is not the only or perhaps the most frequent mechanism that could harm your host. The host damages the reactivated transposon in several ways. First, transposon loci can interfere with the transcription or processing via countless mRNA host mechanisms, including their transcript (Elbarbary et al., 2016; Daniel et al., 2015b). The transcriptional genome-wide repression of TEs in human cell reproduction has been documented (De Cecco et al., 2013) and various mouse tissues like brain, liver, and muscle (De Cecco et al., 2013; Van Meter et al., 2014). Promotors LTR and L1 may cause activation of oncogenic cancer (Babaian and Mager, 2016). Second, TE-encoded proteins can lead to genomic instability and DNA breakdowns as L1 ORF2p endonuclease activity (Hedges and Deininger, 2007). Third, RNA transcript accumulation and copying of extrachromosomal DNA derived by TE can cause an innate immune response that causes autoimmune and sterile inflammation. The activation of interferon response is currently well documented as an endogenous retroviral transcript and can increase the identification and assault of cancer cells via immunotherapy (Chiappinelli et al., 2015). All the mechanisms mentioned above must still make a relative contribution to organism pathologies.

The next step is to translate the coded proteins and reverse transcription in cDNA substrates for translation-suitable retroelements after the transcription and sometimes splintering TEs. The Cytosolic DNA and RNA resulting in the use of the TE-encoded transcriptase protein, DNA hybrids may alert inflammatory tracts. For example, patients with TE-accumulated, cytosensitive DNA in Aicardi-Goutières due to pathway mutations usually block TeD or degrade TE-based DNA (Kassiotis and Stoye, 2016). While not all TE-encodes functional protein, some can make Pol, Gag, or Envelope proteins, including several endogenous retroviruses (Env) (Roulois et al., 2015). Cytotoxicity may be associated with overexpression of the Env-protein, with the multi-sclerosis (Crow and Manel, 2015) and amyotrophic lateral sclerosis of two or more neurodegenerative diseases. The smallest group of human accessory proteins, HERV (HML-2), may play a role in some cancers, although there is no evidence to support this (Vargiu et al., 2016).

8.9 WHEN IS TE MOBILIZATION DONE?

TEs have been considered important drivers of species development due to their ability to move and induce mutations in the genome. Despite their mutational properties, the successful invasion of TEs into genomes is an apparent paradox. In some cases, these elements can also be highly regulated in transposition rates, which have sometimes been associated with environmental conditions changes. TE transformation is generally handled firmly to a low value. Factors that are susceptible to transpositions in natural populations are contributing to the continuity of TE. As causative agents of TE mobilization in various organisms, various factors were suggested: biotic and abiotic stresses, cross-species and intraspecific crosses, and population factors. TEs have also been active under stress, which often proves an adaptive function for TEs to address stress (Negi et al., 2016). TE activation would increase the rate of mutation that results in natural selection variability. In addition, TEs can spread response elements across the genome to help reprogram strain gene networks when stress is activated, as it is known that some TEs contain regulatory strain response sequences. However, in terms of the stress response, this view of TEs has been challenged. It was suggested that experiments on the regulation of TEs are not always vital. The arguments were also made that before assigning these organism-level functions, the unique biological characteristics of TEs should be taken into account (Rey et al., 2016).

8.9.1 TRANSPOSABLE ELEMENTS AND STRESS RESPONSE

Are the TEs under stress conditions always empowered? Recent literature reveals that certain stress-expressed TEs are activated (Ryan et al., 2016; Hummel et al., 2017). The molecular mechanism that causes TEs to stress has been detected in several studies. The longevity regulating protein Sirtuin (SIRT6), for example, silences the nuclear element (LINE)-1s long interspersed under ordinary conditions. SIRT6 links KRAB-Associated Protein (KAP)1 nuclear corepressor protein with the Promotor LINE-1 element and mono-ADP. Ribosylates lead to compacted chromatin and silent LINE-1. The SIRT6 has moved to DNA damage sites so that LINE-1s do not become heterochromatic and therefore transcribed (Ryan et al., 2016). Indirectly, it causes LINE-1 elements to be activated. It has proved possible to suppress TE in stress conditions after initial activation (Shpyleva et al., 2018; Huang et al., 2017). For example, acute stress in the rat hippocampus has demonstrated a tissue-specific increase of H3K9me3 (Lapp and Hunter, 2016). It is associated with a rapid (1 hour after stress) deletion from the TE family [ERV intracisternal-A (IAP), L1 RN, and B2 RN SINE] of the first stress-induced activation. Based on these data, specific TEs targeted by the H3K9me3 response aim to reduce potential genomic instability caused by TE families' activation. UV radiation is a stressor that can accelerate the expression of Hsp genes during embryonic stages in human skin and fish tissues (Vehniäinen et al., 2012). UVC further encouraged the transposition of the Tc1/mariner fungus superfamily *Aspergillus oryzae* and its removal (Ogasawara et al., 2009). Radiation from UVC damages the primary UVC cell chromophores DNA molecules. The absorption spot is known as the photoproducts for DNA; cyclobutane pyrimidine dimers are the most frequent. These spots change the metabolic processes of DNA, as both the replication and transcription machines have a physical block (Costa et al., 2003). Consequently, a cell cycle arrest is observed in phase G1 because the cells cannot progress to phase S and, therefore, lead to cell death (Ortolan and Menck, 2013).

Stress may also induce both TE activation and repression according to context. The TE response to heat stress in 10 different ecotypes has been analyzed (Lapp and Hunter, 2016). The majority (56–80%) of TEs in seven ecotypes showed higher stress expression than nonstress conditions. However, most TEs (63–79%) in the other three ecotypes have shown a lower level of stress expression than in nonstate conditions. Lastly, TE repression was also reported under stress (Barah et al., 2013; Trivedi et al., 2014). One study analyzed the effects of Ty3 transposition in cerevisiae cell cultures by inducing cellular stress responses. The galactose was added to the cell culture to induce Ty3 expression. While Ty3 was transferred by 30°C into a galactose-treated crop, the authors found that no transposition was detected in both stress conditions (Menees and Sandmeyer, 1996). The second example of TE silence was also reported in response to stress (Trivedi et al., 2014). The effect of human neuronal cells was tested in a study. The central opioid for chronic and acute human pain is morphine. While the mechanism is uncertain, oxidative stress can be driven by opioids. Although short-term morphine therapy did not affect global DNA methylation, the authors showed that hypermethylation and LINE-1 expression are reduced. However, after long-term morphine treatment, LINE-1 elements recover to the levels measured in cells that receive morphine (Menees and Sandmeyer, 1996). In general, cumulative data do not demonstrate a uniform topic with studies showing TE activation, removal, and stress repression in TE regulation.

Insertion of TE often involves the upregulation of the neighboring genes in stressful conditions (Hummel et al., 2017; Zovoilis et al., 2016; Forestan et al., 2016; Bouttier et al., 2016). The mouse genome recently identified a complex of the KAP1, which binds to the ERV, as a heat shock protein chaperone (HSP) (Hummel et al., 2017). The EAR binding is used to control the regulatory effect of these ERVs in the nearby gene expression when ERVs are located in genetic regulatory areas. In conditions of stress, the function of the complex HSP90-KAP1 results in neighboring genes. While TE activation is frequently connected with upregulation of the surrounding genes, it sometimes results in the activation and repression of gene expression. An example has shown how the noncoding gene activation and mouse repression from the B2 SINE retrotransposons are molecularly molecular (Zovoilis et al., 2016). B2 RNA links to 50 no translated areas of the stress-efficient genes and intronic regions, reducing the rate of poly-II advance in nonstressed conditions. The expression of B2 is extremely inductive, and the EZH2 protein is recruited in stress genes where B2 RNA is split and allows for more elongation of transcription.

8.10 TRANSPOSABLE ELEMENTS AND EPIGENETIC REGULATION

TEs have often been regarded as a silence of TEs through histone modification and methylation, As the dispersion of TEs evolves along with ways to mitigate expression (Berrens et al., 2017). TEs are also an excellent component of epigenetic marks that may affect genetic expression, leading to the thought that TE can epigenetically alter a particular locus. Epigenetic gene expression regulation remains a complex, unresolved problem in the mobility and distribution of TE as a cause. Evolution favored active intragenic TEs to be placed in the angiogenesis direction, which resulted in developing a trap mechanism that would stop the genome from invading waves that demethylate. To silence the TEs, such overlap and antibiotic transcripts go through a dicer, and Argonaute 2 (AGO2) regulated endo-siRNA pathway, leading to a surge of repressive histone markers when global demethylation activates them. The molecular way to connect repression with the rise of histone repression has still to be verified (Berrens et al., 2017). Recent research on the preimplantation of naive murine embryonic stem cells has shown that a heterochromatin complex of the zinc finger-protein-associated proteins with DNA-methylation marks in the loci center has a vital role in 10–11 translocations (TET) (Coluccio et al., 2018). It was found that two KRAB/ZFP families adapted TE targets to differentiated tissue using histone-based pathways that do not need to affect loci DNA methylation (Ecco et al., 2017).

The study shows that the interactions between TE and their KRAB-ZFP controllers affect the expression of the following gene. The interaction is rather explicit in humans' neural progenitor cells, where primate-specific ERVs were shown to serve as a docking platform for the protein of corepressors, KAP1 (TRIM28). KAP1 binds and suppresses ERV and regulates the expression of the adjacent genes, which are vital to brain development (Brattås et al., 2017). The transcriptional regulators of the human silencing hub (HUCH) and the CW-like Zinfinger 2 (MORC2), which promotes deposition of histone H3K9ME3, are an additional example of TEs' adjacent gene regulation in the transcriptional permissive euchromatic area. The specific L1 effects associated with MORC2/HUSH may extend to nearby genes, decrease mRNA expression, and likely affect POL II elongation rate (RNA Polymerase II) (Liu et al., 2018). In various inbred mouse strains for specific types of TEs, chromatin source variation has been identified (particularly younger LINE). TEs can control specific tissue genes, leading to downstream phenotypical population diversity (Du et al., 2016). TEs can also actively restructure the organization of chromatin that can long regulate gene expression. Overall, approximately 10% of TE families were enriched through different tissues inactive genomic regions. The most enhanced chromatin classes are DNA transposons and SINE, while ERV and L1 LTRs are the expanding classes of epigenetic markings for repressed compounds of H3K9me3 (Trizzino et al., 2018).

8.11 COMBINATION OF TEs MOBILIZATION AND EPIGENETIC LANDSCAPE CHANGES LEAD TO ADAPTION OF PHENOTYPES IN RESPONSE TO CLIMATE CHANGE

The epigenetic components (ECs) contain a network of molecules that can adjust phenotypes according to the requirement, like during development. It can also pass to the next generation without any change in the original DNA sequence (Richards et al., 2012). The ECs connect the genotype and phenotype with the environment and help maintain the response of a living organism according to the changing environmental conditions (Mirouze and Paszkowski, 2011). The activity of TEs can change by the environment, and they create a mutation in regulatory sequences and response to the environment (Oliver and Greene, 2009). ECs and TEs are interconnected and respond to the changing conditions of the environment by influencing the genotype or phenotype (Fedoroff, 2012). The changes in TEs activities have been reported in plants, fungi, and animals. For example, when the fungus was exposed to UV-light and cold stress, the transposon OPHIO2 was activated (Bouvet et al., 2008). The heat stress in Drosophila activates the TEs as well (Jardim et al., 2015). The ECs are also sensitive to drought, heat, and pollutants. For example, the exposure of traffic carbon particles to the human genome has activated DNA methylation (Kim et al., 2015). The changing climatic conditions also affect the genome of wild populations.

8.12 PHENOTYPIC CONSEQUENCES OF TE AND EPIGENETIC ACTIVITIES

TEs affect the phenotype by the formation of new proteins and changes in the existing proteins. The phenotypic changes due to climate change have been studied on pests. The research shows that pests have adapted themselves according to the agricultural chemicals used to destroy the pests. For example, in Drosophila, the retrotransposon has been inserted into the upstream region of the insecticide-resistant gene Cyp6g1, increasing the upregulation of this gene in tissues and creating resistance to insecticide (Rostant et al., 2012). When environmental stress activates the TEs, then they modify the expression of the gene by changing themselves. In maize and rice plants, temperature, salinity, and UV-light exposure started the TEs that created insertions and induced response by regulating genes according to the stress (Makarevitch et al., 2015). The epigenetic response to the environment occurs on demand. When there is a need to respond to climate change, ECs modify the expression of DNA sequences present in somatic cells during development (Feil and Fraga, 2012). The drug resistance produces after several doses of drugs taken is the phenomenon of epigenetic changes. The changes in histone occur to create resistance against drugs. The changes made by DNA methylation can cause heritable changes in the phenotype. The response of toxins present in the environment was tested on the rats. The toxic alters the DNA methylation pattern that was heritable as well. The alteration was associated with male infertility in rats, and this was observed in the next four generations of rats (Anway et al., 2005).

8.12.1 TRANSPOSABLE ELEMENTS IN BIOLOGICAL PROCESSES

Barbara McClintock's career included the discovery and description of TEs, a type of moving genetic factor found in abundance in the genomes of eukaryotic species. Her results were following the famous selfish DNA hypothesis at the time, which suggested that TEs could be thought of as "genomic hitchhikers" or molecular parasites that play zero function in genome evolution and offer no adaptive benefit to the host. While being ahead of her time, Barbara McClintock proposed that TEs would affect genome evolution. It is remarkable how many times this has been proven to be valid since then. The emergence of evolutionary–developmental biology (evo–devo), rapid developments in DNA sequencing technology, and the consequent rise in comparative genomics have all contributed to these discoveries. The modern understanding of TEs is that they can serve as evolutionary agents by extending, reorganizing, and diversifying their host's genetic catalog. TEs benefit as insightful markers to evaluate natural- and stress-induced genetic diversity and improve marker-assisted selection in plant breeding programs due to their high copy number, chromosome coverage, and dynamic arrangement pattern closely related species. Both DNA transposons and retrotransposons may be used to produce markers, with the latter being far more effective. Retrotransposons have been discovered to be the most abundant kind of TE in eukaryotes, accounting for up to 90% of plant genomes. They account for more than half of the maize and cereal genomes and 14% of the Arabidopsis genome. Furthermore, conserved domains at both ends (LTR) can be used to design PCR primers conveniently.

There are some examples of TE exaptation in the kingdom Plantae as well. FHY3 and FAR3 are transcription factors in Arabidopsis that attach to promoter areas and trigger multiple genes associated with far-red light and circadian clock signaling. They are linked to the MuDR family of transposases. Recent research in Arabidopsis showed that injecting the COPIAR7 transposon into the plant disease resistance gene RPP7 increases the host's resistance to a pathogenic organism from a broad group of fungus-like parasites that cause a variety of plant diseases. The Rim2 gene involved in fungal infection protection appears to have been derived directly from a CACTA DNA TE. A copier-like LTR factor selection as a promoter defunctionalized the rice blast disease-resistant gene named Pit. There are tiny details on the direct function of TEs in species domestication operations. The TE Hopscotch gene is introduced within the regulatory locus of the maize domestication gene teosinte branched1 (tb1), which improves its expression and bestows enhanced apical superiority in maize over its progenitor teosinte. The injection of a CACTA-like transposon into the promoter of the photoperiod-sensitive gene ZmCCT will inhibit its expression, allowing maize to extend to long-day temperate regions. The Mustang and Sleeper gene families have sequences extracted from expected transposases from Mutator-like and hAT DNA elements in flowering plants. Mustang genes are only found in angiosperms, and they encode putative transcriptional regulators involved in growth, flower formation, and reproduction.

8.12.2 TES AND REGULATION

Regulation of gene expression may also involve TEs that affect mRNA stability by senseless mediated decay and microRNA activities, circular RNA, and possibly long noncoding RNAs. The Alu sequences of the SINE subclass are frequently integrated into premature mRNAs, regularly in 3″ UTRs or introns (Daniel et al., 2015a; Gong and Maquat, 2011). The Alu elements of mRNA can play a part in the deterioration through the staffing process (SMD). StAU1 can be binding sites of a 3”-UTR SMD target and the Alu element in a cytoplasmic and polyadenylated lncRNA. It was notified. The STAU protein and activate SMD via the mRNA expression profile are available at these sites. The 3′-UTR SINEs can control mRNA levels in various species (including humans) by directing the SMD to show the particular importance of the gene expression profile convergence with lineage-specific SINEs (Lucas et al., 2018). Transposable sequences of elements in RNA transcripts may also be used to regulate mRNA profusion and alternating splicing. TEs insertion within introns is predicted by dispersing elements and considered as exons during transcription in exonization. TEs comprise several donors and receptors splice locations that can help to replace splice. Many of the RNA (RBP) binding proteins can thus interact with composition preferences on sites that transport them to some regions of TE, such as Human Antigen R (HuR) or Fusion Proteins (FUS), preferably binding with purine-rich motifs (Lucas et al., 2018). The depletion of TE sites in the diverse RBPs shows that those sites have a similar effect on the abundance of the transcript and division as in nonrepetitive genetic fields. The RBP binding may vary depending on the family TE in certain specific cases. It was also pointed out. For instance, the RBP HuR provides transcript stability without being attached to an Alu in the U-rich region (Kelley et al., 2014). Stable, structural domains can also be developed, resulting in new biological functions for the Alu RNAs. Kim et al. (2016) and Kralovicova et al. (2016) noted that intronic transposition elements are very similar to a reiterated familial of repeat medium frequencies MER51A. It might affect gene expression, model insertion levels of several NSEs (nonsense-mediated exons), RNA deterioration commuters, pseudo-exons derived from the cryptic splice site stimulation to counteract aluminum-mediated NSE activation.

8.12.3 TES AND MRNA

Circular RNAs are also a class of small RNAs involved in TE regulation, small noncoding RNAs with gene regulatory purposes (circRNas). New studies of mammalian diseases have shown that flanked areas of circRNA can mediate circRNA through reverse complementary transposons (Jeck, 2013; Zhang and Wang, 2014). In humans, in different circRNAs, there are numerous Alu combinations derived by one gene locus in human interfaces that show the competitive role of combinations in creating alternative circulation (Chen and Yang, 2015). An additional maize study shows that sequenced circRNAs are significantly enriched by the LINE1 elements in flanking circRNA regions (Chen et al., 2018), and the other reverse pairs are LLERCPs. Interestingly, the buildup of circRNAs will vary with an increasing LLERCP transcription and a decline in linear transcription. (Jung et al., 2019) explained how TEs could have different effects of long-noncoded cis-natural antisense expressions of another class of RNAs. First of all, NATs may be transcribed from TEs by other promoters. TEs, which complement the exon of a protein-coding gene, are also exonerated. It is also possible to do this exoneration. NATs may also mediate the dissolution of meaning transcripts via dsRNA formations, contributing to RNA interference, or ADAR pathways.

8.12.4 TES AND PROTEIN TRANSLATION

Retrotransposons have evolved with genes, been inserted in different positions across genetic organisms, and have caused many effects. The transposition of the 5′-UTR or 3′-UTR elements of the mRNA influences protein expression for many genes in many aspects. Kitano et al. (2018) disclosed that the upstream open reading framework also includes translations of many genes (uORF). Families like LINEs and SINEs can be prominent retrotransposons to prevent and encourage a translation into eukaryotic mRNAs of downstream canon RFOs. Kitano et al. (2018) showed that approximately 10% of the UORFs for humans are generated and controlled by 5′-UTR in mRNAs and use the human database RefSeq mRNA. Although past studies have shown that retrotransposons are translational regulators, DNA transposons still have a clear role in influencing protein host translation. DNA transposons subfamily, commonly distributable into plant and animal genomes, is the miniature TEs of invert (MITEs). The situation of the 3′-UTRs of rice mRNAs has been reported to be regulated by translational suppression mechanisms.

An example of this is the Ghd2 gene, which regulates essential agricultural characteristically traits in the rice CCT (CO) CO-LIKE and TIMING OF CAB1 family, for example, heading date, plant height, and grain number. [CO] The mechanism underlying the translational MITE GHD 2 repression results in the 3a (OsDCL3a) route that allows the processing of MITE transcripts for short mRNA without poly-A tail. The blocking of translation by mRNA and the repression of mRNA by MITEs remains a mystery. TEs can participate in developing new mRNA isoform splinting alternatives in the gene coding area (Shen et al., 2017). The development of new genes can be viewed as an intermediate step. The splice isoform codes for both genes include the mammal thymopoietin (TMPO) and ZFP 451, both of which relate to the first ORF in a DIRS 1 intermediate retrotransposon splices containing the domains of the 2alpha amine-associable polypeptide. Splice isoform codes in both genes (LAP2alpha). Both mRNAs make the protein that is produced with a new isoform of a noncanonical protein. The crucial role of the cell was particularly coopted with the LAP2alpha-specific TMPO isoform (LAP2a) (Abascal et al., 2015).

The evolutionary introduction of TEs into gene code regions in a mechanism called domestication was also produced. The association of active cytoskeleton with protein (Arc) shows that it comes from the ty3/gypsy vertebrae lines, a well-known example of retrotransposon domestication (Volff, 2009). Arc is an early cellular gene that is especially important for memory and learning, and at a synaptic intersection is the mRNA. Arc mRNA, which is interesting, is similar to viral RNA since Arc has an internal ribosomal input site that allows cap-independent translation. In the subdomains of Arc, the protein structure is identical to the capsid domain of the Gag protein virus (Pastuzyn et al., 2018). Structure of the protein in the subdomains of arc. These results suggest that Gag contains elements that mediate intercellular communication in the developmentally reconstructed nervous system.

8.13 DEFENSE RESPONSE AGAINST TRANSPOSON

8.13.1 PROTECTION AGAINST TES

Since TEs may affect the integrity of genomes, disrupt gene function, and induce conditions of disease, the genome has developed some half-redundant defense mechanisms to decrease TE. The mechanisms are found in silencing transcriptions to transcription editing. Below are the three mechanisms for mammalian genome defense and how they influenced mammalian genome development.

8.13.2 TRANSPOSABLE ELEMENT AND DNA METHYLATION

It maintains CHG methylation mainly by plant-specific chromomethylase protein 3 (CMT3), while CHH is retained by methyltransferase reconfigured domains2 (DRM2). CMT3 is introduced to chromatin through its chromodomain through its interactions with methylated histone H3 tails. In vitro experiments have therefore demonstrated that the CMT3 chromodomain connects directly with H3 tails or trimethylated. In the RNA-directed process of DNA methylation, 24-nucleotide small, interfering RNAs are used to address DNA in the ornithological process of mammalian DNMT3 methyltransferases. In addition, SWI2/SNF2 chromatin-remodeling element, DNA methylation1 (DDM1), corresponds to all sequence contexts, although its exact molecular action mechanism is unknown. Class 2 TE testing in maize presented the first evidence of the DNA-methylphenation function in the regulatory part of TE approximately four decades since B. McClintock detected TEs. According to molecular studies, the methylation of its DNA was due to inhibition of the components of activator (Ac), suppressor-mutator (Spm), and mutator (Mu). Transposable sequences of the Spm and Ac elements promoters are hypomethylated in the active state, resulting in active transcription. Similarly, the transcriptional silence of MuDR's independent Mu-family feature is linked to DNA methylation. Later experiments in Arabidopsis showed that many TE groups were hypomethylated and reconfigured with transcription in a ddm1-induced hypomethylation history.

In addition, it has been found that two DNA transposons, the AtMu1 MULE and CACTA elements, are transposed into ddm1. The characterization of a floral phenotype induced by ddm1 has recently revealed that ddm1 has the mobilization of DN A and ATGP3 gypsum-retrotransposons. Moreover, by hybridizing wild and ddm1 lines on an extremely densely tilling array, the authors could detect and transpose change in copy numbers and thus change many DNA transposon families and retroelements in mutant history. Mutations in LSH1 mouse orthology DDM1 also cause TE hypomethylation and transcriptional reactivation. CHG and CHH methylation are abundant concerning CG methylation, consistent with the favorable interaction with H3K9me2, and successful RdDM pathway.

Consequently, all three kinds of methylation may co-operate or lead to some extent to TE silencing. For example, transcribed and mobilized in the history of hypomethylated mutants CG and non-CG of ddm 1, ATGP3 elements are still transcriptional to single mutants met1 and cmt3. Therefore, in Met1 and Ddm1 mutants such as TEs, loci that lack DNA methylation and epigenetic silencing remain active, and hypomethylation for many generations after mutation is removed. Met1 mutations or ddm1 mutations have hypomethylated DNA in specific genomic regions with epigenetic recombinant inbred lines (epiRILs). CACTA hypomethylation triggers a conversion of these epigenetic allele hypomethylated CACTA ('epialleles) into ddm1-derived epiRILs, as expected. Since CACTA elements are not transposed into a parental mutant context, met1-derived epiRILs should remain stagnant. The mobilization of CACTA by scientists was unexpected in almost 30% of the met1-derived epiRILs. CACTA transposition was likely caused in some epiRIL organisms after many generations of self-pollination. Despite a lack of direct research, CACTA mobilization can be inferred by the descent of met1-derived elements hypomethylated at GC sides and the removal, including non-GC methylation or H3K9me2 adjacent repressive epigenetic markings. Of course, the non-CG methylating patterns and the H 3K9me2 patterns are likely to change and be unstable hereditary, as met1-induced CG methylation is lost.

It is found that EVD retroelements transpose following the inbreeding of certain CACTA identical met1-derived epiRILs. However, the two types of elements were not reconfigured in the same row and emphasized that different epigenetic mechanisms selectively regulated TEs. Scientists have found this selectivity to investigate ddm1-derived epiRILs. It was shown that a subset of TEs, demethylated in ddm1, is related gradually and robust over several generations after being inserted into the wildlife environment. The trend toward demethylation coincides with the preservation of CHH methylation in removable TEs, which seems separate from DDM1. On the other hand, CHH methylation is highly dependent on DDM1 in nonremethylatable TEs. Consequently, CHH methylation is sustained by different pathways at different TEs.

8.13.3 KRAB/KAP1 HISTONE MODIFICATION

However, retrotransposons in premature embryos are silenced by DNA methylation and histone alteration while the discoverer of the mechanism has still been unidentified. The KRAB-ZFPs specifically bind DNA utilizing the C-terminal of ZFP in early embryo retrotransposons. KRAB-ZFPs are now accepted for recruiting protein from KRAB (KAP1s). It can be binding to a single complex of epigenetic regulations, including histone methyltransferases (ESET, HP 1s, nucleosome reproducing, DNMT3A, and

DNMT3B methylated by DNA) (Ecco et al., 2017). The removal of KAP1 from premature mouse embryos causes ERV retrotransposition (Rowe et al., 2010) (Fig. 8.3).

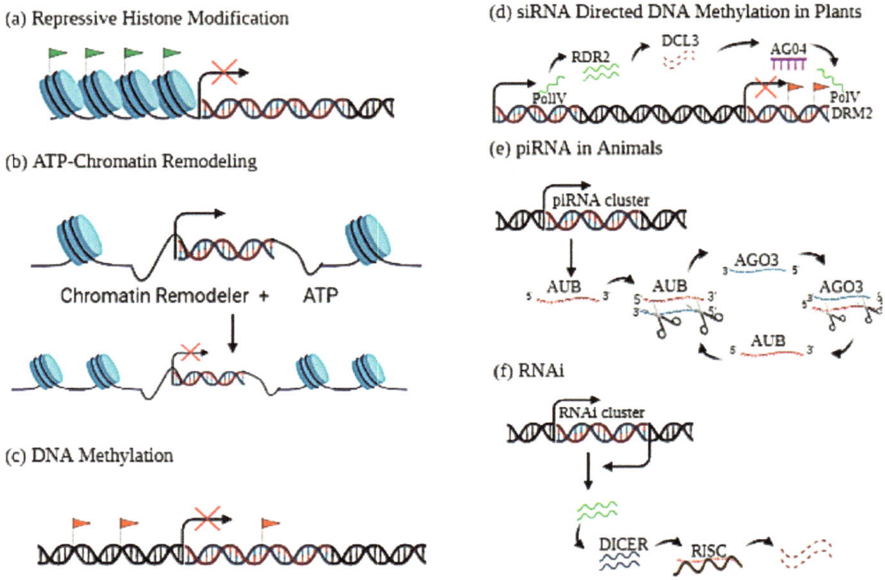

FIGURE 8.3 Mechanisms for silencing the activity of TE. TEs are displayed as DNA strands in red and blue. Blue circles are historical representations. Repressive histone changes are shown as green flags and orange flags are represented by DNA methylation. (a) Repressive histone modifications result in the inactivation of TE. (b) The energy of the ATP hydrolysis is used in ATP chromatin remodeling and to change the nucleosome structure and thus silence the TEs. (c) DNA methylation results in TE silencing. (d) DNA methylation of siRNA-directed in plants. PolIV produces transcripts that are transcribed to dsRNAs by RDR2. These dsRNAs are then divided by DCL3 into 24-nucleotide siRNAs. These siRNAs can be linked to and interact with AGO4. Finally, DRM2 methylates TEs. (e) Drosophila germline pathway for piRNA. A primary piRNA transcript (short red line) is produced from the piRNA (red and blue DNA) cluster and is converted to the 2435-nucleotide piRNAs in the Drosophila germline. These AUB-attached piRNAs are for TE additional mRNA (blue line). A secondary piRNA (short blue line) can be associated with the AGO 3 when cleavage is released from this mRNA. It can then be directly divided into piRNAs and then guided to TE mRNA, to the complementary sequence of the initial piRNA precursor. (f) TE transcripts are dsRNAs that are cleaved by a dicer-family protein into siRNAs. These siRNAs are complementary to SiRNA sequence with a protein of the Argonaute family (RISC). AGO, Argonaute; AUB, Aubergine; DCL3, dicer-like 3; DRM2, domains rearranged methyltransferase 2; piRNA, PIWI-interacting RNA.

8.13.4 SEQUENCE EDITING WITH APOBEC

The cytidine deaminases from the APOBEC family work with reverse transcriptase (Friedli and Trono, 2015). The enzymes mediate cytosine defamation in uracil, and cDNA TE levels are either directly destroyed or weakened. APOBECs have been replaced by genetic duplication in early vertebrates. There are only placental mammals with at least two subfamilies APOBECs (APOBEC3 and APOBEC 1) and APOBEC3. The APOBEC3G virus without the Vif gene and murine leukemia virus was first noticed with viral HIV cDNA. ERVs are structurally similar to these retroviruses, and APOBECs edit them. A POBECs did not impact non-LTR returns, as the APOBEC3G is limited to the cytoplasm, and the reverse transcription of non-LTR is found in the core. However, several LTR and non-LTR feedback were subsequently discovered and inhibited in many ways (Friedli and Trono, 2015; Richardson et al., 2014).

8.14 EFFECTS OF TRANSPOSABLE ELEMENTS

8.14.1 TRANSPOSABLE ELEMENTS AND SOMATIC DISEASES

The expression of TE was considered limited to the germline, but new studies have shown that TE expression is extensive in somatic tissue. If deadly insertions occur, somatic stem-cell insertion is more resistant and more closely connected with aging, neuropathy, and cancer. There are over 100 diseases associated with inserted TE, including various forms of cancer (Hancks and Kazazian, 2016). Cancer TE can occur with the alteration of tumor suppressors or proto-oncogenes. In each case, one ally in a TE must usually be disrupted by deprivation of function on each other allele before the tumorigenesis occurs, for example, tumor-based TE. For example, colorectal cancer is caused by the insertion of the L1 element within the APC suppressor gene (Scott et al., 2016). Splicing of Ks env gene (rec and np9 variants) bind to and suppresses the c-Myc oncogene transcriptional repressor, promyelocytic Zinc finger-protein. Various forms of leukemia are associated with fluctuations in genome structure, including chromosome translocations, recombination, and Alu inserts duplication.

The levels of TE expression in neural tissues are unexpected. In three individuals, only 2200 somatic TE insertions, many of them located inside or near genes that coded protein, were identified. It was estimated that 1 of each 300 neuronal genomes had innovative L1 intakes that could produce thousands of new TE inserted into the typical brain of the mammalian. These and other remarks suggest a certain degree of TE expression is required to develop normal neuronal cells, but other cells can be damaged (Evrony et al., 2016; Reilly et al., 2013). As TEs are inserted in open euchromatin sites, the neural genes can be inserted close or possibly inserted into them. Environmental stimuli of light, heavy metal, flavor, or even physical exercises might boost L1 elements expression over the high levels of the neural tissues. L1 is a fundamental reason for expression (deHaro et al., 2014). The acquisition of TE in stress-related genes such as alcohol or post-traumatic stress disorder has thus increased (Reilly et al., 2013). Alu elements were 16 times in human beings in mitochondria TOMM40 and resulted in significant conformational alterations and shortened proteins that were less efficient (Larsen et al., 2017). A mitochondrial malfunction within neural cells is likely to lead to amplified oxidative stress and consequent inflammation, leading to lower neural and disease functioning in imitate feedback loops.

8.14.2 MOBILE ELEMENTS AND NEURODEGENERATIVE DISORDERS

Endogenous retroelements are being more studied since several neurological disorders have been observed and involved. Such as all three structural genes Gag, Pol, env of HERV-K showed an increased expression of sporadic ALS compared with healthy controls (Li et al., 2015). ALS is neurosubstantial disease-causing damage to neurons. Several pieces of research have recognized the RT activity of serum retroviral in ALS sufferers (Andrews et al., 2000; Steele et al., 2005). Higher expression of HERV K-Env, a powerful immune-pathogenic protein in survivor patients in ALS patients' cortical and spinal neurons after postmortem of the brain. There is no known trigger of HERV-K expression in adult neurons; however, the activation of HERV-K genes showed that motor mouse neurons had decreased dendritic length, ramification, and complexity (Li et al., 2015).

In most sporadic ALS cases, abnormalities are also observed in DNA-binding protein43 (TDP-43). TDP-43 is a nuclear dimeric protein in the heterogeneous family of ribonucleoprotein (hnRNP) (Hergesheimer et al., 2019). The CNS broadly classifies its function as a regulator of pre- and post-transcriptional events because it links with a UG-rich motif in single-stranded RNA/DNA (Hergesheimer et al., 2019; Kuo et al., 2009). Immune precipitation linking (CLIP) and chromatin (ChIP) data indicate that TDP-43 binds widely to human brain tissue and the LTR HERV-K sequence (Li et al., 2012; Li et al., 2015). A study by Krug et al. (2017) investigated whether Drosophila functional abnormal expression of TDP-43 was found and whether there is a more profound depression of retrospective components. hTDP-43 expressions were revealed to induce widespread retreat expression of Drosophila and neuronal glia. Glial expression of dicer-2/Argonaute2 is significantly reduced, while the DNA damage is caused by apoptosis. In a new study, the results of TDP-43 molecular changes are presented for the analysis of diseased neuronal tissue nuclei in patients with frontotemporal ALS degeneration FTd-ALS. Liu et al. (2019) use FACS to divide the diseased neurons and ATAC sequence to access the chromatin. The damage of the TDP-43 is linked with the chromatin decondensation around the elements of LINE-1. While it is uncertain if the loss of protein from TDP-43 is directly responsible for the decondensation of LINE-1 elements, certain specifications or preferences in LINE-1 element decondensation over other repetitive DNA appear. The verdicts together in human and vertebrate models indicate that the unregulated expression of retroelements has little effect on ALS, while no evidence is available that it is the main cause of the syndrome. In ALS clinical trials, potential RTi therapy is also in progress (Alfahad and Nath, 2013; Douville et al., 2011).

8.14.3 AGING

LINE-1 repressive mechanisms are not efficient enough during the aging process. The Line-1 increase in senescent cells was shown in a recent study (De Cecco et al., 2019). Line-1 cDNA accumulation leads to the expression of the (SASP) secret-associated senescence of phenotype. In several tissues with age-related inflammations, IFN (type I interferon) has a typical response. In liver and adipose tissue of 26-month-old mice, expression of LINE-1 mRNA is substantially increased compared to mice of 5 months. In the RT-qPCR evaluation for IFN-I and SASP-responsible genes, the same trend was observed. Reverse transcriptase inhibitors' utilization is contrary to this reaction (RTi). The IFN response and inflammation in this mouse are reduced. Even so, RT is shown that some people believe to be a nonspecific anti-inflammatory element (Fowler et al., 2014). The impact of LINE-1 activities was also tested with mono-ADP-ribosylate/deacetylase in the deficient mice SIRIT6 (Silent Mating Type Information Regel 2 Homolog 6). Mice SIRT6 KO is severely aged for 35 days (Simon et al., 2019). SIRT6 Ko Mice have high LINE-1 levels because of the repressive role of SIRT6 in the ribosylation KAP 1 (Van Meter et al., 2014; Mostoslavsky et al., 2006). The concentration of LINE-1 cytoplasmic DNA was increased without SIRT6, which triggered the IFN response to cGAS (cyclic GMP-AMP synthase).

Treating lamivudine and stavudine with RTi-inhibitors improves mice's life and enhances body mass, mobility, and compatibility (Simon et al., 2019). These results also suggest that LINE-1 is an age-related disease pathologist. Although further studies are required on new L1 cytoplasm copies, this new information encourages the possible therapeutic use of RT in various ages.

8.14.4 TES AND INFLAMMATION

Although the immune system thoroughly studies its contribution to viral protection through innate immunity, the role of TEs in immune responses free from viral infections is little known. It protects the virus by coordinating innate and adaptive immune reactions against infections. If the viral DNA is reached in the cytoplasm, it activates the cGAS-STING mechanism and causes inflammation. The response of interferons by virus cells might be mechanistically connected to the retro deregulation of production (Macia et al., 2017). Research shows that neuroinflammation in endogenous retroelements has a critical role to play. Many autoimmune diseases are at the root of innate endogenous nucleic acid detection (Volkman and Stetson, 2014). How do retroelements work if an incendiary reaction is occurring but no virus infection? How do pathophysiological pathways lead, in particular, to pathological characteristics of disease?

Endogenous retroelements are becoming increasingly attractive as they contribute to a range of inflammatory and neurodegenerative diseases. Retrotransposon intermediates in conditions such as STM and Aicardi-Goutières have been associated as major inflammatory factors (AGS) (Thomas et al., 2017; Treger et al., 2019). In addition, several studies now link older people, TEs, and inflammation.

KEYWORDS

- **transposable elements**
- **transposons**
- **epigenetics**
- **regulation**
- **evolution**

REFERENCES

Abascal, F.; Tress, M. L.; Valencia, A. Alternative Splicing and Co-Option of Transposable Elements: The Case of TMPO/LAP2α and ZNF451 in Mammals. *Bioinformatics* **2015,** *31* (14), 2257–2261.

Alfahad, T.; Nath, A. Retroviruses and Amyotrophic Lateral Sclerosis. *Antiviral Res.* **2013,** *99* (2), 180–187.

Andrews, W.; Tuke, P.; Al-Chalabi, A.; Gaudin, P.; Ijaz, S.; Parton, M.; Garson, J. Detection of Reverse Transcriptase Activity in the Serum of Patients with Motor Neurone Disease. *J. Med. Virol.* **2000,** *61* (4), 527–532.

Anway, M. D.; Cupp, A. S.; Uzumcu, M.; Skinner, M. K. Epigenetic Transgenerational Actions of Endocrine Disruptors and Male Fertility. *Science* **2005,** *308* (5727), 1466–1469.

Babaian, A.; Mager, D. L. Endogenous Retroviral Promoter Exaptation in Human Cancer. *Mobile DNA* **2016,** *7* (1), 1–21.

Barah, P.; Jayavelu, N. D.; Mundy, J.; Bones, A. M. Genome Scale Transcriptional Response Diversity Among Ten Ecotypes of *Arabidopsis thaliana* During Heat Stress. *Front. Plant Sci.* **2013,** *4*, 532.

Berrens, R. V.; Andrews, S.; Spensberger, D.; Santos, F.; Dean, W.; Gould, P.; Sharif, J.; Olova, N.; Chandra, T.; Koseki, H. An EndosiRNA-Based Repression Mechanism Counteracts Transposon Activation During Global DNA Demethylation in Embryonic Stem Cells. *Cell Stem Cell* **2017,** *21* (5), 694–703. e7.

Biscotti, M. A.; Carducci, F.; Olmo, E.; Canapa, A.; Vertebrate Genome Size and the Impact of Transposable Elements in Genome Evolution. In *Evolution, Origin of Life, Concepts and Methods*; Springer, 2019; pp 233–251.

Bourque, G.; Burns, K. H.; Gehring, M.; Gorbunova, V.; Seluanov, A.; Hammell, M.; Imbeault, M.; Izsvák, Z.; Levin, H. L.; Macfarlan, T. S. Ten Things You Should Know About Transposable Elements. *Genome Biol.* **2018,** *19* (1), 1–12.

Bouttier, M.; Laperriere, D.; Memari, B.; Mangiapane, J.; Fiore, A.; Mitchell, E.; Verway, M.; Behr, M. A.; Sladek, R.; Barreiro, L. B.; Alu Repeats as Transcriptional Regulatory Platforms in Macrophage Responses to M. Tuberculosis Infection. *Nucl. Acids Res.* **2016,** *44* (22), 10571–10587.

Bouvet, G. F.; Jacobi, V.; Plourde, K. V.; Bernier, L. Stress-Induced Mobility of OPHIO1 and OPHIO2, DNA Transposons of the Dutch Elm Disease Fungi. *Fungal Genet. Biol.* **2008,** *45* (4), 565–578.

Brattås, P. L.; Jönsson, M. E.; Fasching, L.; Wahlestedt, J. N.; Shahsavani, M.; Falk, R.; Falk, A.; Jern, P.; Parmar, M.; Jakobsson, J. TRIM28 Controls a Gene Regulatory Network Based on Endogenous Retroviruses in Human Neural Progenitor Cells. *Cell reports* **2017,** *18* (1), 1–11.

Canapa, A.; Barucca, M.; Biscotti, M. A.; Forconi, M.; Olmo, E. Transposons, Genome Size, and Evolutionary Insights in Animals. *Cytogenet. Genome Res.* **2015,** *147* (4), 217–239.

Chen, L.; Zhang, P.; Fan, Y.; Lu, Q.; Li, Q.; Yan, J.; Muehlbauer, G. J.; Schnable, P. S.; Dai, M.; Li, L. Circular RNAs Mediated by Transposons Are Associated with Transcriptomic and Phenotypic Variation in Maize. *New Phytol.* **2018,** *217* (3), 1292–1306.

Chen, L.-L.; Yang, L. Regulation of CircRNA Biogenesis. *RNA Biol.* **2015,** *12* (4), 381–388.

Chiappinelli, K. B.; Strissel, P. L.; Desrichard, A.; Li, H.; Henke, C.; Akman, B.; Hein, A.; Rote, N. S.; Cope, L. M.; Snyder, A. Inhibiting DNA Methylation Causes an Interferon

Response in Cancer via dsRNA Including Endogenous Retroviruses. *Cell* **2015,** *162* (5), 974–986.

Coluccio, A.; Ecco, G.; Duc, J.; Offner, S.; Turelli, P.; Trono, D. Individual Retrotransposon Integrants Are Differentially Controlled by KZFP/KAP1-Dependent Histone Methylation, DNA Methylation and TET-Mediated Hydroxymethylation in Naïve Embryonic Stem Cells. *Epigenet. Chromatin* **2018,** *11* (1), 1–18.

Costa, R. M.; Chigançgas, V.; da Silva Galhardo, R.; Carvalho, H.; Menck, C. F. The Eukaryotic Nucleotide Excision Repair Pathway. *Biochimie* **2003,** *85* (11), 1083–1099.

Crow, Y. J.; Manel, N. Aicardi–Goutières Syndrome and the Type I Interferonopathies. *Nat. Rev. Immunol.* **2015,** *15* (7), 429–440.

Daniel, C.; Behm, M.; Öhman, M. The Role of Alu Elements in the CIS-Regulation of RNA Processing. *Cell. Mol. Life Sci.* **2015,** *72* (21), 4063–4076.

Daniel, C.; Lagergren, J.; Öhman, M. RNA Editing of Non-Coding RNA and Its Role in Gene Regulation. *Biochimie* **2015,** *117*, 22–27.

De Cecco, M.; Criscione, S. W.; Peterson, A. L.; Neretti, N.; Sedivy, J. M.; Kreiling, J. A. Transposable Elements Become Active and Mobile in the Genomes of Aging Mammalian Somatic Tissues. *Aging (Albany NY)* **2013,** *5* (12), 867.

De Cecco, M.; Ito, T.; Petrashen, A. P.; Elias, A. E.; Skvir, N. J.; Criscione, S. W.; Caligiana, A.; Brocculi, G.; Adney, E. M.; Boeke, J. D. LINE-1 Derepression in Senescent Cells Triggers Interferon and Inflammaging. *Nature* **2019,** *566* (7742), 73.

deHaro, D.; Kines, K. J.; Sokolowski, M.; Dauchy, R. T.; Streva, V. A.; Hill, S. M.; Hanifin, J. P.; Brainard, G. C.; Blask, D. E.; Belancio, V. P.; Regulation of L1 Expression and Retrotransposition by Melatonin and Its Receptor: Implications for Cancer Risk Associated with Light Exposure at Night. *Nucl. Acids Res.* **2014,** *42* (12), 7694–7707.

Denli, A. M.; Narvaiza, I.; Kerman, B. E.; Pena, M.; Benner, C.; Marchetto, M. C.; Diedrich, J. K.; Aslanian, A.; Ma, J.; Moresco, J. J. Primate-Specific ORF0 Contributes to Retrotransposon-Mediated Diversity. *Cell* **2015,** *163* (3), 583–593.

Douville, R.; Liu, J.; Rothstein, J.; Nath, A. Identification of Active Loci of a Human Endogenous Retrovirus in Neurons of Patients with Amyotrophic Lateral Sclerosis. *Ann. Neurol.* **2011,** *69* (1), 141–151.

Du, J.; Leung, A.; Trac, C.; Lee, M.; Parks, B. W.; Lusis, A. J.; Natarajan, R.; Schones, D. E. Chromatin Variation Associated with Liver Metabolism Is Mediated by Transposable Elements. *Epigenet. Chromatin* **2016,** *9* (1), 1–16.

Ecco, G.; Imbeault, M.; Trono, D. KRAB Zinc Finger Proteins. *Development* **2017,** *144* (15), 2719–2729.

Elbarbary, R. A.; Lucas, B. A.; Maquat, L. E. Retrotransposons as Regulators of Gene Expression. *Science* **2016,** *351* (6274).

Evrony, G. D.; Lee, E.; Park, P. J.; Walsh, C. A. Resolving Rates of Mutation in the Brain Using Single-Neuron Genomics. *Elife* **2016,** *5*, e12966.

Fedoroff, N. V. Transposable Elements, Epigenetics, and Genome Evolution. *Science* **2012,** *338* (6108), 758–767.

Feil, R.; Fraga, M. F. Epigenetics and the Environment: Emerging Patterns and Implications. *Nat. Rev. Genet.* **2012,** *13* (2), 97–109.

Forestan, C.; Cigliano, R. A.; Farinati, S.; Lunardon, A.; Sanseverino, W.; Varotto, S. Stress-Induced and Epigenetic-Mediated Maize Transcriptome Regulation Study by Means of Transcriptome Reannotation and Differential Expression Analysis. *Sci. Rep.* **2016,** *6* (1), 1–20.

Fowler, B. J.; Gelfand, B. D.; Kim, Y.; Kerur, N.; Tarallo, V.; Hirano, Y.; Amarnath, S.; Fowler, D. H.; Radwan, M.; Young, M. T. Nucleoside Reverse Transcriptase Inhibitors Possess Intrinsic Anti-Inflammatory Activity. *Science* **2014,** *346* (6212), 1000–1003.

Friedli, M.; Trono, D. The Developmental Control of Transposable Elements and the Evolution of Higher Species. *Annu. Rev. Cell. Dev. Biol.* **2015,** *31*, 429–451.

Gong, C.; Maquat, L. E. lncRNAs Transactivate STAU1-Mediated mRNA Decay by Duplexing with 3′ UTRs via Alu Elements. *Nature* **2011,** *470* (7333), 284–288.

Hancks, D. C.; Kazazian, H. H. Roles for Retrotransposon Insertions in Human Disease. *Mobile DNA* **2016,** *7* (1), 1–28.

Hedges, D.; Deininger, P. Inviting Instability: Transposable Elements, Double-Strand Breaks, and the Maintenance of Genome Integrity. *Mutat. Res./Fundamental Mol. Mech. Mutagen.* **2007,** *616* (1–2), 46–59.

Hergesheimer, R. C.; Chami, A. A.; de Assis, D. R.; Vourc'h, P.; Andres, C. R.; Corcia, P.; Lanznaster, D.; Blasco, H. The Debated Toxic Role of Aggregated TDP-43 in Amyotrophic Lateral Sclerosis: A Resolution in Sight? *Brain* **2019,** *142* (5), 1176–1194.

Hermant, C.; Torres-Padilla, M.-E. TFs for TEs: The Transcription Factor Repertoire of Mammalian Transposable Elements. *Genes Dev.* **2021,** *35* (1–2), 22–39.

Huang, J.; Wang, Y.; Liu, W.; Shen, X.; Fan, Q.; Jian, S.; Tang, T.; EARE-1, a Transcriptionally Active Ty1/Copia-Like Retrotransposon Has Colonized the Genome of Excoecaria Agallocha Through Horizontal Transfer. *Front. Plant Sci.* **2017,** *8*, 45.

Hummel, B.; Hansen, E. C.; Yoveva, A.; Aprile-Garcia, F.; Hussong, R.; Sawarkar, R. The Evolutionary Capacitor HSP90 Buffers the Regulatory Effects of Mammalian Endogenous Retroviruses. *Nat. Struct. Mol. Biol.* **2017,** *24* (3), 234.

Jardim, S. S.; Schuch, A. P.; Pereira, C. M.; Loreto, E. L. S. Effects of Heat and UV Radiation on the Mobilization of Transposon Mariner-Mos1. *Cell Stress Chaperones* **2015,** *20* (5), 843–851.

Jeck, R.; Sorrentino, A. J.; Wang, K.; Slevin, K. M.; Burd, E. C.; Liu, J.; William, F.; Marzluff, F. W.; Sharpless, E. N. **2013**, 141–157.

Jung, J.; Lee, S.; Cho, H.-S.; Park, K.; Ryu, J.-W.; Jung, M.; Kim, J.; Kim, H.; Kim, D.-S. Bioinformatic Analysis of Regulation of Natural Anti-Sense Transcripts by Transposable Elements in Human mRNA. *Genomics* **2019,** *111* (2), 159–166.

Jurka, J.; Bao, W.; Kojima, K. K. Families of Transposable Elements, Population Structure and the Origin of Species. *Biol. Direct* **2011,** *6* (1), 1–16.

Kassiotis, G.; Stoye, J. P. Immune Responses to Endogenous Retroelements: Taking the Bad with the Good. *Nat. Rev. Immunol.* **2016,** *16* (4), 207–219.

Kelley, D. R.; Hendrickson, D. G.; Tenen, D.; Rinn, J. L. Transposable Elements Modulate Human RNA Abundance and Splicing via Specific RNA-Protein Interactions. *Genome Biol.* **2014,** *15* (12), 1–16.

Kim, E. Z.; Wespiser, A. R.; Caffrey, D. R. The Domain Structure and Distribution of Alu Elements in Long Non-Coding RNAs and mRNAs. *RNA* **2016,** *22* (2), 254–264.

Kim, J. H.; Jun, S.-A.; Kwon, Y.; Ha, S.; Sang, B.-I.; Kim, J. Enhanced Electrochemical Sensitivity of Enzyme Precipitate Coating (EPC)-Based Glucose Oxidase Biosensors with Increased Free CNT Loadings. *Bioelectrochemistry* **2015,** *101*, 114–119.

Kitano, S.; Kurasawa, H.; Aizawa, Y. Transposable Elements Shape the Human Proteome Landscape via Formation of CIS-Acting Upstream Open Reading Frames. *Genes Cells* **2018,** *23* (4), 274–284.

Kralovicova, J.; Moreno, P. M.; Cross, N. C.; Pêgo, A. P.; Vorechovsky, I.; Antisense Oligonucleotides Modulating Activation of a Nonsense-Mediated RNA Decay Switch Exon in the ATM Gene. *Nucl. Acid Therap.* **2016,** *26* (6), 392–400.

Krug, L.; Chatterjee, N.; Borges-Monroy, R.; Hearn, S.; Liao, W.-W.; Morrill, K.; Prazak, L.; Rozhkov, N.; Theodorou, D.; Hammell, M. Retrotransposon Activation Contributes to Neurodegeneration in a Drosophila TDP-43 Model of ALS. *PLoS Genet.* **2017,** *13* (3), e1006635.

Kuo, P.-H.; Doudeva, L. G.; Wang, Y.-T.; Shen, C.-K. J.; Yuan, H. S.; Structural Insights Into TDP-43 in Nucleic-Acid Binding and Domain Interactions. *Nucl. Acids Res.* **2009,** *37* (6), 1799–1808.

Lapp, H. E.; Hunter, R. G.; The Dynamic Genome: Transposons and Environmental Adaptation in the Nervous System. *Epigenomics* **2016,** *8* (2), 237–249.

Larsen, P. A.; Lutz, M. W.; Hunnicutt, K. E.; Mihovilovic, M.; Saunders, A. M.; Yoder, A. D.; Roses, A. D.; The Alu Neurodegeneration Hypothesis: A Primate-Specific Mechanism for Neuronal Transcription Noise, Mitochondrial Dysfunction, and Manifestation of Neurodegenerative Disease. *Alzheimer's Dementia* **2017,** *13* (7), 828–838.

Lee, S.-I.; Kim, N.-S. Transposable Elements and Genome Size Variations in Plants. *Genom. Inform.* **2014,** *12* (3), 87.

Li, W.; Jin, Y.; Prazak, L.; Hammell, M.; Dubnau, J. Transposable Elements in TDP-43-Mediated Neurodegenerative Disorders. *PloS One* **2012,** *7* (9), e44099.

Li, W.; Lee, M.-H.; Henderson, L.; Tyagi, R.; Bachani, M.; Steiner, J.; Campanac, E.; Hoffman, D. A.; Von Geldern, G.; Johnson, K. Human Endogenous Retrovirus-K Contributes to Motor Neuron Disease. *Sci. Transl. Med.* **2015,** *7* (307), 307ra153–307ra153.

Liang, Z.; Anderson, S. N.; Noshay, J. M.; Crisp, P. A.; Enders, T. A.; Springer, N. M. Genetic and Epigenetic Variation in Transposable Element Expression Responses to Abiotic Stress in Maize. *Plant Physiol.* **2021**.

Liu, E. Y.; Russ, J.; Cali, C. P.; Phan, J. M.; Amlie-Wolf, A.; Lee, E. B. Loss of Nuclear TDP-43 Is Associated with De-Condensation of LINE Retrotransposons. *Cell Rep.* **2019,** *27* (5), 1409–1421. e6.

Liu, N.; Lee, C. H.; Swigut, T.; Grow, E.; Gu, B.; Bassik, M. C.; Wysocka, J. Selective Silencing of Euchromatic L1s Revealed by Genome-Wide Screens for L1 Regulators. *Nature* **2018,** *553* (7687), 228–232.

Lucas, B. A.; Lavi, E.; Shiue, L.; Cho, H.; Katzman, S.; Miyoshi, K.; Siomi, M. C.; Carmel, L.; Ares, M.; Maquat, L. E. Evidence for Convergent Evolution of SINE-Directed Staufen-Mediated mRNA Decay. *Proc. Natl. Acad. Sci.* **2018,** *115* (5), 968–973.

Macia, A.; Tejwani, L.; Mesci, P.; Muotri, A.; Garcia-Perez, J. L. Activity of Retrotransposons in Stem Cells and Differentiated Cells. In *Human Retrotransposons in Health and Disease*; Springer, 2017; pp 127–156.

Makarevitch, I.; Waters, A. J.; West, P. T.; Stitzer, M.; Hirsch, C. N.; Ross-Ibarra, J.; Springer, N. M. Transposable Elements Contribute to Activation of Maize Genes in Response to Abiotic Stress. *PLoS Genet.* **2015,** *11* (1), e1004915.

McClintock, B. The Origin and Behavior of Mutable Loci in Maize. *Proc. Natl. Acad. Sci. USA* **1950,** *36* (6), 344–355.

McCullers, T. J.; Steiniger, M. Transposable Elements in Drosophila. *Mob. Genet. Elements* **2017,** *7* (3), 1–18.

Menees, T. M.; Sandmeyer, S. B. Cellular Stress Inhibits Transposition of the Yeast Retrovirus-Like Element Ty3 by a Ubiquitin-Dependent Block of Virus-Like Particle Formation. *Proc. Natl. Acad. Sci.* **1996,** *93* (11), 5629–5634.

Mirouze, M.; Paszkowski, J. Epigenetic Contribution to Stress Adaptation in Plants. *Curr. Opin. Plant Biol.* **2011,** *14* (3), 267–274.

Mostoslavsky, R.; Chua, K. F.; Lombard, D. B.; Pang, W. W.; Fischer, M. R.; Gellon, L.; Liu, P.; Mostoslavsky, G.; Franco, S.; Murphy, M. M. Genomic Instability and Aging-Like Phenotype in the Absence of Mammalian SIRT6. *Cell* **2006,** *124* (2), 315–329.

Negi, P.; Rai, A. N.; Suprasanna, P. Moving Through the Stressed Genome: Emerging Regulatory Roles for Transposons in Plant Stress Response. *Front. Plant Sci.* **2016,** *7,* 1448.

Ogasawara, H.; Obata, H.; Hata, Y.; Takahashi, S.; Gomi, K.; Crawler, a novel Tc1/Mariner-Type Transposable Element in *Aspergillus oryzae* Transposes Under Stress Conditions. *Fungal Genet. Biol.* **2009,** *46* (6–7), 441–449.

Oliver, K. R.; Greene, W. K. Transposable Elements: Powerful Facilitators of Evolution. *Bioessays* **2009,** *31* (7), 703–714.

Ortolan, T. G.; Menck, C. F. M. UVB-Induced Cell Death Signaling Is Associated with G1-S Progression and Transcription Inhibition in Primary Human Fibroblasts. *PLoS One* **2013,** *8* (10), e76936.

Pastuzyn, E. D.; Day, C. E.; Kearns, R. B.; Kyrke-Smith, M.; Taibi, A. V.; McCormick, J.; Yoder, N.; Belnap, D. M.; Erlendsson, S.; Morado, D. R. The Neuronal Gene Arc Encodes a Repurposed Retrotransposon Gag Protein That Mediates Intercellular RNA Transfer. *Cell* **2018,** *172* (1–2), 275–288. e18.

Platt, R. N.; Vandewege, M. W.; Ray, D. A. Mammalian Transposable Elements and Their Impacts on Genome Evolution. *Chromosome Res.* **2018,** *26* (1), 25–43.

Pray, L. A. Transposons: The Jumping Genes. *Nat. Educ.* **2008,** *1* (1), 204.

Reilly, M. T.; Faulkner, G. J.; Dubnau, J.; Ponomarev, I.; Gage, F. H. The Role of Transposable Elements in Health and Diseases of the Central Nervous System. *J. Neurosci.* **2013,** *33* (45), 17577–17586.

Rey, O.; Danchin, E.; Mirouze, M.; Loot, C.; Blanchet, S. Adaptation to Global Change: A Transposable Element–Epigenetics Perspective. *Trends Ecol. Evol.* **2016,** *31* (7), 514–526.

Richards, C. L.; Verhoeven, K. J.; Bossdorf, O. Evolutionary Significance of Epigenetic Variation. In *Plant Genome Diversity Volume 1*; Springer, 2012; pp 257–274.

Richardson, S. R.; Narvaiza, I.; Planegger, R. A.; Weitzman, M. D.; Moran, J. V. APOBEC3A Deaminates Transiently Exposed Single-Strand DNA During LINE-1 Retrotransposition. *Elife* **2014,** *3,* e02008.

Rogers, R. L.; Slatkin, M. Excess of Genomic Defects in a Woolly Mammoth on Wrangel Island. *PLoS Genet.* **2017,** *13* (3), e1006601.

Rostant, W. G.; Wedell, N.; Hosken, D. J. Transposable Elements and Insecticide Resistance. *Adv. Genet.* **2012,** *78,* 169–201.

Roulois, D.; Yau, H. L.; Singhania, R.; Wang, Y.; Danesh, A.; Shen, S. Y.; Han, H.; Liang, G.; Jones, P. A.; Pugh, T. J. DNA-Demethylating Agents Target Colorectal Cancer Cells by Inducing Viral Mimicry by Endogenous Transcripts. *Cell* **2015,** *162* (5), 961–973.

Rowe, H. M.; Jakobsson, J.; Mesnard, D.; Rougemont, J.; Reynard, S.; Aktas, T.; Maillard, P. V.; Layard-Liesching, H.; Verp, S.; Marquis, J. KAP1 Controls Endogenous Retroviruses in Embryonic Stem Cells. *Nature* **2010,** *463* (7278), 237–240.

Ryan, C. P.; Brownlie, J. C.; Whyard, S. Hsp90 and Physiological Stress Are Linked to Autonomous Transposon Mobility and Heritable Genetic Change in Nematodes. *Genome Biol. Evol.* **2016,** *8* (12), 3794–3805.

Scott, E. C.; Gardner, E. J.; Masood, A.; Chuang, N. T.; Vertino, P. M.; Devine, S. E. A Hot L1 Retrotransposon Evades Somatic Repression and Initiates Human Colorectal Cancer. *Genome Res.* **2016,** *26* (6), 745–755.

Shen, J.; Liu, J.; Xie, K.; Xing, F.; Xiong, F.; Xiao, J.; Li, X.; Xiong, L. Translational Repression by a Miniature Inverted-Repeat Transposable Element in the 3′ Untranslated Region. *Nat. Commun.* **2017,** *8* (1), 1–10.

Shpyleva, S.; Melnyk, S.; Pavliv, O.; Pogribny, I.; James, S. J. Overexpression of LINE-1 Retrotransposons in Autism Brain. *Mol. Neurobiol.* **2018,** *55* (2), 1740–1749.

Simon, M.; Van Meter, M.; Ablaeva, J.; Ke, Z.; Gonzalez, R. S.; Taguchi, T.; De Cecco, M.; Leonova, K. I.; Kogan, V.; Helfand, S. L. LINE1 Derepression in Aged Wild-Type and SIRT6-Deficient Mice Drives Inflammation. *Cell Metabol.* **2019,** *29* (4), 871–885. e5.

Staunstrup, N. H.; Madsen, J.; Primo, M. N.; Li, J.; Liu, Y.; Kragh, P. M.; Li, R.; Schmidt, M.; Purup, S.; Dagnæs-Hansen, F. Development of Transgenic Cloned Pig Models of Skin Inflammation by DNA Transposon-Directed Ectopic Expression of Human β1 and α2 Integrin. *PLoS One* **2012,** *7* (5), e36658.

Steele, A.; Al-Chalabi, A.; Ferrante, K.; Cudkowicz, M.; Brown, R.; Garson, J. Detection of Serum Reverse Transcriptase Activity in Patients with ALS and Unaffected Blood Relatives. *Neurology* **2005,** *64* (3), 454–458.

Thomas, C. A.; Tejwani, L.; Trujillo, C. A.; Negraes, P. D.; Herai, R. H.; Mesci, P.; Macia, A.; Crow, Y. J.; Muotri, A. R. Modeling of TREX1-Dependent Autoimmune Disease Using Human Stem Cells Highlights L1 Accumulation as a Source of Neuroinflammation. *Cell Stem Cell* **2017,** *21* (3), 319–331. e8.

Treger, R. S.; Pope, S. D.; Kong, Y.; Tokuyama, M.; Taura, M.; Iwasaki, A. The Lupus Susceptibility Locus Sgp3 Encodes the Suppressor of Endogenous Retrovirus Expression SNERV. *Immunity* **2019,** *50* (2), 334–347. e9.

Trivedi, M.; Shah, J.; Hodgson, N.; Byun, H.-M.; Deth, R. Morphine Induces Redox-Based Changes in Global DNA Methylation and Retrotransposon Transcription by Inhibition of Excitatory Amino Acid Transporter Type 3–Mediated Cysteine Uptake. *Mol. Pharmacol.* **2014,** *85* (5), 747–757.

Trizzino, M.; Kapusta, A.; Brown, C. D. Transposable Elements Generate Regulatory Novelty in a Tissue-Specific Fashion. *BMC Genom.* **2018,** *19* (1), 1–12.

Van Meter, M.; Kashyap, M.; Rezazadeh, S.; Geneva, A. J.; Morello, T. D.; Seluanov, A.; Gorbunova, V. SIRT6 Represses LINE1 Retrotransposons by Ribosylating KAP1 But This Repression Fails with Stress and Age. *Nat. Commun.* **2014,** *5* (1), 1–10.

Vargiu, L.; Rodriguez-Tomé, P.; Sperber, G. O.; Cadeddu, M.; Grandi, N.; Blikstad, V.; Tramontano, E.; Blomberg, J. Classification and Characterization of Human Endogenous Retroviruses; Mosaic Forms Are Common. *Retrovirology* **2016,** *13* (1), 1–29.

Vehniäinen, E. R.; Vähäkangas, K.; Oikari, A.; UV-B Exposure Causes DNA Damage and Changes in Protein Expression in Northern Pike (Esox lucius) Posthatched Embryos. *Photochem. Photobiol.* **2012,** *88* (2), 363–370.

Volff, J. N.; Cellular Genes Derived from Gypsy/Ty3 Retrotransposons in Mammalian Genomes. *Ann. NY Acad. Sci.* **2009,** *1178* (1), 233–243.

Volkman, H. E.; Stetson, D. B.; The Enemy Within: Endogenous Retroelements and Autoimmune Disease. *Nat. Immunol.* **2014,** *15* (5), 415–422.

Zhang, F.; Wang, J.; Xu, J.; Zhang, Z.; Koppetsch, B. S.; Schultz, N.; Vreven, T.; Meignin, C.; Davis, I.; Zamore, P. D.; UAP56 Couples piRNA Clusters to the Perinuclear Transposon Silencing Machinery. *Cell* **2012,** *151* (4), 871–884.

Zhang, W.; Wang, H.; Bin. *Zhou, KG* 2014.

Zovoilis, A.; Cifuentes-Rojas, C.; Chu, H.-P.; Hernandez, A. J.; Lee, J. T.; Destabilization of B2 RNA by EZH2 Activates the Stress Response. *Cell* **2016,** *167* (7), 1788–1802. e13.

CHAPTER 9

Transposable Elements and Polyploidy in Plants

PARUL TYAGI, AYUSHI SINGH, POOJA SARASWAT, AMBIKA CHATURVEDI, and RAJIV RANAJAN*

Dayalbagh Educational Institute, Department of Botany, Dayalbagh, Agra, Uttar Pradesh 282005, India

**Corresponding author. E-mail: rajivranjanbt@gmail.com*

ABSTRACT

Besides polyploidization, transposable elements (TEs) contribute majorly to plant evolution as they serve to be a source for coding new regulatory genetic sequences and thereby causing changes in size of the genome. These changes largely involve epigenetic restructuring. Together with polyploidization events (hybridization between relatively divergent genomes), whole-genome duplication results in drastic reorganizations in the genome; thus, TEs are acknowledged as the pre-eminent participant in the plant evolution. Plant genomes and TEs evolve simultaneously and their coexistence has led to controlled governance of TEs activity, and growing evidence suggests their relationship is mutual. Distinct mechanisms are provided by polyploid systems to check TE activities and also affect the process of evolution by genome reorganization induced by TE and hence help to combat various biotic or environmental stresses and different genomic conditions like allo-polyploidization and hybridization conditions. In shorter run, TE insertion onto the different gene locations might induce variations in gene regulation, and also might increase the pace of gene mutation as well in a shorter run. Because of interelement recombination, TE burst-induced genome

Plant Transposable Elements: Biology and Biotechnology, Deepu Pandita, Anu Pandita, (Eds.)

restructuring occurs. These include larger loss of a genome region and rearrangement of chromosomes and therefore causing the reduction of genome size, as well as the number of chromosomes as being part of diploidization process.

9.1 INTRODUCTION

Nearly about 50 years ago, Waring and Britten used reassociation studies and described that the genome of most of eukaryote carry a larger amount of repetitive nucleotide fragments, major portion of the repetitive sequence originate in TEs (Consortium, 1998; Stein et al., 2003). The portion of these often repeating sequences varies accordingly, for instance, about 12% and 50% in nematodes and in mammals, respectively, and much higher in plants, that is, 80% (Waterston et al., 2002). TEs significantly influence the organization of the genome and evolution due to their large numbers. Much advancement has been achieved in the knowledge regarding the role of TEs but the real mystery behind the role of TEs in host genome is still unsolved. TEs are also responsible for a variety of challenges to the organism genome community regarding their detection aspects, classification, genomic assembly, annotation, comparison, and genome mapping. Here, this chapter contributes to add some content regarding TE heterogeneity and evolution, and the extensive techniques related to their analysis are also discussed (Lander et al., 2001).

Discrete fragments of DNA that can migrate in new locations within the genome of the hosts are termed as transposable elements (TEs). Transposons and jumping genes are the alternate terms that are employed for these DNA fragments. Apparently, these mobile elements are prevalent in almost every eukaryotic organism and constitute a large portion of the genome. In humans, they occupy 45% of the total genes. At first, Barbara McClintock acknowledged the concept of jumping genes while working on *Zea mays* (maize). TEs are nonautonomous in nature, that is, they cannot shift themselves and get introduced into a different region in the genome of the host organism. They require a set of enzymes that would facilitate their movement. While relocating it, they make their duplicate copies as they carry their own origin of replication (SanMiguel et al., 1998; Carlos et al., 2017). In 1944, Barbara McClinckton conducted an experiment in maize that contributed to the discovery of TEs and named them as controlling elements because of their characteristics to control the phenotypic traits (SanMiguel et al., 1998).

But her discovery remains unappreciated for several years. In this spare time, a new discovery blew the mind, that is, the same of insertion sequences (IS) in bacteria was discovered around the 1970s by Szybalski and colleagues (Kalendar et al., 2000). That was a time when McClintock discovery came to light in which it was mentioned that "Genetic Elements" are present in almost all higher organisms and can easily rearrange themselves from one locus to the other within a genome. It was expected that they might resemble the present study related to IS insertions. Then in 1983, McClinktock was appreciated by the Nobel Prize for the discovery of these mobile elements. Subsequently, Szybalski rediscovered TEs and Susumu Ohno named them as junk DNA (Jordan et al., 1968) because these fragments are not involved in any for any of protein synthesis. Few researchers observed evolutionary changes in TEs in the early 1990s (Lerman et al., 2005). Jumping genes have an active involvement in the process of evolution to regulate a gene and researchers have now started to explore more in this direction (Moran et al., 1999; Sayah et al., 2004).

In his intriguing critique on macroevolution, gene silencing and TEs, a significant participant has been overlooked by John McDonald that has an active role in the genome advancement of both plants and animals together with TEs, that is, polyploidy. The upheaved expansions in the number of genes corresponding to milestone transformative occasions have regularly come about because of the entire duplication of genomes. Two rounds of polyploidization were seen for the evolution from invertebrates to vertebrates (Elliot and Gregory, 2015). Also, it is found that about 70% of angiosperms show polyploidy in the history. It is seen that diploid genomes are at more risk of insertional mutagenesis than those which have polyploid genomes. As they contain copies and generally kept up with the harmful results of transposition. Transposable components favor to duplicate and remain in polyploid genomes since, additional duplicates make up for modifications in articulation caused through TE insertions (Chalopin et al., 2015). Concerning quality silencing in the event, it is accepted that TEs are the essential focus in reference to the epigenetic alterations, for example, methylation. As polyploid genomes accommodate a large number of TEs, they are exceptionally more methylated than genomes of diploid organisms. Surely, an unpleasant connection exists; methylation is prevalent in vertebrates and genomes of plants that contain a considerable amount of TEs, though partial methylation is common in few invertebrates and *Arabidopsis thaliana* (which is a diploid plant and has little quantity of TEs) (Fujino et al., 2010; Grandbastien et al., 2005).

Obviously, factors affecting polyploidy, for example, the inclusion site inclinations of various TEs can likewise pose an impact on appropriation and plenitude of TEs in genomes. A component that especially incorporates to duplicate itself, for example, *Tdd-1*, in the sludge form *Dictyostelium discoideum* (Thomas et al., 2020), might actually achieve a large number of duplicate copies in a genome of diploid organisms. Epigenetic silencing framework results in expansions in gene number prior protective epigenetic changes, directed against TEs or infections. These perceptions propose that qualities whose suppression is associated with modulation contain TE remainders or even complete TE in their promoters. Considering another aspect, epigenetic silencing induced by these mobile components can be suggested for development if genetically transmitted silencing occurs meiotically (Benoit et al., 2012). We are just barely starting to comprehend the various ways by which TEs could collaborate in cis- and in trans-relationships for initiating quality silencing and epigenetic modifications. In plants and maybe in different living beings, non-Mendelian legacy can result from this cooperation in light of the fact that interfacing alleles or loci display adjusted articulation in the wake of isolating in descendants. At times, epigenetic silencing initiated by unfamiliar arrangements can endure and affect aggregates for some sexual generations and in this way be vague from hereditary transformations. In rundown, the hereditary and epigenetic outcomes of interpretation in polyploids are a significant part of the imaginative force credited to redundancy (Petrov et al., 1996; Wang et al., 2014).

Polyploidy, being a repetitive cycle in nearly all organic entities could be considered a system for significant speciation (Wood et al., 2009). It is evident in plants, furthermore, regularly happens in a few taxa (Mable et al., 2011). Specifically, in angiosperms, the entire genome is seen to undergo at least one round of duplication (Jiao et al., 2011) and therefore plant genomes have extensive hereditary repetitions. Traditionally, two basic types of polyploids have been perceived, addressing extreme examples of a continuum. Autopolyploids are seen in individuals who contain two complimentary copies of the ancestral species and are described by polysomic inheritance that is dominating in nature (Parisod et al., 2010b). Variations in allopolyploid chromosomes come about because of the convergence of disparate genome (i.e., homologous in nature; AABB) and for the most part show disomic legacy (Leitch and Leitch, 2008). The distinction between homologous genomes and homologous genomes is not really obvious, and also it shows continuity among autopolyploidy as well as allopolyploidy. Hence, it is crucial to understand that the transformative production of every normal polyploid includes hybridization between differently related genomes.

9.2 CATEGORIZATION OF TRANSPOSABLE ELEMENTS

Finnegan (1989) proposed TEs classification system and categorized it into two classes, namely; RNA or retrotransposons (class I transposons) and DNA transposons (class II transposons). The mechanism of transposition is commonly termed as "copy and paste" as well as "cut and paste." Basis of TE hierarchical classification structural characteristics along with the mode of replication (Wicker et al., 2007). Eukaryotic TEs belong to class II in accordance with the hierarchical classification.

9.2.1 MOBILE ELEMENTS (CLASS II OR DNA TRANSPOSONS) OF TES

Unlike subclass I, TE subclass II is not concerned with the production of RNA intermediates and is composed of two orders. Ordinarily, DNA TEs replicate with double-strand cleavage but those constituting subclass II follow different mechanisms. Rolling-circle mechanism is employed in helitron replication, where insertion does not result in duplication (Kapitonov and Jurka, 2001). Tyrosine recombinase along with some other related proteins like DNA polymerase Bare integrated to encode TEs. Firstly, helitrons were discovered in fungi, plants, and mammals (Hood, 2005; Pritham et al., 2007). Excluding plants, mavericks are the largest group of TEs that are present in many eukaryotic lineages (Pritham et al., 2007). Out of several TE types in the genome of plants, the long terminal repeat retrotransposons (LTRs) and miniature inverted transposable elements (MITEs) are more bountiful TEs (Casacuberta and Santiago, 2003).

The larger shape and size of LTR retrotransposons make them more commonly sequenced among the plant genomes, including 2.5% in *Utricularia gibba* (humped or floating bladderwort) (Ibarra-Laclette et al., 2013), whereas *Fritillaria* species contain up to 90% of the same (Ambrozová et al., 2011) and few other examples are shown in Table 9.1. Along with polyploidization, TEs enhancement is viewed as the principal mechanism in expansion of the plant genome and its development (Wendel et al., 2016). In reality, as examined beneath, polyploidizations as well as TE amplification are not totally free systems. Actually, these two phenomena incredibly impact each other, building up their capability resulting in plant genome advancement. TEs play a major role in the evolution of plant genes. Genome is not just essential for prolonged plant advancement, but also is of vital significance for late yield taming and breeding (Olsen and Wendel, 2013).

TABLE 9.1 Comparative Study on Changes in Gene Expression Associated with the Percent Transposable Elements in Few Plant Species.

Common name	Scientific name	Chromosome number and ploidy level	Genomic size	Percent of transposon	Auto/ allo	Type of transposon	Functional effect	References
Barley	*Hordeumvulgare*	7Chr, diploid	5100 Mb	80%		LTR retrotransposons	Specifically target promoters and downstream regions and also associated with decreased methylation levels	Wicker et al., (2017)
Rice	*Oryza sativa*	48Chr, tetraploid	385 Mb	–	Auto	Diverse	Associated with methylation and change in Si RNA	Li et al., (2014)
Maize	*Zea mays*	10Chr	2198.23 Mb	85%	Allo	Diverse	Regulation of transcription	Vicient et al., (2010)
Bread Wheat	*Triticumaestivum*	42Chr, allohexaploid	14439 Mb	90%	Allo	Diverse, LTR retrotransposons	Increased siRNA density for TEs Genomic, centromere rearrangements after polyploidizationevents. Differential amplification in the subgenomes.	Salentijn et al., (2009); Li et al., (2014); Li et al., (2013)
Rape	*Brassica napus*	38Chr, allopolyploid	848.3 Mb	56.8–58.2%	Allo	Diverse	Known for dominant expres-sion among subgenomes.	Yin et al. (2020); Song et al., (2020)
Coffee	*Coffea Arabica*	44Chr, allotetraploid	1094.45 Mb	–	Allo	LTR transposons	Distinctive insertions in subgenomes	Hernandez et al. (2017) Sanglard et al., (2018)
Cotton	*Gossypiumhirsutum*	52Chr, allotetraploid	2296.55 Mb	67.2%	Allo	LTR retrotransposon	Deletions in the TE genome fractions	Wang et al., (2016)
	Gossypiumraimondii	13Chr, diploid	761.565 Mb	57%	Allo	LTR retrotransposon	–	Wang et al., (2016)
	Gossypiumarboreum	52Cr, allotetraploid	1778.42 Mb	68.5%	Allo	LTR retrotransposon,	Most of maternal origin sequence loss	Wang et al., (2016); Grover et al., (2007)

TABLE 9.1 *(Continued)*

Common name	Scientific name	Chromosome number and ploidy level	Genomic size	Percent of transposon	Auto/ allo	Type of transposon	Functional effect	References
Thale cress	*Arabidopsis thaliana*	10Cr, diploid	119.709 Mb	14%	Auto	En-Spm-like transposon	Activation and initiation of transcription, remodeling of CG methylation as well as epigenetic modifications.	Madlung et al. (2005); Ma et al., (2019)
Tobacco	*Nicotianatabacum*	48Chr, allotetraploid	3734.23 Mb	–	Allo	Diverse	Parental origin sequence loss	Renny-Byfield et al., (2011)
Sunflower	*Helianthus annuus*	17Chr, diploid	3010.05 Mb	More than 81% in 25% genome	Allo	LTR-retrotransposons	Shaping the chromatin and DNA landscape of the sunflower genome.	Giordani et al., (2014)
Sorghum	*Sorghum bicolor*	20Chr, Diploid	687.75 Mb	–	Allo	Diverse	–	Hodnett et al., (2019)

Later in the chapter, we discuss about the connections among TEs dynamics and polyploidization, as well as the functional role of TEs in the development of plant genomes (wild and domestic breeding).

9.3 EXPANSION OF LTR RETRO-TRANSPOSONS AND CONTRACTION OF PLANT GENOMES

Albeit all plant genomes possess a significant part of TEs, and LTR retrotransposons are most bountiful. The pervasiveness of specific families among species also among assortments of similar species is highly variable. Copy numbers have been expanded by various TE families in one heredity as seen in various cases. A solitary kind of LTR retrotransposon clarifies the greater portion of the *Capsicum annuum* (Shimla mirch) genome extension, and a solitary Ty3/vagabond like retrotransposon, Monster, comprises approximately 38% of Hungarian vetch genome (Neumann, 2006). At times, capacity enhancement of a family is pooled by a various plant species that are related; however, it is noticed that a TE family in one animal group might possess a high copy number, whereas a low duplicate number might be present in a nearby family member (Hawkins et al., 2009). Also, significant contrasts are seen among various similar species. As, for example, 1450 copies of Grande LTRs are seen in innate line B73 of maize, whereas "Palomero Toluque~no" contains 3500 copies of the same.

Though, the existence of a solitary or a couple of exceptionally dull TE families in a genome is regular; genomes with a few TE families with comparative copy numbers have additionally been noticed. The genomes of conferring mostly contain repetitive sequences in which LTRs occupy a considerable quantity (Wegrzyn et al., 2014). These LTR retrotransposon studies regarding the limit with respect to TEs recommend that both components and the genomes direct TE to strike that specific genome considering its limit of multiplication. Few of its components have the capability to escape their regulation from the controlling system of genome, whereas few genomes are more tolerant and permissive toward TE multiplication. TE activity is tightly controlled by the epigenetic mechanisms (Bennetzen and Wang, 2014). The tolerance of certain genomes to TEs can be due to lower silencing effectiveness. Also, climatic change and a momentary drop in silencing are one of the reasons responsible for TEs multiplication blasts (Willing et al., 2015). The limit of certain TEs to neutralize genome silencing accounts for characteristic actions of specific TE. For sure, it is seen that plant retrotransposons can escape from silencing by interacting with silencing factors. Also, the transcription of TEs, their amplification and transposition can get reactivated under specific conditions, like DNA methylation or genome rearrangement (Ito et al., 2014).

For instance, the initiation of the expression of some TEs is found in the pollen vegetative nurse cell which surrounds the sperm cells by triggering the formation of small interfering RNAs (siRNAs) to maintain the epigenetic silencing of TEs in the coming generations. Likewise, a few TEs are activated under various stress situations. It has been seen that the biotic (living) and abiotic (nonliving) stresses can activate the transcription of tobacco retrotransposon (*Tnt1*), as reported by Grandbastien and co-workers. Many stresses have been reported that cause activation TEs such as rice miniature inverted-repeat transposable elements (MITE) mPing under cold and salt stress, *A. thaliana* retrotransposon ONSEN under heat stress, and in vitro culture initiated the activation of *Oryza sativa* (rice) and maize TEs. TE promoters suggest a transcriptional proposes mechanism, with the presence of stress-associated transcription factor-binding sites (TFBSs); however, a reduction in silencing related to stress could likewise represent the broad relationship of stress and reactivation of TEs (Tittel-Elmer et al., 2010). The activation of TEs during stress may cause an increase in the mutation related to TEs and some of that may bring versatile changes to the stressful conditions, as reported for the ONSEN retrotransposon in *Arabidopsis*. Changes like interspecific crosses and polyploidization in the genome have been seen to cause worldwide epigenetic changes and activation of TEs transcription; therefore, considered as "genome stresses" (Yaakov and Kashkush, 2012).

9.4 TRANSPOSONS AND MALDEVELOPMENT IN PLANTS

TEs are considered the most dynamic portion of the plant genome (Tenaillon et al., 2010). The divergence in taxa causes different arrangements of TEs, which reduces recombination and result in incompatibilities of chromosomes during hybridization (He et al., 2009; Abbott et al., 2013). This type of dynamic arrangement of TEs may lead to different copies in the diverging genomes which may later conflict during genome merging because such hybrid genomes fail in their regulatory activities. The small interfering RNA (siRNA) controls the TE activities as the inserted homologous copies are targeted and silenced by DNA methylation (Bourehis et al., 2010). The siRNA and TEs are matched in somatic cells that ensure genome stability and appropriate suppression (Martienssen, 2010). At the time of gametogenesis, the TEs get reactivated and increase siRNAs in the neighboring cells and assure to maintain repression of copies along the generation. It is evident that siRNA in the cytoplasm could fail in repressing all copies after the genomes with diverge TEs are merged (Parisod et al., 2012). In endosperm, such conflicts might even be more evident because maternal and paternal genomes are merged together.

The activated TEs, when the divergent genomes are merged undergo rapid transposition which may result in the failure of endosperm development and mutations in the zygote, which is called hybrid dysgenesis. Accordingly, as the divergent genomes are merged, polyploidy induces reconstruction in the genome more specifically affecting the TEs (Parisod et al., 2010b). Despite all the advancements made in the recent years (reviewed in Parisod and Senerchia, 2012), there is much more that could be elucidated about TEs.

9.5 RELATIONSHIP BETWEEN TRANSPOSONS AND POLYPLOIDY DYNAMICS IN PLANTS

Some plant genes are carried from transcriptional elements in a process known as exaptation, and TEs are additionally added to the advancement of promoters, introns, and exons. Genes can be modified and also diversified by the mechanism involving transcriptional elements (Contreras et al., 2015). Insertional inactivation is the phenomenon for the coding or regulation of the gene. However, it is seen that the insertion of TEs in a gene can create more unobtrusive transformations, for example, alterations in the encoded protein arrangement or modification in their expressions (Huang et al., 2015). TEs have their own promoters and/or potentially enhancers, empowering the regulatory components. TEs can intensify and rearrange TFBSs, making new administrative organizations or overhauling new genes into the currently present genes (He'naff et al., 2014). Components regulating transcription serve genomes to be more transcriptionally flexible and, thus, is valuable for quick variation to evolving conditions.

TEs transposition burst and their results can be seen in several generations to generations. Indeed, even without a new transposition process, it is expected to be more continuous. Because of their greater abundance, they might neutralize genome development and might also bring about gene loss, changes in the gene (various types of mutation), and genome restructuring. TEs play a significant role in restoring harmony after the genome is duplicated. It has been reported that allopolyploidization instigates a TEs transcription in various allopolyploid and diploid parental plants (Kashkush et al., 2003). In a study of engineered polyploids of Arabidopsis and other species like *A. thaliana*, an increment in the RNA level of three En-Spm-like components and a *Ty-1*copia-like retrotransposon was identified (Madlung et al., 2005). Also, high transcriptional activity of *Wis2-1* are trotransposon was seen in freshly combined wheat amphidiploid and its other diploid members, and higher expression of Tip100 was observed in allopolyploid espresso, *Coffea arabica*, in comparison with the other members like *C. eugenioides* and *C. canephora* (Lopes et al., 2013).

Copy numbers of TEs are generally greater in diploid species and are more or less related. *Tnt1* retrotransposons in *Nicotiana tabacum* (tobacco) allotetraploid and Au SINE in *Triticum aestivum*, that is, wheat polyploids are present. After polyploidization, some of the TEs play a vital role in proliferation, for examples, Tekay families multiply after *Orobanche gracilis* and the Stowaway-like Bugs translate subsequent allopolyploidization events in species of wheat and Brassica (Yaakov and Kashkush, 2012). Additionally, in Helianthus species, hybridization of the three diploids resulted in atrocious TE derepression.

However, polyploidization should not generally be joined by an expansion in TEs. For example, there was no increase in the copy number of Au SINE in allopolyploid wheat lines (Ben-David et al., 2013), *Spartina anglica* and *Brassica napus* allotetraploids. Likewise, the variations are seen in the activation of various TE families inside a genome. For example, in Aegilops allotetraploid, some gypsy-like retrotransposons multiply while other remains inactive. Nonetheless, the impact on a specific family of TE may depend upon the parental species, as seen for the Sabine retrotransposon which multiplies specifically in polyploids of wheat and is hugely wiped out in others (Senerchia et al., 2014). It appears that the mechanism of polyploidization is different among TE families and genomes. Furthermore, unique TE families can be regulated differently within a solitary genome, depending on the type of TEs, their respective copy number, promoter arrangements, and chromosome limitation.

Likewise, then again, various genomes vary in their TEs control productivity due, among others, in contrast to siRNA and methylation status. An expanding measure of information shows that polyploidization may trigger epigenetic changes like alteration of TEs during DNA methylation (Zhang et al., 2015). DNA methylated TEs were seen in autotetraploid rice accompanied by variations of 24 nucleotide in siRNA. It has been seen that the allopolyploids may get their TE methylation site after demethylation (Zhang et al., 2015). The methylation, either hyper- or hypo-methylation, depends upon the TE family and is found reproducible. Although retrotransposons show hypomethylation of most parts in the newly formed allopolyploids and hypermethylation of class II DNA components studied in rice and wheat (Yaakov et al., 2011; Zhang et al., 2015). Polyploidization depends upon the different processes (auto- or allo-polyploidization) and the genome and may influence some groups of TEs that more inclined to activation. Likewise, the casual determination in polyploids may consider higher maintenance of TE inclusions which will add up to an expansion in TE copy number because of the increment of gene duplicates.

Events in duplication of whole-genome leading to polyploidy are typical subjects in plant advancement. Polyploidy is outstretched in almost all plants, except for gymnosperms. Whether regular or trained, it is perceived as a significant speciation process (Soltis et al., 2015; Shimizu-Inatsugi et al., 2017). Polyploidizations significantly affect genomes. For better understanding, refer to Figure 9.1. Few potential outcomes of polyploidy include heterosis, regenerative isolation, heterosis, gene redundancy, mating framework rearrangements, modification in cell engineering, issues in cell cycle, quality administrative changes, etc. (Soltis et al., 2015). Duplication genes might be abolished, held, or kept up, frequently obtaining new capacities. Thus, polyploids regularly show various disparities, and their diploid ancestors may add their utility for farming. Polyploidization is oftentimes joined by an increment in TEs content (Adams and Wendel, 2005). This can be the consequence of an actuated explosion of interpretation. Nonetheless, then again, quality duplication permits genomes to adapt to a higher TEs action, as the limit of mutagenicity for TE is supported by the duplication of fundamental qualities. The increment in TE additions might modify gene capacities and does not affect inactivation of duplicated genes (Gaeta et al., 2007).

9.6 ORIGEN OF NATURAL PATHWAYS OF POLYPLOIDS

9.6.1 MERGING OF COMPLETE GAMETES IN CONTRAST TO SOMATIC DOUBLING

In plants, new polyploid genealogies originate by fusion of unreduced gametes or through uncontrolled multiplication of chromosomes in substantial cells (Ramsey et al., 1998). The pathways including unrestricted multiplication of chromosome sets do not appear to be continuous in nature. Mitotic nondisjunction is possible in plants in its the life cycle that supports multiplication of chromosomes unrestrictedly and may give rise to mixoploid living beings

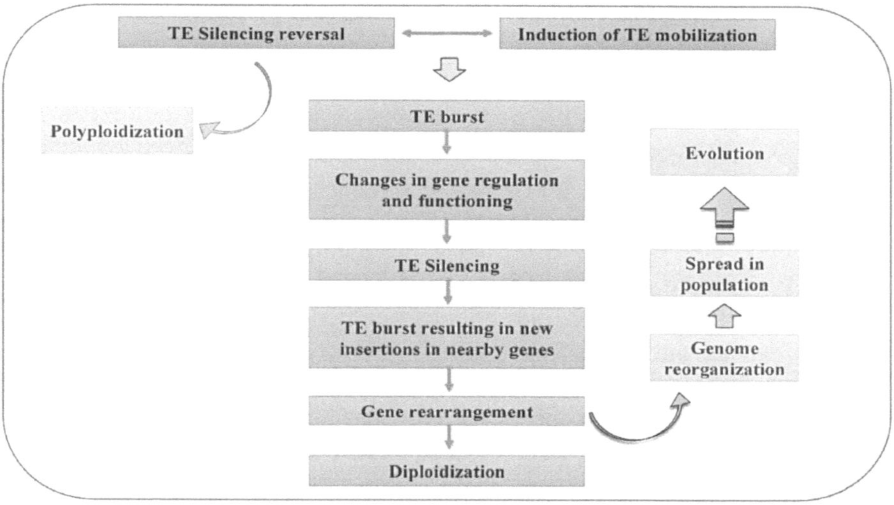

FIGURE 9.1 Effect of transportable elements on polyploidy and their association. Connection between TE and polyploidy.

probably at the root of meristematic cells, which is polyploid in nature that eventually form another polyploid organic entity [e.g., *Primula kewensis* (Kew primrose)] (Grant, 1981; Christoph et al., 2017). Little is understood about the consistency of uncontrolled chromosome multiplying.

Nowadays, synthetic polyploidy is utilized for somatic doubling that closely resembles and imitates the traditional ones. Delivering unreduced gametes (e.g., diplogametes) at the rate much higher than accepted brings about polyploidy. Different processes responsible to generate unreduced gametes have been distinguished in a variety of plant taxa. Diplogametes production is exceptional among species and is assessed to be 0.56% in nonhybrid flowering plants. Diplogametes seem to be more regular under natural stresses (biotic and abiotic), like ice, injury, herbivore, water, or supplement lack (Nogler, 1984; Bicknell et al., 2003). Discernibly, mixtures between disparate genomes display near about 50-overlay expansion to form unreduced gamete as distinguished from nonhybrid frameworks (Zhang et al., 2015). Furthermore, the formation of diplogametes has all the earmarks of being heritable and represented by genes (Zhang et al., 2015). The formation of unreduced gamete starts to be disentangled; however, much remaining parts have to be perceived. Due to the production of unreduced gamete at a higher pace, their association probably upholds the initiation of both allo- and auto-polyploids under regular conditions.

9.6.2 UNILATERAL VERSUS BILATERAL POLYPLOIDIZATION OF UNREDUCED GAMETES

The probability of meeting two unreduced gametes is quite low in a single-step process. The development of tetraploids has been expected to be uncertain, including triploid intermediates (triploid stage supporting one-sided polyploidization) (Harlan et al., 1975). Triploid individuals can be obtained conveniently using an experimental setup whereas in nature, they are observed at a lower frequency (Ramsey et al., 1998). The fusion of unreduced gamete with the reduced one results in triploid formation. It can be carried out by selfing or crossing with different triploids or even with diploid progenitors. Gametes' ploidy level that is attained by triploids, fitness, and rate of formation of triploids defines the development of tetraploids (Husband, 2004). Triploid seeds abort early due to a phenomenon known as "triploid block" and hence triploidy is identified to be fatal. The endosperm cannot develop in polyploids properly as they lack balanced set of chromosomes (Ramsey et al., 1998; Husband, 2004).

Various models are designed to produce of triploid seeds based on the relative level of seed tissues, in spite of triploid block. In 1999, failure in the endosperm development was detected by Soltis and Soltis in both interploidy hybrids as well as homoploid hybrids. This might be because of different imprinted genes of polycomb group. Failure of endosperm in triploids results from uneven interactions between targets of polycomb groups and their global transcriptional repressors. The unbalanced interaction in turn originates from parental v/s maternal dosage ratios and/or paternal divergence. However, there are cases where few fertile viable triploids have been reported unexpectedly. Ramsey and Schemske (1998) reported 31.9% mean pollen fertility in triploid angiosperms.

Comparatively large number of euploid viable gametes (where $n = x$, $2x$, and $3x$) are produced by triploids without considering their origin, be it allo- or auto-polyploid. Therefore, they strongly affect the development of polyploids (Husband, 2004). Tetraploids can perhaps be formed either through bilateral polyploidization specifically, direct fusion of two diplogametes or through homoploid hybrids without involving any triploid intermediate. Diplogametes production is higher in hybrid systems (i.e., 27%) as compared to nonhybrid systems (i.e., 5.6%) (Ramsey et al., 1998). Homoploid hybrids are expected to withstand polyploidization. When polyploidization occurs bilaterally, hybrids can be directly subjected to polyploidy regardless of their origin (Ramsey et al., 1998; Husband, 2004).

In the case of individuals that are self-fertilizing, plants with larger amount of diplogametes might promote bilateral polyploidization (Bretagnolle et al., 1995). In 1998, Ramsey and Schemske in 1998 analyzed that rate of production of polyploids by either bilateral polyploidization or unilateral polyploidization is almost similar. Though, they did not examine the prevalence and rate of occurrence of somatic doubling arising spontaneously. In accordance to their concluding remark, "thorough analysis of the mechanisms, pathways, and rate of formation of polyploids in nature" continue to be relevant.

9.6.3 PATHWAYS AND ORIGINS FOR VARIABLE POLYPLOID PRODUCTION

Different sources of polyploid species are right now considered as the standard rather than the exclusion (Soltis et al., 1999). It has appeared for a couple of allopolyploids and autopolyploids (Parisod and Besnard, 2007), neglecting special case exceptions, for instance, *Arabidopsis suecica* (Swedish thale-cress) or *Draba ladina*. Conspicuously, different origins (multiple) may happen over restricted topographical distributions.

Various elements of polyploids may incorporate innately. Morphologically isolated individuals might achieve critical phenotypic and genetic changes similarly as atomic or nucleo-cytoplasmic associations at the polyploid level (Soltis et al., 1999; Mable, 2003). Adding to the complexity, various divergent pathways might have resulted to independent polyploid ancestries, leading to combination products among various genomes. Individuals of polyploidy linages formed autonomously might hybridize (Modliszewski and Willis, 2012). But apparently, the hypothesis is not been experimented yet. Correspondingly, polyploid taxon history might be more complicated than typically expected.

9.7 GENE REGULATION OF TRANSPOSABLE ELEMENT IN POLYPLOIDS

The expression of nearby gene can be reduced by epigenetic silencing of transposable components thus TE silencing related to polyploidization induces modifications in the expression of genes. After polyploidization, gene arranged in the nearby region of reactivated TEs could then be influenced by different TEs and therefore, their transcriptional activity is altered and chromatin is rearranged. Also, reactivated TEs are capable to reduplicate themselves. These changes in TE induce transcriptional modifications in the neighboring genes. As seen in wheat polyploids, decreased control in TEs silencing following polyploidization, restores the genome in functionally working order (Kashkush et al., 2003). Oddly, replicated characteristics confirmed in allopolyploid progenies typically show disparities with their parental origin. This phenomenon is called genome strength. This can be seen in morphological traits that are differentially controlled (Feldman et al., 2012). It is seen that genome strength is greater in former polyploids than in newly engineered individuals, depicting that it requires a few generations to establish. Moreover, the oldest allopolyploids exhibit genome predominance, whereas autopolyploids do not exhibit the same (Woodhouse et al., 2014).

Various processes are intended for masking intergenomic gene movement, including alterations in chromatin and the characteristic masking of genes in neighboring TEs regions (Feldman et al., 2012). The gene suppression cycle in regions adjoining TEs in a polyploid genome is initiated by methylation and is higher in one of the two parental genomes. Cytoplasmic TE-controlling factors (e.g., siRNAs) are contributed by female parent and, therefore, TEs of maternal genome are subjected to higher suppression in the early stages of polyploidy (Zhang et al., 2015). A possibility of having diverse suppression efficiency of TE possessed by two parental genomes also exists. It is probable that if either of the genomes is comprised of more prominent TE content and also if genes are in close vicinity with the transposons, it behaves as latent subgenome in the allotetraploid (Garsmeur et al., 2014). In *Brassica rapa,* transposon-inferred 24 nucleotide RNAs focus on the upstream region of genes specially situated in the latent subgenome. This gave rise to the theory that the genome, whether paternal or maternal, having minimal TE substance might turn into the prevailing polyploidy genome (Woodhouse et al., 2014). Whatever the underlying explanation is, this distinction starts a course of cycles depending on the way that a gene that is less translated is a gene that can be changed or adjusted more effectively without phenotypic results. The impacts will be significant as the disparity among the parental species increases (Cheng et al., 2016).

9.8 TE CONTROL THROUGH SILENCING

Large numbers of sequences existing in plant genomes are remainders of transposons exposing the presence of effective transposon inactivating apparatus. Among the various mechanisms in eukaryotic genomes to diminish the mobility of transposons and therefore ensuring to regulate their mutagenic behavior, silencing is universal as well as a powerful process (Vance et al., 2002). Silencing mechanism was first time performed in transgenically modified plants but nowadays, silencing-related phenomena is found in a wide spectrum of organisms. Post-transcriptional gene silencing (TGS), involving mechanism of sequence-specific RNA degradation, has a vital role in plants providing immunity against antiviral diseases. On the other hand, TGS is a promoter inactivation system required to counteract TE transcription (Vaucheret et al., 2001). It is responsible for suppression of promotors found within the repeated sequences. Various variables affect the recognition of TGS, yet the existence of several copies of the target sequence overall enhances gene silencing. Over the recent couple of years, examples of TE silencing initiated by high copy number is notified. For instance, the activity of a LINE retrotransposon of Drosophila is restricted as various duplicates of the transgene are introduced in the region close to the element (Jensen et al., 1999). Increasing the copy number of *Tto1* retrotransposon found in tobacco seems to get silenced in Arabidopsis following a few cycles of retrotransposition (Hirochika et al., 2000). Additionally, existence of short interfering RNA (siRNA), that is, an intermediatory element of silencing, assists genome silencing machinery in reducing these components. Silenced promoters are tolerably protected from DNase I as they are hypermethylated, proposing the DNA structures, and heterochromatin regions are rearranged in terms of methylation (Vaucheret and Fagard, 2001).

It seems that transformations influence diverse chromatin rebuilding factors which in turn reactivate silenced transposons (Miura et al., 2001; Wright and Voytas, 2002). As discussed earlier, tobacco *Tto1* retrotransposon is inactivated in *Arabidopsis*, it is linked with its hypermethylation and it could be reverted back around the nonmethylated sites (Hirochika et al., 2000). This implies that the rate of hypermethylation speed up the inefficiency of silenced components. In yeast, it is observed that the high turnover of *Ty1* retrotransposons is the result of recombination among LTRs, as well as the support of few elements of dynamic components (Jordan et al., 1999).

In plants, retrotransposons reverse transcriptase gene showed phylogenetic analysis in which refined selection of species with few copy numbers of *Ty1*-copia components is established, indicating higher turn-over of lower copy number of retrotransposons (Navarro-Quezada et al., 2002). Plants additionally carry retrotransposon families in high copy numbers which are largely inert. Also, plant genomes can uphold massive variations in their genome regardless of significant outcomes. The specific type of TEs called nonautonomous transposons contains unique sequences that code for the protein required for translation. TE-encoding components in reference to

MITEs are present in not many duplicates, while the translating and multiplying unit, that is, the actual MITE attain high copy number.

The mechanism of silencing coordinated with the MITE cannot be inactivated by TGS as it is unable to get transcribed and it is not intended to silence the transposase-encoding components that are low in copy number. MITE coordinated TGS affects the transposase release because transposase-encoding component carry the terminal sequences to its reversal derivative, that is, the MITE. It is suggested that MITE is activated by transposases encoded by related components (Feschotte et al., 2002b). Despite the fact that MITEs cannot be inactivated by TGS; silencing-related cycles like methylation could affect its capacity. Holomethylated Ac/Ds components are incapable to interact with Ac transposase while hemimethylated components bind the transposase with high affinity and translate effectively. The relationship between replication of DNA and Ac/Ds translation could be well explained by the tendency of Ac transposase to prefer DNA that is hemimethylated. It is enticing to guess that a comparative replication-dependent translation could assist MITEs for attaining the very high copy number they manifest in eukaryotic genomes. Short junk components called terminal-repeat retrotransposons in miniature (TRIMs) have additionally been depicted (Witte et al., 2001), though, immediate connection to an autonomous component has not been found at this point.

Furthermore, the TRIMs and retrotransposons themselves can achieve extremely high copy numbers. Dynamic retrotransposons accordingly avoid silencing by different mechanisms. Individual components situated at specified areas within a chromosome might be less susceptible to silencing and therefore they could increase their number. LTR-retrotransposons can efficiently circumvent silencing because of their high sequence variability. Curiously, inconsistency in sequences in tobacco *Tnt1* retrotransposon is not homogeneously dispersed; promoter, being the target of the TGS components (Vernhettes et al., 1998). It appears to be troublesome that the sequence inconstancy showed by retrotransposons is sufficient to bypass the effect of incredibly effective TGS mechanisms, which is capable to inactivate and distinguish short repeated sequences as short stretch of sequence as 90 nt. Not many retrotransposons in plants keep up their capacity to translate and are dynamic under stress (biotic or abiotic) conditions only (Grandbastien, 1998). Various illustrations can explain this relationship of stress environment with transposon mobility, yet genomic silencing systems are unregulated under these circumstances and therefore permitting TEs to transiently sidestep the genomic control. Strangely, it has been shown in

Zea mays (maize) that reduction in pressure can be caused methylation and consequently, retrotransposon-related arrangements are initiated (Steward et al., 2002).

9.9 CONCLUSION AND FUTURE PERSPECTIVES

Equilibrium between production and eradication system contribute to various TEs ratios ranging from 25% to 90% in a specific plant genome. The expanding of information on whole-genome sequences for diverse plant species highlights the significance of TEs in the advancement of domestic as well as wild plants. In plant genomes, TEs directly pose an impact on genes as they offer new sequences, and consequently coding and regulation are affected; indirectly causing epigenetic changes in the chromatin near genes. TEs and plant polyploidy play significant roles in the development of plant genes. Although, they cannot be considered as autonomous sources of variation because polyploidy promotes TE activity and TE explains the variability related to polyploidy. Moreover, polyploidization leads to diploid genomes. The degree of contribution added by TEs to diploidization has more to explore. Yet, a favorable model to explore the different variables regulating TEs could be polyploid speciation; hence the knowledge regarding the activity of TEs could serve to deepen the understanding related to the elements of polyploid genomes.

KEYWORDS

- **plant genome**
- **transposable elements**
- **epigenetic changes**
- **polyploidization**
- **genome reorganization**
- **whole genome doubling**
- **chromosomal rearrangement**
- **genome stress**
- **silencing**
- **chromosomal rearrangement**
- **diploidization**

REFERENCES

Adams, K. L.; Wendel, J. F. Polyploidy and Genome Evolution in Plants. *Curr. Opin. Plant Biol.* **2005,** *8*, 135–141.

Ambrozová, K.; Mandáková, T.; Bures, P.; Neumann, P.; Leitch, I. J.; Koblízková, A.; Macas, J.; Lysak, M. A. Diverse Retrotransposon Families and an AT-Rich Satellite DNA Revealed in Giant Genomes of Fritillaria Lilies. *Ann Bot.* **2011,** *107* (2), 255–268.

Ben-David, S.; Yaakov, B.; Kashkush, K. Genome-Wide Analysis of Short Interspersed Nuclear Elements SINES Revealed High Sequence Conservation, Gene Association and Retrotranspositional Activity in Wheat. *Plant J.* **2013,** *76*, 201–210.

Bennetzen, J. L.; Wang, H. The Contributions of Transposable Elements to the Structure, Function, and Evolution of Plant Genomes. *Annu. Rev. Plant Biol.* **2014,** *65*, 505–530.

Bicknell, R. A.; Lambie, S. C.; Butler, R. C. Quantification of Progeny Classes in Two Facultatively Apomictic Accessions of Hieracium. *Hereditas* **2003,** *138* (1), 11–20.

Bretagnolle, F.; Thompson, J. Gametes with the Stomatic Chromosome Number: Mechanisms of Their Formation and Role in the Evolution of Autopolypoid Plants. *New Phytol.* **1995,** *129*, 1–22.

Chalopin, D.; Naville, M.; Plard, F.; Galiana, D.; Volff, J. -N. Comparative Analysis of Transposable Elements Highlights Mobilome Diversity and Evolution in Vertebrates. *Genome Biol. Evol.* **2015,** *7* (2), 567–580. DOI: 10.1093/gbe/evv005.

Cheng, F.; Sun, C.; Wu, J. Epigenetic Regulation of Subgenome Dominance Following Whole Genome Triplication in *Brassica rapa. New Phytol.* **2016,** *211*, 288–299.

Christoph, C. F. S.; Kirchheimer, B.; Dullinger, S.; Geelen, D.; De Storme, N.; Horandl, E. Pathways to Polyploidy: Indications of a Female Triploid Bridge in the Alpine Species *Ranunculus kuepferi* (Ranunculaceae). *Plant Syst. Evol.* **2017,** *303* (8), 1093–1108.

Consortium, C. e. S. Genome Sequence of the Nematode *C. elegans*: A Platform for Investigating Biology. *Science* **1998,** *282* (5396), 2012–2018.

Contreras, B.; Vives, C.; Castells, R.; Casacuberta, J. M. The Impact of Transposable Elements in the Evolution of Plant Genomes: From Selfish Elements to Key Players. In *Evolutionary Biology: Biodiversification from Genotype to Phenotype*; Pontarotti, P., Ed. Springer International Publishing: Cham, 2015; pp 93–105.

Elliot, T. A.; Gregory, T. R. What's in a Genome? The C-Value Enigma and the Evolution of Eukaryotic Genome Content. *Phil. Trans R. Soc. B.* **2015,** *370*, 20140331.

Feldman, M.; Levy, A. A.; Fahima, T.; Korol, A. Genomic Asymmetry in Allopolyploid Plants: Wheat as a Model. *J. Exp. Bot.* **2012,** *63*, 5045–5059.

Feschotte, C.; Zhang, X.; Wessler, S. R. In *Miniature Inverted-repeat Transposable Elements (MITEs) and Their Relationship with Established DNA Transposons, Mobile DNA II*; Craig, N., Craigie, R., Gellert, M., Lambowitz, A., Eds.; American Society of Microbiology Press: Washington, DC, 2002b; pp 1147–1158.

Finnegan, D. J. Eukaryotic Transposable Elements and Genome Evolution. *Trends Genet.* **1989,** *5* (4), 103–107.

Fujino, K.; Hashida, S.; Ogawa, T.; Natsume, T.; Uchiyama, T.; Mikami, T.; Kishima, Y. Temperature Controls Nuclear Import of Tam3 Transposase in Antirrhinum. *Plant J.* **2010,** *65* (1), 146–155. DOI: 10.1111/j.1365-313x.2010.04405.x.

Gaeta, R. T.; Pires, J. C.; Iniguez-Luy, F.; Leon, E.; Osborn, T. C. Genomic Changes in Resynthesized Brassica Napus and Their Effect on Gene Expression and Phenotype. *Plant Cell.* **2007,** *19*, 3403–3417.

Garsmeur, O.; Schnable, J. C.; Almeida, A.; Jourda, C.; D'Hont, A.; Freeling, M. Two Evolutionarily Distinct Classes of Paleopolyploidy. *Mol. Biol. Evol.* **2014,** *31*, 448–454.

Giordani, T.; Cavallini, A.; Natali, L. The Repetitive Component of the Sunflower Genome. *Curr. Plant Biol.* **2014,** *1*, 45–54. DOI: 10.1016/j.cpb.2014.05.001.

Grandbastien, M. A. Activation of Plant Retrotransposons Under Stress Conditions. *Trends Plant Sci.* **1998,** *3*, 181–187.

Grandbastien, M. -A.; Audeon, C.; Bonnivard, E.; Casacuberta, J. M. Chalhoub, B.; Costa, A. -P. P.; Mhiri, C. Stress Activation and Genomic Impact of Tnt1 Retrotransposons in Solanaceae. *Cytogenet. Genome Res.* **2005,** *110* (1–4), 229–241. DOI: 10.1159/000084957.

Grant, V. *Plant Speciation*; Columbia University Press: New York, 1981.

Grover, C. E.; Kim, H.; Wing, R. A.; Paterson, A. H.; Wendel, J. F. Microcolinearity and Genome Evolution in the AdhA Region of Diploid and Polyploid Cotton (Gossypium). *Plant J.* **2007,** *50* (6), 995–1006.

Harlan, J.; DeWet, J.; On, O. Winge and a Prayer: The Origins of Polyploidy. *Bot. Rev.* **1975,** *41*, 361–390.

Hawkins, J. S.; Proulx, S. R.; Rapp, R. A.; Wendel, J. F. Rapid DNA Loss as a Counterbalance to Genome Expansion Through Retrotransposon Proliferation in Plants. *Proc. Natl. Acad. Sci. U.S.A.* **2009,** *106*, 17811–17816.

Hernandez-Hernandez, E. M.; Fernández-Medina, R. D.; Navarro-Escalante, L.; Nuñez, J.; Benavides-Machado, P.; Carareto, C. M. Genome-Wide Analysis of Transposable Elements in the Coffee Berry Borer *Hypothenemushampei* (Coleoptera: Curculionidae): Description of Novel Families. *Mol. Genet. Genom.* **2017,** *292* (3), 565–583.

Hirochika, H.; Okamoto, H.; Kakutani, T. Silencing of Retrotransposons in Arabidopsis and Reactivation by the ddm1 Mutation. *Plant Cell.* **2000,** *12*, 357–369.

Hodnett, G. L.; Ohadi, S.; Pugh, N. A. Sorghum Bicolor × S. Halepense Interspecific Hybridization Is Influenced by the Frequency of 2n Gametes in S. Bicolor. *Sci. Rep* **2019,** *9*, 17901. DOI: https://doi.org/10.1038/s41598-019-53193-3.

Hood, M. E. Repetitive DNA in the Automictic Fungus Microbotryumviolaceum. *Genetica* **2005,** *124* (1), 1–10.

Huang, J.; Gao, Y.; Jia, H.; Liu, L.; Zhang, D.; Zhang, Z. Comparative Transcriptomics Uncovers Alternative Splicing Changes and Signatures of Selection from Maize Improvement. *BMC Genomi.* **2015,** *16*, 363. DOI: 10.1186/s12864-015-1582-5.

Husband, B. The Role of Triploid Hybrids in the Evolutionary Dynamics of Mixed-Ploidy Populations. *Biol. J. Linn. Soc.* **2004,** *82*, 537–546.

Ibarra-Laclette, E.; Lyons, E.; Hernández-Guzmán, G.; Pérez-Torres, C. A.; Carretero-Paulet, L. Architecture and Evolution of a Minute Plant Genome. *Nature* **2013,** *498* (7452), 94–98.

Ito, H.; Kakutani, T. Control of Transposable Elements in Arabidopsis Thaliana. *Chromosome Res.* **2014,** *22*, 217–223.

Jensen, S.; Gassama, M. P.; Heidmann, T. Taming of Transposable Elements by Homology-Dependent Gene Silencing. *Nat. Genet.* **1999,** *21*, 209–212.

Jordan, E.; Saedler, H.; Starlinger, P. O° and strong-polar mutations in the gal operon are insertions. *Mol. Gen. Genet.* **1968,** *102* (4), 353–363.

Jordan, I. K.; McDonald, J. F. Tempo and Mode of Ty Element Evolution in Saccharomyces Cerevisiae. *Genetics* **1999,** *151*, 1341–1351.

Kalendar, R.; Tanskanen, J.; Immonen, S.; Nevo, E.; Schulman, A. H. Genome Evolution of Wild Barley (Hordeumspontaneum) by BARE-1 Retrotransposon Dynamics in Response to Sharp Microclimatic Divergence. *Proc. Natl. Acad. Sci. U.S.A.* **2000,** *97* (12), 6603–6607.

Kapitonov, V. V.; Jurka, J. Rolling-Circle Transposons in Eukaryotes. *Proc. Natl. Acad. Sci. U.S.A.* **2001,** *98,* 8714–8719.

Kashkush, K.; Feldman, M.; Levy, A. A. Transcriptional Activation of Retrotransposons Alters the Expression of Adjacent Genes in Wheat. *Nat. Genet.* **2003,** *33,* 102–106.

Lander, E. S.; Linton, L. M.; Birren, B.; Nusbaum, C.; Zody, M. C.; Baldwin, J. Initial Sequencing and Analysis of the Human Genome. *Nature* **2001,** *409* (6822), 860–921.

Leitch, A. R.; Leitch, I. J. Genomic Plasticity and the Diversity of Polyploid Plants. *Science* **2008,** *320,* 481–483.

Lopes, F. R.; Jjingo, D.; da Silva C. R. Transcriptional Activity, Chromosomal Distribution and Expression Effects of Transposable Elements in Coffea Genomes. *PLoS One.* **2013,** *8,* e78931. DOI: 10.1371/journal.pone.0078931.

Ma, B.; Xin, Y.; Kuang, L.; He, N. Distribution and Characteristics of Transposable Elements in the Mulberry Genome. *Plant Genome* **2019,** *0* (0), 0. DOI: 10.3835/plantgenome2018.12.0094.

Mable, B. K. Breaking Down Taxonomic Barriers in Polyploidy Research. *Trends Plant Sci.* **2003,** *8,* 582–590.

Mable, B. K.; Alexandrou, M. A.; Taylor, M. I. Genome Duplication in Amphibians and Fish: An Extended Synthesis. *J. Zool.* **2011,** *284,* 151–182.

Madlung, A.; Tyagi, A. P.; Watson, B.; Jiang, H.; Kagochi, T.; Doerge, R. W.; Martienssen, R.; Comai, L. Genomic Changes in Synthetic Arabidopsis Polyploids. *Plant J.* **2005,** *41,* 221–230.

Miura, A.; Yonebayashi, S.; Watanabe, K.; Toyama, T.; Shimada, H.; Kakutani, T. Mobilization of Transposons by a Mutation Abolishing Full DNA Methylation in Arabidopsis. *Nature* **2001,** *411,* 212–214.

Modliszewski, J. L.; Willis, J. H. Allotetraploid Mimulussookensis Are Highly Interfertile Despite Independent Origins. *Mol. Ecol.* **2012,** *21,* 5280–5298.

Navarro-Quezada, A.; Schoen, D. J. Sequence Evolution and Copy Number of Ty1-Copia Retrotransposons in Diverse Plant Genomes. *Proc. Natl. Acad. Sci. U.S.A.* **2002,** *99,* 268–273.

Neumann, P. Significant Expansion of *Viciapannonica* Genome Size Mediated by Amplification of a Single Type of Giant Retroelement. *Genetics* **2006,** *173* (2), 1047–*1056.* DOI: 10.1534/genetics.106.056259.

Nogler, G. A. Gametophyticapomixis. In: *Embryology of Angiosperms*; Johri, B. M., Ed., 1; Springer: Berlin, 1984; pp 475–518.

Olsen, K. M.; Wendel, J. F. A Bountiful Harvest: Genomic Insights into Crop Domestication Phenotypes. *Annu. Rev. Plant Biol.* **2013,** *64,* 47–70.

Parisod, C.; Besnard, G. Glacial In Situ Survival in the Western Alps and Polytopicautopolyploidy in *Biscutellalaevigata* L. (Brassicaceae). *Mol. Ecol.* **2007,** *16,* 2755–2767.

Parisod, C.; Senerchia, N. Responses of Transposable Elements to Polyploidy. In *Plant Transposable Elements*; Grandbastien, M. A., Casacuberta, J. M., Eds.; Springer: Berlin, 2012; pp 147–168.

Petrov, D. A.; Lozovskaya, E. R.; Hartl, D. L. High Intrinsic Rate of DNA Loss in Drosophila. *Nature* **1996,** *384* (6607), 346–349. DOI: 10.1038/384346a0.

Pritham, E. J.; Feschotte, C. Massive Amplification of Rolling-Circle Transposons in the Lineage of the Bat Myotislucifugus. *Proc. Natl. Acad. Sci. U.S.A.* **2007,** *104,* 1895–1900.

Pritham, E. J.; Putliwala, T.; Feschotte, C. Mavericks, a Novel Class of Giant Transposable Elements Widespread in Eukaryotes and Related to DNA Viruses. *Gene* **2007,** *390* (1–2), 3–17. DOI: https://doi.org/10.1016/j.gene.2006.08.008.S0378-1119(06)00537-3.

Ramsey, J.; Schemske, D. W. Pathways, Mechanisms, and Rates of Polyploid Formation in Flowering Plants. *Annu. Rev. Ecol. Syst.* **1998,** *29*, 467–501.

Renny-Byfield, S.; Chester, M.; Kovařík, A.; Le Comber, S. C.; Grandbastien, M. A.; Deloger, M.; Leitch, A. R. Next Generation Sequencing Reveals Genome Downsizing in Allotetraploid Nicotianatabacum, Predominantly Through the Elimination of Paternally Derived Repetitive DNAs. *Mol. Biol. Evol.* **2011,** *28* (10), 2843–2854.

Salentijn, E. M.; Goryunova, S. V.; Bas, N.; van der Meer, I. M.; van den Broeck, H. C.; Bastien, T.; Smulders, M. J. Tetraploid and Hexaploid Wheat Varieties Reveal Large Differences in Expression of Alpha-Gliadins from Homoeologous Gli-2 Loci. *BMC Genomi.* **2009,** *10* (1), 48. DOI: 10.1186/1471-2164-10-48.

Sanglard, N. A.; Amaral-Silva, P. M.; Sattler, M. C.; de Oliveira, S. C.; Cesário, L. M.; Ferreira, A.; Clarindo, W. R. Indirect Somatic Embryogenesis in Coffea with Different Ploidy Levels: A Revisiting and Updating Study. *Plant Cell Tissue Organ Culture (PCTOC).* **2018.** DOI: 10.1007/s11240-018-1511-9.

SanMiguel, P.; Gaut, B. S.; Tikhonov, A.; Nakajima, Y.; Bennetzen, J. L. The Paleontology of Intergeneretrotransposons of Maize. *Nat. Genet.* **1998,** *20* (1), 43–45.

Sarilar, V.; Marmagne, A.; Brabant, P.; Joets, J.; Alix, K. BraSto, a Stowaway MITE from Brassica: Recently Active Copies Preferentially Accumulate in the Gene Space. *Plant Mol. Biol.* **2011,** *77*, 59–75.

Sarilar, V.; Palacios, P. M.; Rousselet, A. Allopolyploidy Has a Moderate Impact on Restructuring at Three Contrasting Transposable Element Insertion Sites in Resynthesized Brassica Napusallotetraploids. *New Phytol.* **2013,** *198*, 593–604.

Senerchia, N.; Felber, F.; Parisod, C. Contrasting Evolutionary Trajectories of Multiple Retrotransposons Following Independent Allopolyploidy in Wild Wheats. *New Phytol.* **2014,** *202*, 975–985.

Shimizu-Inatsugi, R.; Terada, A.; Hirose, K.; Kudoh, H.; Sese, J.; Shimizu, K. K. Plant Adaptive Radiation Mediated by Polyploid Plasticity in Transcriptomes. *Mol. Ecol.* **2017,** *26*, 193–207.

Soltis, D. E.; Soltis, P. S. Polyploidy: Recurrent Formation and Genome Evolution. *Trends Ecol. Evol.* **1999,** *14*, 348–352.

Soltis, P. S.; Marchant, D. B.; Van de Peer, Y.; Soltis, D. E. Polyploidy and Genome Evolution in Plants. *Curr. Opin. Genet. Dev.* **2015,** *35*, 119–125.

Song, J. M.; Guan, Z.; Hu, J.; Guo, C.; Yang, Z.; Wang, S.; Guo, L. Eight High-Quality Genomes Reveal Pan-Genome Architecture and Ecotype Differentiation of Brassica Napus. *Nat. Plants* **2020,** *6* (1), 34–45.

Stein, L. D.; Bao, Z.; Blasiar, D.; Blumenthal, T.; Brent, M. R.; Chen, N.; Chinwalla, A.; Clarke, L.; Clee, C. The Genome Sequence of *Caenorhabditisbriggsae*: A Platform for Comparative Genomics. *PLoS Biol.* **2003,** *1* (2), E45.

Steward, N.; Ito, M.; Yamaguchi, Y.; Koizumi, N.; Sano, H. Periodic DNA Methylation in Maize Nucleosomes and Demethylation by Environmental Stress. *J. Biol. Chem.* **2002,** *277*, 37741–37746.

Tittel-Elmer, M.; Bucher, E.; Broger, L.; Mathieu, O.; Paszkowski, J.; Vaillant, I. Stress-Induced Activation of Heterochromatic Transcription. *PLoS Genet.* **2010,** *6*, e1001175. DOI: 10.1371/journal.pgen.1001175.

Vance, V.; Vaucheret, H. RNA Silencing in Plants—Defense and Counter Defense. *Science* **2002,** *292*, 2277–2280.

Vaucheret, H.; Fagard, M. Transcriptional Gene Silencing in Plants: Targets, Inducers and Regulators. *Trends Genet*. **2001,** *17*, 29–35.

Vernhettes, S.; Grandbastien, M. A.; Casacuberta, J. M. The Evolutionary Analysis of the Tnt1 Retrotransposon in Nicotiana Species Reveals the High Variability of Its Regulatory Sequences. *Mol. Biol. Evol.* **1998,** *15*, 827–836.

Vicient, C. M.; Casacuberta, J. M. Impact of Transposable Elements on Polyploid Plant Genomes. *Ann. Bot.* **2017,** *120* (2), 195–207. DOI: 10.1093/aob/mcx078.

Wang, K.; Huang, G.; Zhu, Y. Transposable Elements Play an Important Role During Cotton Genome Evolution and Fiber Cell Development. *Sci. China Life Sci.* **2016,** *59* (2), 112–121.

Wang, X.; Fang, X.; Yang, P.; Jiang, X.; Jiang, F.; Zhao, D.; Kang, L. The Locust Genome Provides Insight into Swarm Formation and Long-Distance Flight. *Nat. Commun.* **2014,** *5* (1). DOI: 10.1038/ncomms3957.

Waterston, R. H.; Lindblad-Toh, K.; Birney, E.; Rogers, J.; Abril, J. F.; Agarwal, P.; Agarwala, R.; Ainscough, R.; Alexandersson, M.; An, P.; Antonarakis, S. E.; Attwood, J. Initial Sequencing and Comparative Analysis of the Mouse Genome. *Nature* **2002,** *420* (6915), 520–562.

Wegrzyn, J. L.; Liechty, J. D.; Stevens, K. A. Unique Features of the Loblolly Pine (Pinustaeda L.) Megagenome Revealed Through Sequence Annotation. *Genetics* **2014,** *196*, 891–909.

Wendel, J. F.; Jackson, S. A.; Meyers, B. C.; Wing, R. A. Evolution of Plant Genome Architecture. *Genome Biol.* **2016,** *17*, 37.

Wicker, T.; Sabot, F.; Hua-Van, A.; Bennetzen, J. L.; Capy, P.; Chalhoub, B.; Flavell, A.; Leroy, P.; Morgante, M.; Panaud, O.; Paux, E.; SanMiguel, P.; Schulman, A. H. A Unified Classification System for Eukaryotic Transposable Elements. *Nat. Rev. Genet.* **2007,** *8* (12), 973–982. DOI: https://doi.org/10.1038/nrg2165.

Wicker, T.; Schulman, A. H.; Tanskanen, J.; Spannagl, M.; Twardziok, S.; Mascher, M.; Gundlach, H. The Repetitive Landscape of the 5100 Mbp Barley Genome. *Mob. DNA* **2017,** *8* (1). DOI: 10.1186/s13100-017-0102-3.

Willing, E. M.; Rawat, V.; Mandáková, T.; Maumus, F.; James, G. V.; Nordström, K. J. V.; Schneeberger, K. Genome Expansion of Arabisalpina Linked with Retrotransposition and Reduced Symmetric DNA Methylation. *Nat. Plants* **2015,** *1* (2), 14023. DOI: 10.1038/nplants.2014.23.

Witte, C. P.; Le, Q. H.; Bureau, T.; Kumar, A. Terminal-Repeat Retrotransposons in Miniature (TRIM) are Involved in Restructuring Plant Genomes. *Proc. Natl. Acad. Sci. U.S.A.* **2001,** *98*, 13778–13783.

Wood, T. E.; Takebayashi, N.; Barker, M. S.; Mayrose, I.; Greenspoon, P. B.; Rieseberg, L. H. The Frequency of Polyploid Speciation in Vascular Plants. *Proc. Natl. Acad. Sci. U.S.A.* **2009,** *106*, 13875–13879.

Woodhouse, M. R.; Cheng, F.; Pires, J. C.; Lisch, D.; Freeling, M.; Wang, X. Origin, Inheritance, and Gene Regulatory Consequences of Genome Dominance in Polyploids. *Proc. Natl. Acad. Sci. U.S.A.* **2014,** *111*, 5283–5288.

Wright, D. A.; Voytas, D. F. Athila4 of Arabidopsis and Calypso of Soybean Define a Lineage of Endogenous Plant Retroviruses. *Genome Res.* **2002,** *12*, 122–131.

Yaakov, B.; Kashkush, K. Massive Alterations of the Methylation Patterns Around DNA Transposons in the First Four Generations of a Newly Formed Wheat Allohexaploid. *Genome* **2011,** *54*, 42–49.

Yaakov, B.; Kashkush, K. Mobilization of Stowaway-Like MITEs in Newly Formed Allohexaploid Wheat Species. *Plant Mol. Biol.* **2012,** *80*, 419–427.

Yin, L.; Zhu, Z.; Luo, X.; Huang, L.; Li, Y.; Mason, A. S.; Fu, S. Genome-Wide Duplication of Allotetraploid Brassica Napus Produces Novel Characteristics and Extensive Ploidy Variation in Self-Pollinated Progeny. *G3: Genes, Genomes, Genet.* **2020,** *10* (10), 3687–3699.

Zhang, J.; Liu, Y.; Xia, E. H.; Yao, Q. Y.; Liu, X. D.; Gao, L. Z. Autotetraploid Rice Methylome Analysis Reveals Methylation Variation of Transposable Elements and Their Effects on Gene Expression. *Proc. Natl. Acad. Sci. U.S.A.* **2015,** *112*, E7022–E7029.

CHAPTER 10

Transposable Elements and Plant Architecture

SHILPY SINGH[1], SR ASSUMI[2], and POOJA BHADRECHA[3*]

[1]*Department of Biotechnology and Microbiology, School of Sciences, Noida International University, Greater Noida, U.P., India*

[2]*Division of System Research and Engineering, ICAR Research Complex for NEH Region, Umiam, Meghalaya, India*

[3]*Department of Biotechnology (UIBT), Chandigarh University, Mohali, Punjab, India*

[*]*Corresponding author. E-mail: pbhadrecha.pb@gmail.com*

ABSTRACT

Transposable elements (TEs) are mobile genetic components or DNA segments present in both prokaryotes and eukaryotes that are capable of altering their position within the genome. TEs occupy a large proportion of genomes in plants and it is of clear significance for understanding genome regulation and phenotypic variations to consider their impact on the structure of the genome and the expression of genes when mobilized. TEs have important functions in chromosome structural organization, gene regulation, genetic polymorphism, dynamic shifts, and mutations. This can range from simple insertional polymorphisms to complex rearrangements with profound effects on gene evolution, dosage, and regulation, resulting in gene knockouts. While TEs are considered essentially parasitic to the host and their behavior at the level of the individual organism is often deleterious, they can provide valuable genetic diversity as a basis for natural- or human-driven

Plant Transposable Elements: Biology and Biotechnology, Deepu Pandita, Anu Pandita, (Eds.)

selection at the population level. Plant architect is the arrangement of each part and relies on growth processes that are significantly driven by internal and external factors. While environmental conditions and various biotic and abiotic circumstances constitute the external factors, gene regulation is the key internal factor that directs growth patterns. TEs, being capable of reprogramming gene expressions, are well known for altering plant genome, eventually leading to changes in plant phenotype. Hence, the use of TEs is widely recommended to bring about desired transposon-mediated diversity in plants. The following chapter discusses genetic diversities brought about by TEs and their effects on the plant's architect.

10.1 INTRODUCTION

Some genetic elements located in the main chromosomes have been shown to migrate from one position to another in previous studies. These are mobile elements that occur naturally in large quantities. These genetic elements are mentioned by several titles like controlling elements, jumping genes, mobile genes, mobile genetic elements, and transposons. "Transposable elements" (TEs) may change locations on the same chromosome or to various chromosomes altogether The mutations TEs induce to inactivate genes at insertion sites, have been used to classify TEs genetically through model organisms such as *Escherichia coli*, corn, yeast, and Drosophila (Bennetzen and Wang, 2014; Gregory et al., 2007; Wicker et al. 2007). As per the genome sequencing data from different organisms, like bacteria, plants, and animals, TEs can be found in nearly all organisms. Eukaryotic TEs are categorized as class I and class II: in the class I elements, the DNA component of the genome is transcribed into an RNA copy, and hence they are also mentioned as "RNA elements." The RNA copy is then changed to DNA and intruded at a new desired location in the genome of the host. Class I elements are also known as "retroelements" since their mobilization (called retrotransposition) is determined based on the inverted coding of the genome from RNA to DNA (Kapitonov and Jurka, 2008; Wicker et al., 2007). The host genome may incorporate the immense quantity of class I elements. A single class I element can transcribe a large number of RNAs, each of which can potentially lead to intrusion of a new DNA sequence in the host's genome, allowing for such high copy numbers. Class I element insertions into the genome are, therefore, basically irreversible since the RNA copy acts as a transposition intermediate. In other words, they cannot be deleted at the donor site. Despite this, they are also referred to as "mobile" since they are capable of adding their transcripts at the desired location on the DNA. Class II elements, on the other hand, are called "DNA elements" because they move from one part of the genome to another. "Class II elements," unlike class I elements, may be removed from the donor site, hence ensuring that if injection into a gene causes a mutation, the original mutation can be reversed by excision of the component. The first TEs discovered genetically in maize were Class II elements, which were extracted from kernel pigmentation, resulting in unusual spotted kernels (McClintock, 1950). Surprisingly, the prevalent class of TEs may differ significantly among taxa and organisms, as can their genomic frequency, position, and effectiveness, also throughout the population (Lockton and Gaut, 2010; Bennett et al., 2004). TEs are found all over nature (Aziz et al., 2010), and their influence on their hosts is generally regarded as negative. They are often regarded as selfish elements. However, since TEs transpose in the genome, gene disruption is not the only effect they have. By addition of "inversions," TEs often are responsible for altered regulation, expansions of the genome, and the generation of novel variants of the chromosome. Scientists have utilized TE's ability of intruding at new sites on the genome in their study. TEs engineered in the test tube are useful tools for genetic mapping, producing mutations, copying genes, and even creating transgenic species in both prokaryotes and eukaryotes. Let us take a look at some of the assessments that contributed to our current knowledge of TEs.

10.2 DISCOVERY OF PLANT TRANSPOSABLE ELEMENTS

"Barbara McClintock" identified TEs in maize as genetic agents responsible for mutant kernel pigmentation changes more than half a century ago (McClintock, 1950). The 1983 Nobel Prize was awarded to her for the discovery and subsequent characterization of the genetic properties of TEs in the genomes of *E. coli*, *Caenorhabditis elegans*, *Drosophila melanogaster*, *Saccharomyces cerevisiae,* and humans (Berg and Howe, 1989).

10.2.1 BARBARA MC CLINTOCK'S EXPERIMENTS

While researching the colored kernels of maize in the 1940s, she made an amazing discovery. McClintock looked at the chromosome breakdown in maize, which has 10 chromosomes. Chromosome breakage happens at random and only once in a lifetime for every organism. However, McClintock discovered that in one maize strain and at one location, chromosome 9 had broken quite frequently (locus). She discovered that the existence of two genetic factors caused the chromosome to break at this location. At the site of the split, she discovered a factor she called *Ds* (Dissociation). To "trigger" chromosome 9 breakage at the *Ds* locus, another unrelated genetic factor was needed. So, McClintock termed the second factor as "*Ac*" (which stands for "Activator") (McClintock, 1950; Fedoroff, 1984).

The hint of *Ac* and *Ds* being mobile in nature was highlighted when mapping of *Ac* became challenging. It was traced to a single spot in some plants and multiple spots in others from the same line. As if the mapping was not interesting enough, the original strain, with prevalent breaks in chromosome 9, might have developed odd kernels with vastly different phenotypes. The results were colorless kernel with pigmented spots. In Figure 10.1, the phenotype of the chromosome-breaking strain is compared to one of these derivative strains. When the chromosome splits at or close *Ds* in the chromosome-breaking strain, "wild-type alleles" of the *C*, *Sh*, and *Wx* genes are lost, for example, as depicted in Figure 10.1, a single cell split caused a large section of mutant tissue to divide mitotically (*c, sh, wx*). Breaking may occur repeatedly on a single kernel, but all three genes are lost in each tissue field. Each new derivative, on the other hand, affected only one gene's expression. Figure 10.1 shows a derivative that only had an effect on the *C* gene's expression. Pigmented spots appeared on the colorless kernel's context in this scenario. Even when the expressions of *C* gene were modified in unusual ways, expression of Sh and Wx genes was consistent. Elucidation of these observations was considered as the location shifts of *Ds* from near the centromere to a *C* gene near the telomeric end. *Ds* directed some unique alterations in the *C* gene's expression at the new location. Here, an example of dysfunctional phenotype is exhibited by a spotted kernel. According to McClintock, such dysfunctional phenotypes are caused by the motility of *Ds* from the *C* gene. Pigmented spots appeared on the colorless kernel's context in this scenario. Unusual alterations in the expression of *C* gene were recorded, though the expression of *Sh* and *Wx* genes was unaltered. To clarify the new derivatives, McClintock believed *Ds* had shifted from a centromere neighboring position to a *C* gene closer to the telomeric end. At the new position, *Ds* has led to oddly altered expression of the *C* gene. According to McClintock, such dysfunctional phenotypes are caused by the mobility or transposition of *Ds* from the *C* gene.

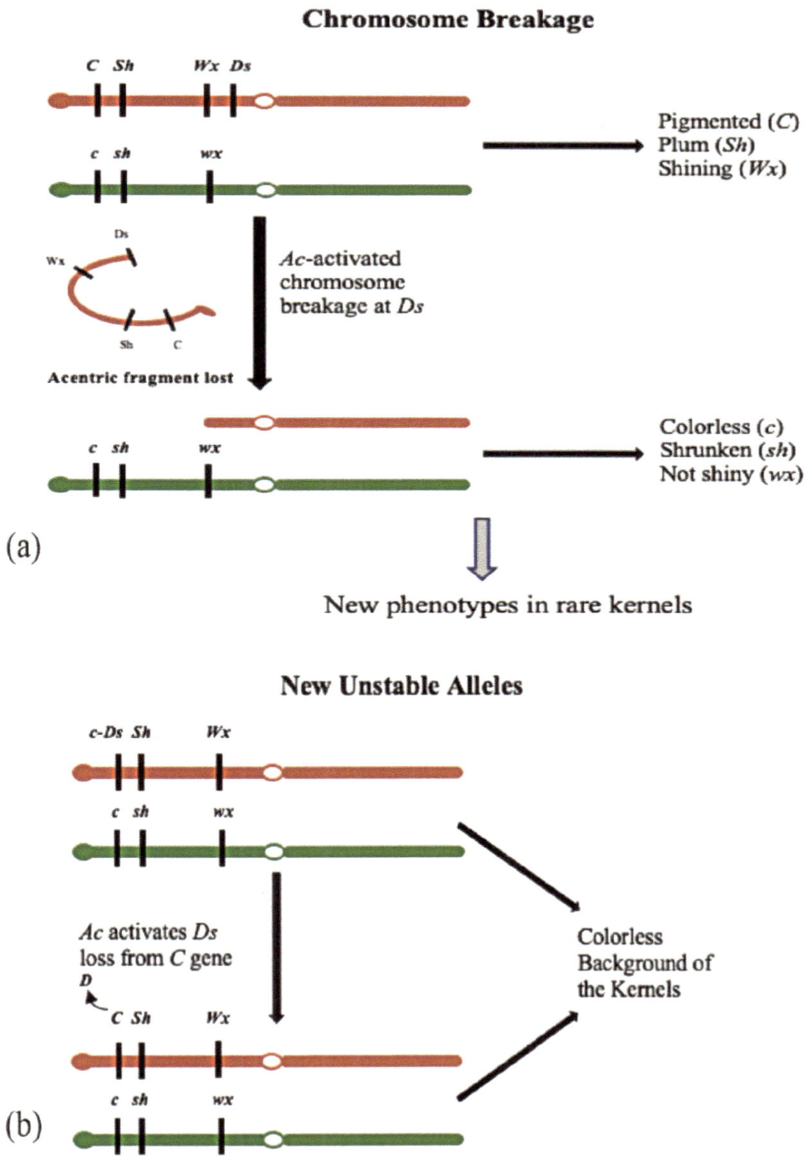

FIGURE 10.1 (a) Chromosome fragment lost due to breakage of *Ds* locus and resulted colorless kernels because of recessive alleles. (b) Insertion of *Ds* into *C* gene and production of colorless kernels. Replacement of *Ac* at the location of *Ds* in *C* gene, producing spotted phenotype.

Source: Modified from McClintock 1954.

As an example, Figure 10.2 depicts interactions of the genes *Ds*, *Ac*, and *C* (McClintock, 1950). *Ds* is a DNA fragment that is inserted into the coding region of the *C* gene to render it inactive [the allele is known as "c-mutable (*Ds*)" or "c-m(*Ds*)"]. The line with c-m(*Ds*) and without *Ac* showed colorless kernels since *Ds* was perplexed within the *C* gene. Spotted kernels are also recorded in strains with c-m(*DS*) since the *Ds* may get activated by *Ac* in a few cells, allowing it to expel from the *C* gene (also referred as "excision" or "transpose"). Other strains [dubbed c-m (*Ac*)] have been discovered with the *Ac* element incorporated at the *C* gene. The *c-m* (*Ds*) allele is unstable only in the case of presence of *Ac* in the genome, whereas the *c-m* (*Ac*) allele is always unstable. In addition, McClintock discovered that on rare occasions, an *Ac*-type allele could be transformed to a *Ds*-type allele. Such transformations were recorded when the *Ac* element got intruded on the *Ds* element (Feschotte et al., 2002a, 2002b; Yu et al., 2011).

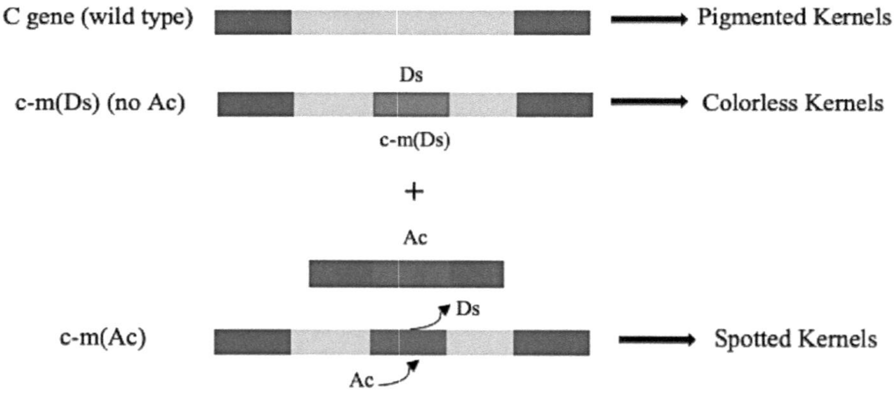

FIGURE 10.2 Example of *Ac* and *Ds* genes that act on the *C* gene, capable of controlling pigmentation.

Source: Modified from McClintock 1954.

To put it another way, *Ds* is almost definitely a mutated and incomplete variant of *Ac*. Many structures, such as "*Ac*/*Ds*," were discovered by McClintock and other researchers who worked on maize plant (Du et al., 2011). In 1938, Marcus Rhoades discovered a pathway termed as "Dotted" (*Dt*) and another pathway, that is, suppressor/mutator (*Spm*) discovered by Peterson in 1953 and McClintock in 1954, who called it "Enhancer/Inhibitor" (*En*/*In*), have been documented. Also, elements from bacteria, plants, and animals have been isolated with the same genetic behavior. Based on their common genetic behavior, geneticists suggested a new classification for all of the elements. *Ac* and a few other elements alike common genetic characters are now referred to as autonomous elements since they do not need the help of other elements to move. *Ds* and elements with similar genetic properties, on the other hand, are nonautonomous elements. These elements are distinguished as part of families, which contains the autonomous components and the nonautonomous elements. "Nonautonomous elements" cannot move, until an "autonomous element" from a similar family is available at any other location on the genome.

10.3 CLASSIFICATION OF TRANSPOSABLE ELEMENTS

In 1989, Finnegan and team introduced the classification pattern for TEs, with the help of which different variants of TEs could be characterized as per their mechanism of movement. These were the "DNA transposons" (also referred to as "type II elements") and the "retrotransposons" (also termed as type II elements). Generally, the transposons are DNA-based TEs that can relocate from one site to another location on the parent's genome. Retrotransposons are TEs that are initially converted to RNAs, which are later reverse-transcribed as DNA (Fig. 10.3). Within each type, TEs are also subdivided as orders, superfamilies, and families depending on the replication strategy and the way they conserve sequences (Kapitonov and Jurka, 2008; Wicker et al., 2007) (Fig. 10.4). Families of both the "retrotransposons" and "transposons" comprise of autonomous and nonautonomous components. Transposons with autonomous elements have everything they need, including a functional gene for an active transposase enzyme and a terminal inverted repeat (TIR), to promote their own transposition. Nonautonomous transposons, on the other hand, only have inverted repeats at the ends and do not encode a transposase. As a result, autonomous transposons encode transposase, which is needed for transposition. Transposition is a major source of genome variation and transformation because it can induce mutation. An actively working gene can also be turned into an inactive gene when a transposon is added into the gene's coding region. Transcription and translation termination signals can be found in transposable components, preventing other genes from expressing downstream of the insertion site. The one-way mutational effect is referred to as polar mutations.

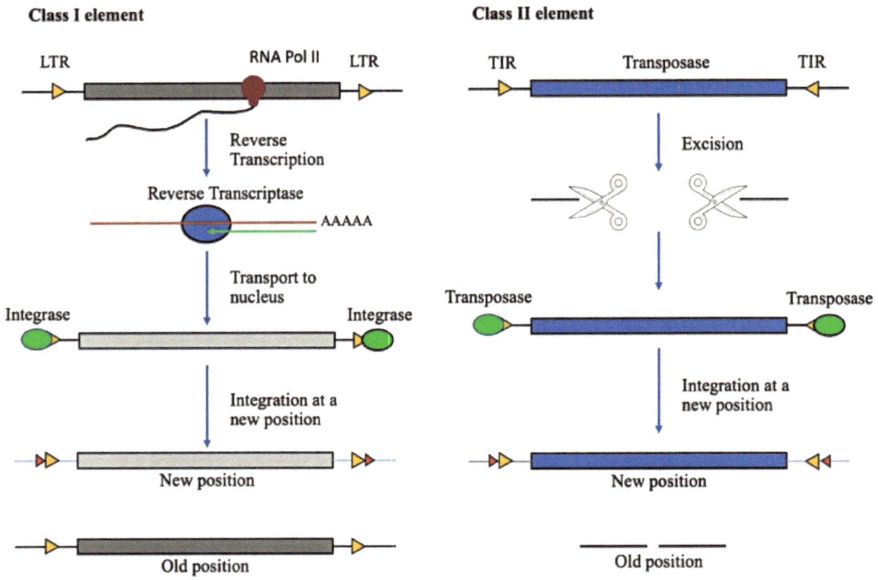

FIGURE 10.3 Class I and Class II TEs in plants. Class I TEs, though permanent after insertion, have a mobile characteristic because RNA is reversed into DNA and added at different sites in the genome. Class II TEs have a different mechanism for changing locations, that is, the DNA is removed from the current location and inserted at a new one.

Source: Modified/redrawn from Lisch, 2013.

10.3.1 CLASS I ELEMENTS

The "Class II elements" are also termed as "retrotransposons." They are the most common form of a retroelement in plants, and they transpose using a "copy-and-paste" process where the RNA polymerase II enzyme transcribes mRNA, reverse-transcribed to a "cDNA," followed by insertion at a novel location at the genome with the help of enzyme "integrase." According to the whole genome sequences of many plant species, retrotransposons compose a slightly higher proportion of higher-plant genomes. For example, TEs account for about 84% of the genome of barley (*Hordeum vulgare*), with retrotransposons accounting for 76% (Mayer et al., 2012). Furthermore, retrotransposons comprise to more than 75% of the maize genome, whereas 8.6% genome contains DNA transposons (Schnable et al., 2009). The most popular retrotransposons are "long terminal repeats (LTR) retrotransposons," "DIRS-like elements," "Penelope-like elements," long interspersed elements (LINEs), and short interspersed elements (SINEs), all of which have different integration mechanisms. "Non-LTR" retrotransposons are

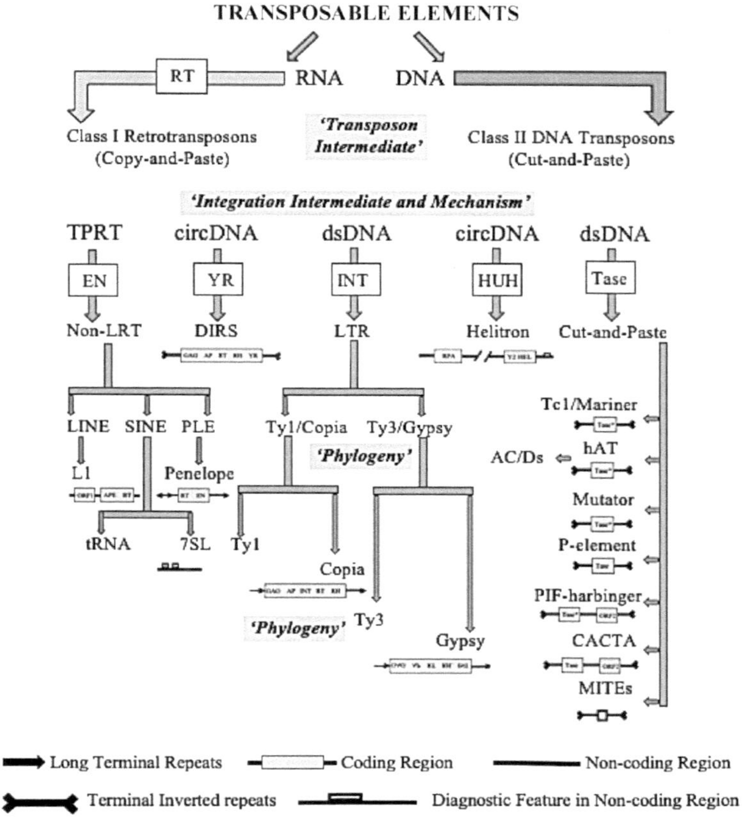

Where: circDNA-circular DNA intermediate, DIRS-Dictyostelium repetitive sequence, dsDNA- linear double-stranded DNA intermediate, PLEs-Penelope-like elements, HUH-Rep/Helicase protein with HUH endonuclease activity, RT- reverse transcriptase, TPRT- target primed reverse transcription, AP-Aspartic proteinase, APE-Apurinic endonuclease, EN-Endonuclease, ENV-Envelope protein, GAG-Capsid protein, HEL-Helicase, INT-Integrase, ORF-Open reading frame of unknown function, RH-RNase H, RPA-Replication protein A (found only in plants), RT-Reverse Transcriptase, Tase-Transposase (* with DDE motif), YR-Tyrosine recombinase and Y2, YR with YY motif

FIGURE 10.4 Classification and structures of plant transposable elements.

Source: Modified from Wicker et al. 2007

divided into "LINEs" and "short interspersed nuclear elements" (SINEs), while LTR retrotransposons are bifurcated as "Ty1-copia" and "Ty3-gypsy" groups (SINEs). The LTR retrotransposons are differentiated on the basis of the availability of LTRs differing in length from fewer to various thousand base pairs. The *gag* genes are well known for coding the "capsid protein" which is necessary for the whole enzymatic mechanism of reverse transcription, as well as insertion inside the host's genome, "reverse transcriptase," ribonuclease H and integrase enzymes (coded by *pol* gene) (Hughes, 2015). Several proteins are deciphered by Ty1-copia and Ty3-gypsy (genes *gag* and *pol*), which are complexed as a "polyprotein" prior to cleavage as functioning peptides by protease (which is element-encoded). The packaging of retrotransposon RNA requires structural proteins which are deciphered by the *gag* gene and the required enzymes for life cycle of retrotransposon are decoded from the *pol* gene. The sequence in where these enzymes get deciphered (in the pol gene) varies amongst the Ty1-copia and Ty2-gypsy elements. The integrase is located prior to reverse transcriptase and RNAseH in the Ty1-copia elements, which is established downstream protease encoding sequence, whereas, among the Ty3-gypsy elements, it is found at the end of the *pol* gene. Although some LTR retrotransposons have traces of an *env* gene, their ability to insert is limited to the genome from which they came. LTR retrotransposons, for example, are target-specific genomic sites for reinsertion, usually around chromosomes, with potentially important functional consequences for the host gene (Kazazian, 2004). LTR retrotransposons have been discovered in various large genomes, for example, corn, wheat, and barley comprise thousands of LTR retrotransposon families. In spite of such huge diverseness, few families like Angela (wheat, BARE21 (barley), Opie (maize), and Retrosort6 (sorghum) constitute for the majority of DNA repetitions in these varied genomes (Peterson et al., 2002; Wicker et al., 2001; Vicient et al., 1999; SanMiguel et al., 1998).

The DIRS order groups structurally distinct transposons that lack duplication of the desired target site and have *YR* gene that codes for tyrosine recombinase, in spite of the gene *INT*, which codes for integrase (TSDs). These genes have some inverted repeats as their termini, which are also mentioned as "split direct repeats." Such properties point to another method for integration, then the other class I mobile components. DIRS were originally identified during the initial years of the 1980s, in the genome of the slime mold *Dictyostelium discoideum* (Zuker and Lodish, 1981), and they can now be found in all major phylogenetic lineages, including vertebrates (Goodwin and Poulter, 2001). They have been discovered to be widespread in hydrothermal vent species (Piednoel and Bonnivard, 2009).

In comparison to LTR retrotransposons, plant LINEs such as Ta11 in *Arabidopsis thaliana* and Cin4 in maize tend to be uncommon. The binding sites and activities for enzymes like endonucleases, reverse transciptases, and the nucleic acids are requisite for the retroposition of LINE (Martin, 2010). LINEs are a component of almost all the eukaryotes' genomes, including plants, advising that they appeared early in evolution (Moran and Gilbert,

2002). The length of LINE elements ranges from 4 to 9 kbp, though most LINEs contained in the genome are of small size since they are 5-truncated. Cin-4 has been documented as the first LINE in plants to be revealed as an insertion in the maize *A1* gene's 3′-untranslated region (Schwarz-Sommer et al., 1987a, 1987b). The presence of LINEs is reported as a huge variety of organisms throughout the kingdom of plants, including monocots and dicots, which are established on the basis of the conserved regions of the gene that codes for the enzyme reverse transcriptase (Schmidt, 1999). Except in the case of the del2 family in *Lilium* species, LINEs in plants are documented from fewer (100) to average (up to 1000) transcripts (Leeton and Smyth, 1993). Del2 is noteworthy for its high abundance (>2.5×105 copies in *Lilium speciosum*, accounting for almost 4% of the complete genome) also, not just abundant, mostly are full-size, unlike LINEs. In rice plants, these LINEs are also introduced as the "Karma Element" and the "LIb Element" from sweet potatoes (Yamashita and Tahara, 2006; Komatsu et al., 2003). The evolution of SINEs is believed to be from the genes of RNA, such as *7SL* and tRNA genes (Kramerov and Vassetzky, 2011a, 2011b; Wicker et al., 2007). By definition, they are small, with a length of up to 1000 base pairs. They are called nonautonomous components since they do not have their own retrotranslation machinery and are usually enabled by the L1 machinery (Kajikawa and Okada, 2002). In the plant kingdom, SINEs are the most usual nonautonomous retroelements. In case of the *Waxy* gene of rice plants, the very first SINE in plants, that is, the "p-SINE" was discovered (Umeda et al., 1991).

10.3.2 CLASS II ELEMENTS

Class II elements can be passed around the genome using a conventional cut-and-paste method where the donor sequence is removed, and later reinserted at a novel location in the genome. The total amount of DNA strands cleaved in the process of transposition divide DNA transposons into two subclasses (Wicker et al., 2007). The classification of traditional "cut and paste" transposons are based on the TIRs and group I. The TIRs require transposase enzymes, which are capable of binding to the inverted repeats and facilitates motility, which generally is not a reoccurring procedure, until the gap created via excision is restored by the sister chromatid. As soon a transposon is intruded into a new site, it leaves small gaps that are repaired by the host enzymes, allowing the sequence to replicate at the desired site. Length-wise, the size of these TSDs varies depending on the transposon. Tc1/Mariner, Mutator, and CACTA, among others, are TIR order superfamilies (Gao et al., 2012). A nonautonomous group of elements that is heterogeneous, small, and nonautonomous. TIRs (Bureau and Wessler, 1994) are MITEs that have thousands of copies in certain genomes, such as Stowaway in the genome of rice (Feschotte et al., 2003) and tourist in most bamboo genomes (Zhou et al., 2016). Two orders of TEs in subclass II do not form RNA intermediates like those in subclass I. Opposite to the regular DNA transposons, they are not subject to the cleavage of double-strands. The mechanism of "rolling-circle" is employed by Helitrons and their addition has not been reported to cause any duplications at the target sites (Gozukirmizi et al., 2016; Lisch, 2013; Wicker et al., 2007). Tyrosine recombinase and a few other proteins are also encoded by these genes. Helitrons were found in plants first, but they have since been discovered in fungi and mammals (Pritham and Feschotte, 2007; Hood, 2005). With the exception of plants, "mavericks" are known as huge transposons located in a number of eukaryotic lineages (Pritham et al., 2007). They encode proteins such as the integrase and DNA polymerase B, among others. Their life cycle steps include deletion of a single strand, followed by extrachromosomal replication, and finally the reintegration at a fresh site (Kapitonov and Jurka, 2006). The activator (Ac) of maize is a prototype transposable factor from the superfamily of hAT transposon, which has members in plants, fungi, and animals. Ac encodes a single transposase protein that mobilizes the autonomous Ac and nonautonomous dissociation (Ds) components. Ac/Ds transposons are known to be active in several species of plants, and they are the most widely employed TEs in plants to achieve tagging of genes and for functional genomics (Lazarow et al., 2013). The structure of MITEs appears like that of faulty class II transposons when the coding strength is not available, but the TIRs are present. The MITEs can be strongly replicated because of the higher copy number of each MITE subfamily, preserving the sequences, and conservation of size, even when a limited amount of primogenitor is available (which refers to be a function of class I elements), and hence, the MITEs were uncategorized for a longer time period. MITEs being a form of "defective" class II transposons have been recently discovered by the reporting of connection among the MITE family with a possible autonomous element. A MITE-family-linked transposase-encoding factor has been discovered in *Arabidopsis* (Feschotte and Mouches, 2000). From that time, various elements that encode transposons, coreferential to the majority of MITEs families, were identified among plants and various other organisms, leading to the hypothesis of MITEs being a form of "defective class II" entity mobilized with transposases that are deciphered by some related autonomous elements, via transposases which are deciphered from related autonomous elements (Feschotte et al., 2002a, 2002b; Feschotte and Wessler, 2002; Turcotte and Bureau, 2002). The mechanism that amplifies these elements, on the other hand, is unknown.

10.4 EFFECT OF TRANSPOSABLE ELEMENTS ON PLANT ARCHITECT

The insertion and expression of TEs in the host plant can bring about both beneficial and harmful effects on the plant's architect. Several such examples are discussed in the following examples:

The addition of TEs leads to TE-mediated insertional mutations which do not just alter the plant's genome but also bring about various phenotypic changes and hence variance in the plant's architecture. Examples include changes in the color of the seed's coat and that of the flowers of *Ipomoea purpurea*, altered coloration of the coats of *Brassica rapa*'s seeds, differences in the hull and internode color in rice, variations in the skin color of the grape, altered skin color of potato tuber skin, purple colored cauliflower, soybean with the unusual coloration of the flowers and differently colored coats of the seeds, distinguishable changes in tomato fruits, variant petal colors in *Antirrhinum* plant, etc. (Wei and Cao, 2016; Uchiyama et al., 2013; Hong et al., 2012; Li et al., 2012; Chiu et al., 2010; Momose et al., 2010; Xiao et al., 2008; Park et al., 2007; Zabala and Vodkin, 2007; Zabala and Vodkin, 2005; Kobayashi et al., 2004; Clegg and Durbin, 2000). *Ghd2* genes act as the suppressor of the timings of the rice plant's flower. But, overexpression of gene *Ghd2* not only delays the flowering time, it also leads to bigger size of the panicle, more spikelets, and better yield of the grains, just like the phenotypes of the *Ghd7* overexpression transgenic rice (Shen et al., 2017). Earlier research has documented that the "CCT domain-containing genes" are capable of regulating various biological mechanisms, like the control over growth, reception and overall development related of the plants, responsive to the environmental changes (Xue et al., 2008; Yan et al., 2004; Kaczorowski and Quail, 2003; Strayer et al., 2000). Among known plants' developmental relaxation of TE silencing (referred to as "DRTS"), the DRTS observed in the shoots of rice by Tamakia et al. (2015) was found to be ample. Many of the registered TEs (4379 out of 7549) were noted to be decrypted at the shoot apex of rice, as well as the majority among them were the "retrotransposons." Most of the reproductive shoot apexes were transcriptionally operational, though ~5% were silenced (Tamakia et al., 2015). Mirouze and Paszkowski (2011) reported that the TEs in spite of being generally silent can be widely expressed exclusively in organs at particular stages of development. Due to the regulation "post-transcription," majority of the TEs do not transpose, though the amount of TEs expressed is very huge. Although the possibility of TEs being transposed to new sites is limited, addition of recently activated retrotransposons provides possibility for a bigger genome size of a species, for example, *Oryza australiensis*'s genome size has been doubled by increasing the number of retrotransposons (Piegu et al., 2006). A general consideration is that the explosion of transposition is a result of biotic and/or abiotic stress factors because living being which is able to evolve genetically in the conditions of stress have higher chances of survival and reproduction (Wessler, 1996). The amplification and proliferation of the TEs can also be brought about by the process of "hybridization" and *Gossypium* being known for hybridization activity on regular basis and flow of genes at interspecific levels can be employed for explicating the eruption of LTR retrotransposon activities in *Gossypium* (Hawkins et al., 2008; Ungerer et al., 2006; Cronn et al., 2003; Liu and Wendel, 2000). In addition to their significant role as a controller and influencer of the genome size, the TEs present in plants are crucial for the "global genome architecture." TE insertions in rice and maize have been documented to direct the evolution of the chromosome's centromeric regions which are well known to be very significant in the process of cell's division (Gao et al., 2015; Sharma et al., 2013; Wolfgruber et al., 2009). Researchers have documented their findings about the capturing and amplification of active genes in the genome of maize and rice, due to mobile Helitrons and MULE-elements (also known as PackMULEs) (Jameson et al., 2008; Jiang et al., 2004). TEs are also considered as essential factor in the domestication of cereals. Plants of wheat and sorghum were the first studied to report that the TE-induced mutations may impart aluminum resistance in plants. There are reports suggesting that when plants are exposed to toxic aluminum concentrations, some regions of the TEs show higher activities, and the TEs are also observed in the transcribed portions of a few genes that are associated with aluminum resistance (Guo et al., 2017; Mattiello et al., 2010). Hundreds of genes are credited for evolution and adaptations in sunflower plants, providing a wide range for TEs to intrude and express. Along with bringing about developmental changes in flower development, various TEs have been documented to affect several processes during plant domestication

(Baute et al., 2015; Mandel et al., 2014; Mandel et al., 2013; Chapman and Burke, 2012; Blackman et al., 2011). Sunflower plants have *HaCYC2d* and *HaCYC2c* genes, whose major role is the control, regulation, and evolution of radiated inflorescence and zygomorphism of ray flowers (Tähtiharju et al., 2012; Chapman et al., 2008). But, on insertion of partial CACTA transposons or retrotransposons, these ray flowers get transmuted from zygomorphs to actinomorphs, hence acquiring similarity to the flowers of disk, a phenotypic characteristic of the tubular ray flower (*turf*) and the mutants "tubular-rayed (*tub*)" (Chapman et al., 2012; Fambrini et al., 2011). Roccaro et al. (2005) isolated and identified ROSINA (RSI), which is a DNA-binding factor from *Antirrhinum majus* genome, capable of binding to CArG-box at the promoter of the *MADS*-box gene *DEFICIENS* (*DEF*). *RSI*, known to be aggregated in the genome of *A. majus*, was found to be capable of binding exclusively to peculiar promoter region at *DEF* gene, influencing the development of petal and stamen (Roccaro et al., 2005). Consecutive studies evidenced *RSI* as a component of the CACTA TE and were coreferential to epigenetic mechanisms for the regulation of genes which consorted the command for flower development (Roccaro et al., 2007). Changes in the tier of expression have been reported to produce aberrant petals, unchanged with the functioning of *DEF*, generally expressed following initiation of floral meristem following sepal primordia. Subsequent stages of growth and development exhibit lower levels of transcripts of *DEF* in the bases of sepals. Studies have confirmed early detection of DEF in the growing carpels, at the fourth whorl; and the mutants of *def* recorded homeotic alterations of petals to sepals in the second whorl and stamens to carpels in the third whorl (Schwarz-Sommer et al., 1992; Jack et al., 1992; Sommer et al., 1990). Insertion of TEs has been reported to result in null mutations of anthocyanin biosynthesis genes, for example, MYB-transcription factor, coded by *Y* gene in sorghum ortholog of maize p1, is essential for red-colored grain pericarp. The DNA transposon *Cs1* (from the CACTA family), when intruded in the second intron of *Y*, leads to the production of differently colored sorghum grain pericarps. Niu et al. (2019) designed a study on *Capsella rubella* certified TEs as an essential factor for genetic variations, resulting in altered parent genome and variant phenotypes. They reported that the insertion of TEs on the FLOWERING LOCUS C (FLC) yielded plants with altered flowering timings.

The following section discusses architectural modifications in plants brought about by the insertions of TEs in the host genome, based on the regulatory mechanisms involved.

10.4.1 INSERTIONAL MUTAGENESIS

The TEs in maize plant may be active or inactive at several steps of the plant's life, and the inactive TEs become incapable of transposing, and suppressing/enhancing the mutants produced by their addition. Martienssen et al. (1990) reported that the *hcf106* mutant's phenotypic characters were repressed in mutants without *Mu* activity, that is, the homozygous hcf106 seedlings were dark green colored and found to be feasible for insertion of TEs from the Robertson's Mutator family (modified at terminal IRs). Their research team considered this characteristic to precede clonal leaf sectors with altered *Mu* elements, highlighting progression of epigenetic transposition including *Mu* DNA, on the aging of meristem. The host plants morphology is affected by insertion of TEs because they are also known to be "long-distance enhancers" which govern the decryption of genes, for example, it has been reported in maize, regulated "Hopscotch," an LTR retrotransposon (Studer et al., 2011). It is placed at a location 60 kb upstream the gene for domestication (*tb1*, that is, *teosinte branched1*) which works like a "long-distance enhancer," known for heightening the transcription of *tb1* in *Zea mays*. This gene tb1 deciphers a member of the TCP family which suppresses the growth of branches and is significant for plant's apical dominance (Cubas et al., 1999; Doebley et al., 1995). Heightened deciphering of gene *tb1* leads to enhanced dominance of apical and suppression of branching in the maize plants, when compared with the wild-type *teosinte*, which is known to have various branches of long size. Hence, the long-distance enhancer in TEs arbitrates the architecture of branches in the maize plants (Studer et al., 2011).

10.4.2 DOMESTICATED TRANSPOSASE GENES

Sunlight is known as the major element for plant's development and growth, whereas the photoreceptor phytochrome A (phyA) acts as intermediate for different far-red light-induced responses. Lin et al. (2007) demonstrated that enactment of the *FHY3* and *FAR1* genes activate transcription of *FHY1* and *FHL* genes and regulates phyA signaling. Genes *FHY1* and *FHL* are necessary for nuclear accumulation induced by light and light responses; whereas the genes *FHY3* and *FAR1* have separate domains for the binding and transcription activation, exclusive for the Mutator-like transposases, and the phyA signaling negatively affects the regulation of the expression of *FHY3* and *FAR1*. Their study suggested that the genes *FHY3* and *FAR1* symbolize transcription factors that were chosen out of an ancient "mutator-like transposase" which regulates homeostasis of phyA-signaling. Significant demotions in development and reproduction, like smaller size of the plant, delay in flowering process, altered growth of flowers, and diminished fertility, were recorded due to mutations in *Arabidopsis MUSTANG* genes (Joly-Lopez et al., 2012; Cowan et al., 2005). The BONSAI gene is located upstream to LINE element in plant *Arabidopsis* Col (Columbia), though it is absent n Cvieco-type plants. In the case of the mutants "ddm1" (*decrease in DNA methylation1*), desuppression of LINE element leads to the production of antisense transcripts and the 24-nt small RNAs in Col plants, which lead to spread of DNA methylation, silencing *BONSAI* and dwarf plants (Saze et al., 2008; Saze and Kakutani, 2007). TEs have been known to influence some key factors in plants, namely, growth, development, and reproduction. Among the various changes brought about by TEs, there are examples of fitness benefits in flowering plants, induced by "*Mustang* and *Sleeper* genes," which include sequences derived of transposases from "*Mutator*-like DNA-TEs" as well as "*hAT* DNA-TEs" (Joly-Lopez et al., 2012; Knip et al., 2012; Cowan et al., 2005; Bundock and Hooykaas, 2005). Expression of *Mustang* genes at the angiosperm linage and code for putative transcriptional regulators, known for playing significant character for the plant's growth, development of flowers, and reproduction. These have been essential as a fitness point of view since the expression of *Mustang* gene in plants produce defected development of floral organs, fertility, and the timing for reproduction (Joly-Lopez et al., 2012). Sleepers genes have also been documented for such alterations (Knip et al., 2012; Bundock and Hooykaas, 2005). *Mustang* and *Sleeper* genes are supposed play significant roles in the "phyletic differentiation" of the angiosperms, as well as supposedly essential evolutionary potential due to "TE-Thrust" (Oliver et al., 2013). TEs are virile factors in evolution and adaptation because they get replicated in the genome and bring about major editions for transforming the host's parent genes into novel genes, essential for exaptation. Such TEs are known as exapted transposable elements (ETEs). Joly-Lopez et al. (2016) designed their study to understand ETE diversification by unveiling the phenotypic context and pattern of ETE at different timings. Their study concluded that the ETEs MUG and FHY3 plants were paraphyletic and they proposed that such pattern exhibited their occurrence in various exaptation events. Angiosperm evolution resulted in such ETE variegation and exaptations. They also discovered evidences of various novel ETE families.

10.4.3 EPIGENETIC REGULATION

Tsugane et al. (2006) reported a mutable *virscent* allele (*pyl-v*) produced due to interrupting a significant gene for chloroplast proteases (*OsClpP5*), by inserting a nonautonomous DNA transposon of 607-bp, "nonautonomous DNA-based active rice transposon one (*nDart1*)," from the superfamily *hAT*. This transformation can be stimulated via a crossing with a host of autonomous element, *aDart*, and can be braced by sequestration of *aDart*. They also discovered a new dwarf allele "*thi-m*," which was mutable on addition of "*nDart1*." Some major phenotypic changes documented in the autotetraploid rice as compared with the diploid parent are: a prominent feature in rice plants' architecture is "tillering" and the TEs comprise more than 35% of rice plant's genome and repeatedly occupy the euchromatic regions (Tan et al., 2016; Wei et al., 2014; Zemach et al., 2010; International Rice Genome Sequencing Project, 2005). Chances of gene expression and hence of a supple development by "RdDm" at TEs have been reported to be heightened, and also, *OsDCL3a*'s knocking-down results in 24-nt siRNAs. *OsDRM2*'s knock-down is essential for most of the CHH methylation in rice, resulting in major defects in development, like dwarfism and small-sized panicle (Tan et al., 2016; Wei et al., 2014). The leaf angles of the *OsDCL3a* knockdown lines are enlarged, though the osdrm2 mutants are reported as sterile (Tan et al., 2016; Wei et al., 2014). The *OsNRPD1a* and *OsNRPD1b* genes present in rice code for Pol IV holoenzyme's biggest subunits. Mutations in these genes lead to a reduction in 24-nt siRNAs and DNA methylation, particularly CHH methylation. A noticeable change in the phenotype is high tillering, which is imparted by lowered methylation of CHH at MITEs in the promoter region of *OsMIR156d* and *OsMIR156j* suppresses *OsMIR156d* and *OsMIR156j* expression. High-tillering phenotype is also the result of a decrease in "CHH methylation" of the MITE downstream of D14. The explanation for suppression of rice tillering by RdDM directs the deciphering of essential genes *OsMIR156d/j* and *D14*. Though the RdDM acts on two MITEs in the promoters of *OsMIR156d/j*, repressing their expression, it acts on the MITE downstream to D14, activating the expression of D14. In spite of being negatively affected by RdDM, both of the signaling pathways result in lower tillers, which is a significant property of ideal plant architecture (Xu et al., 2020). "Epigenetic silencing" of DWARF1 (*Epi-d1*) in rice plants results in "metastable dwarf" phenotype, while on the contrary Epi-df is a benefit of FERTILIZATION INDEPENDENT ENDOSPERM 1 (FIE1) epiallele, that causes pleiotropic defects (Zhang et al., 2012; Miura et al., 2009). Epialleles are believed to be widespread and are of essential values for agricultural traits in rice plants. Hence, the observations prove "24-nt siRNA associated MITEs," as well as several other TEs are responsible for altering the deciphering of neighboring genes as well as the control of other plant characteristics in rice, which are of huge significance from agricultural point of view. Future upgrading of epignenomics and phenomics will make it easier to analyze the epigenetic phase alterations brought about by TEs are elected for agriculturally important characters, especially the crop plants (Wei et al., 2014). Autotetraploid rice plant lines were formulated by Zhang et al. (2015a, 2015b) with its diploid donor, *Oryza sativa* ssp. *indica* cv. *Aijiaonante*, which can both pollinate individually, for more than 48 generations. The transcriptomes of these new plants were compared, along with the equivalence analysis of "siRNAomes" and base-pair resolution methylomes. They recorded widespread diversions in the methylation of TEs, in the autotetraploid rice, by reporting predominant hypermethylation of class II DNA transposons at "CHG" and "CHH." Even though the pattern of methylation of genes has been identical among diploid and autotetraploid rice, TEs (in particular the class II TEs) show essential hypermethylation variations in autotetraploid rice. Noticeable variations in the morphological traits, for example, bigger leaf size, lower fertility, decreased branching, lower spikelet count, and bigger grain size, were discovered in the autotetraploid rice. Whereas, since it merely considered a subset of methylation-sensitive amplified polymorphism, similarity was noticed in the locations of DNA methylation in diploids and autotetraploids in *Paspalum* sp. and watermelon (Wang et al., 2009; Martelotto et al., 2007).

10.5 CONCLUSION

TEs are well known for their potential of integrating in the host genome and getting expressed with the parent genome, leading to some unpredictable changes in the genotype and phenotype, which can be both harmful and beneficial for the host. Nowadays, these mischievous elements are very well employed by researchers to bring about desired, magnificent modifications in the host organisms, and resulting in wonderful final products. A wide range of TEs and their beneficial effects on plants have been well studied and documented. Targeted insertion of the competent TEs in host genome enables the fulfillment of craved necessities, making it possible to get desired architectural changes in the plant and plant's yields. Betterment of the current technologies and discoveries of more efficient methods will enable desired mutations, higher productivity, and longevity in plant species.

KEYWORDS

- **polymorphism**
- **jumping genes**
- **transposons**
- **retrotransposons**
- **chromosome breakage**
- **dysfunctional phenotypes**
- **autonomous and nonautonomous elements**
- **helitrons**
- **insertional mutagenesis**
- **epigenetic regulation**

REFERENCES

Aziz, R. K.; Breitbart, M.; Edwards, R. A. Transposases Are the Most Abundant, Most Ubiquitous Genes in Nature. *Nucl. Acids Res.* **2010,** *38*, 4207–4217.

Baute, G. J.; Kane, N. C.; Grassa, C.; Lai, Z.; Rieseberg, L. H. Genome Scans Reveal Candidate Domestication and Improvement Genes in Cultivated Sunflower, As Well As Post-Domestication Introgression with Wild Relatives. *New Phytol.* **2015,** *206*, 830–838.

Bennett, E. A.; Coleman, L. E.; Tsui, C.; Pittard, W. S.; Devine, S. E. Natural Genetic Variation Caused by Transposable Elements in Humans. *Genetics* **2004,** *168*, 933–951.

Bennetzen, J. L.; Wang, H. The Contributions of Transposable Elements to the Structure, Function, and Evolution of Plant Genomes. *Annu. Rev. Plant Biol.* **2014,** *65*, 505–530.

Berg, D. E.; Howe, M. H. *Mobile DNA*; American Society for Microbiology Press: Washington, DC, 1989.

Blackman, B. K.; Rasmussen, D. A.; Strasburg, J. L.; Ruduski, A. R.; Burke, J. M.; Knapp, S. J.; Michaels, S. D.; Rieseberg, L. H. Contributions of Flowering Time Genes to Sunflower Domestication and Improvement. *Genetics* **2011,** *187*, 271–287.

Bundock, P.; Hooykaas, P. An *Arabidopsis hAT*-Like Transposase Is Essential for Plant Development. *Nature* **2005,** *436*, 282–284.

Bureau, T. E.; Wessler, S. R. Stowaway: A New Family of Inverted Repeat Elements Associated with the Genes of Both Monocotyledonous and Dicotyledonous Plants. *Plant Cell.* **1994,** *6* (6), 907–916.

Chapman, M. A.; Burke, J. M. Evidence of Selection on Fatty Acid Biosynthetic Genes During the Evolution of Cultivated Sunflower. *Theor. Appl. Genet.* **2012,** *125*, 897–907.

Chapman, M. A.; Leebens-Mack, J. H.; Burke, J. M. Positive Selection and Expression Divergence Following Gene Duplication in the Sunflower *CYCLOIDEA* Gene Family. *Mol. Biol. Evol.* **2008,** *25*, 1260–1273.

Chapman, M. A.; Tang, S.; Draeger, D.; Nambeesan, S.; Shaffer, H.; Barb, J. G. Genetic Analysis of Floral Symmetry in Van Gogh's Sunflowers Reveals Independent Recruitment of *CYCLOIDEA* Genes in the *Asteraceae*. *PLoS Genet.* **2012,** *8* (3), e1002628. DOI: https://doi.org/10.1371/journal.pgen.1002628.

Chiu, L. W.; Zhou, X.; Burke, S.; Wu, X.; Prior, R. L.; Li, L. The Purple Cauliflower Arises from Activation of a MYB Transcription Factor. *Plant Physiol.* **2010,** *154*, 1470–1480.

Chopra, S.; Brendel, V.; Zhang, J.; Axtell, J. D.; Peterson, T. Molecular Characterization of a Mutable Pigmentation Phenotype and Isolation of the First Active Transposable Element from Sorghum Bicolor. *Proc. Natl. Acad. Sci. U.S.A.* **1999,** *96*, 15330–15335.

Clegg, M. T.; Durbin, M. L. Flower Color Variation: A Model for the Experimental Study of Evolution. *Proc. Natl. Acad. Sci. U.S.A.* **2000,** *97*, 7016–7023.

Cowan, R. K.; Hoen, D. R.; Schoen, D. J.; Bureau, T. E. *Mustang* Is a Novel Family of Domesticated Transposase Genes Found in Diverse Angiosperms. *Mol. Biol. Evol.* **2005,** *22*, 2084–2089.

Cronn, R. C.; Small, R. L.; Haselkorn, T.; Wendel, J. F. Cryptic Repeated Genomic Recombination During Speciation in *Gossypium gossypioides*. *Evolution* **2003,** *57*, 2475–2489.

Cubas, P.; Lauter, N.; Doebley, J.; Coen, E. The TCP Domain: A Motif Found in Proteins Regulating Plant Growth and Development. *Plant J.* **1999,** *18*, 215–222.

Doebley, J.; Stec, A.; Gustus, C. Teosinte Branched1 and the Origin of Maize: Evidence for Epistasis and the Evolution of Dominance. *Genetics* **1995,** *141*, 333–346.

Du, C.; Hoffman, A.; He, L. The Complete Ac/Ds Transposon Family of Maize. *BMC Genom.* **2011,** *12* (588).

Fambrini, M.; Salvini, M.; Pugliesi, C. A Transposon-Mediate Inactivation of a *CYCLOIDEA*-Like Gene Originates Polysymmetric and Androgynous Ray Flowers in *Helianthus annuus*. *Genetica* **2011,** *139*, 1521–1529.

Fedoroff, N. V. Transposable Genetic Elements in Maize. *Sci. Am.* **1984,** *250* (6), 84–99.

Feschotte, C.; Jiang, N.; Wessler, S. R. Plant Transposable Elements: Where Genetics Meets Genomics. *Nat. Rev. Genet.* **2002a,** *3*, 329–341.

Feschotte, C.; Mouches, C. Evidence that a Family of Miniature Inverted-Repeat Transposable Elements (MITEs) from the *Arabidopsis thaliana* Genome Has Arisen from a Pogo-Like DNA Transposon. *Mol. Biol. Evol.* **2000,** 17, 730–737.

Feschotte, C.; Swamy, L.; Wessler, S. R. Genome-Wide Analysis of Mariner-Like Transposable Elements in Rice Reveals Complex Relationships with Stowaway Miniature Inverted Repeat Transposable Elements (MITEs). *Genetics* **2003,** *163* (2), 747–758.

Feschotte, C.; Wessler, S. R. Mariner-Like Transposases Are Widespread and Diverse in flowering Plants. *Proc. Natl. Acad. Sci.* **2002,** *99*, 280–285.

Feschotte, C.; Zhang, X.; Wessler, S. R. *Miniature Inverted-Repeat Transposable Elements (MITEs) and their relationship with established DNA transposons*. In *Mobile DNA II*; Craig, N., Craigie, R., Gellert, M., Lambowitz, A., Eds.; American Society of Microbiology Press: Washington, DC, 2002b; pp 1147–1158.

Finnegan, D. J. Eukaryotic Transposable Elements and Genome Evolution. *Trends Genet.* **1989,** *5*, 103–107.

Gao, D.; Chen, J.; Chen, M.; Meyers, B. C.; Jackson, S. A Highly Conserved, Small LTR Retrotransposon That Preferentially Targets Genes in Grass Genomes. *PLoS One.* **2012,** *7* (2), e32010.

Gao, D.; Jiang, N.; Wing, R. A.; Jiang, J.; Jackson, S. A. Transposons Play an Important Role in the Evolution and Diversification of Centromeres Among Closely Related Species. *Front. Plant Sci.* **2015,** *6*, 216.

Goodwin, T. J.; Poulter, R. T. The DIRS1 Group of Retrotransposons. *Mol. Biol. Evol.* **2001,** *18* (11), 2067–2082.

Gozukirmizi, N.; Temel, A.; Marakli, S.; Yilmaz, S. Transposon Activity in Plant Genomes. *Plant Omics Trends Appl.* **2016,** 83–108.

Gregory, T. R.; Nicol, J. A.; Tamm, H.; Kullman, B.; Kullman, K.; Leitch, I. J.; Murray, B. G.; Kapraun, D. F.; Greilhuber, J.; Bennett, M. D. Eukaryotic Genome Size Databases. *Nucl. Acids Res.* **2007,** *35* (1), 332–338.

Guo, P.; Qi, Y. P.; Yang, L. T.; Lai, N. W.; Ye, X.; Yang, Y.; Chen, L. S. Root Adaptive Responses to Aluminum-Treatment Revealed by RNA-Seq in Two Citrus Species with Different Aluminum-Tolerance. *Front. Plant Sci.* **2017,** *8*, 330.

Hawkins, J. S.; Hu, G.; Rapp, R. A.; Grafenberg, J. L.; Wendel, J. F. Phylogenetic Determination of the Pace of Transposable Element Proliferation in Plants: *copia* and LINE-Like Elements in *Gossypium*. *Genome* **2008,** *51*, 11–18.

Hong, L.; Qian, Q.; Tang, D.; Wang, K.; Li, M.; Cheng, Z. A Mutation in the Rice Chalcone Isomerase Gene Causes the Golden Hull and Internode 1 Phenotype. *Planta* **2012,** *236*, 141–151.

Hood, M. E. Repetitive DNA in the Automictic Fungus *Microbotryum violaceum*. *Genetica* **2005,** *124* (1), 1–10.

Hughes, S. H. Reverse Transcription of Retroviruses and LTR Retrotransposons. *Mob. DNA III.* **2015,** *30*, 1051–1077.

International Rice Genome Sequencing Project. The Map-Based Sequence of the Rice Genome. *Nature* **2005,** *436* (7052), 793–800.

Jack, T.; Brockman, L. L.; Meyerowitz, E. M. The Homeotic Gene *APETALA3* of *Arabidopsis thaliana* Encodes a MADS Box and Is Expressed in Petals and Stamens. *Cell* **1992,** *68*, 683–697.

Jameson, N.; Georgelis, N.; Fouladbash, E.; Martens, S.; Hannah, L. C.; Lal, S. Helitron Mediated Amplification of Cytochrome P450 Monooxygenase Gene in Maize. *Plant Mol. Biol.* **2008,** *67*, 295–304.

Jiang, N.; Bao, Z.; Zhang, X.; Eddy, S. R.; Wessler, S. R. Pack-MULE Transposable Elements Mediate Gene Evolution in Plants. *Nature* **2004,** *431*, 569–573.

Joly-Lopez, Z.; Forczek, E.; Hoen, D. R.; Juretic, N.; Bureau, T. E. A Gene Family Derived from Transposable Elements During Early Angiosperm Evolution Has Reproductive Fitness Benefits in *Arabidopsis thaliana*. *PLoS Genet.* **2012,** *8* (9), e1002931.

Joly-Lopez, Z.; Hoen, D. R.; Blanchette, M.; Bureau, T. E. Phylogenetic and Genomic Analyses Resolve the Origin of Important Plant Genes Derived from Transposable Elements. *Mol. Biol. Evol.* **2016,** *33*, 1937–1956.

Kaczorowski, K. A.; Quail, P. H. *Arabidopsis* PSEUDO-RESPONSE REGULATOR7 Is a Signaling Intermediate in Phytochrome-Regulated Seedling Deetiolation and Phasing of the Circadian Clock. *Plant Cell.* **2003,** *15*, 2654–2665.

Kajikawa, M.; Okada, N. LINEs Mobilize SINEs in the Eel Through a Shared 30 Sequence. *Cell* **2002,** *111* (3), 433–444.

Kapitonov, V. V.; Jurka, J. A Universal Classification of Eukaryotic Transposable Elements Implemented in Repbase. *Nat. Rev. Genet.* **2008,** *9*, 411–412.

Kapitonov, V. V.; Jurka, J. Rolling-Circle Transposons in Eukaryotes. *Proc. Natl. Acad. Sci.* **2001,** *98*, 8714–8719.

Kapitonov, V. V.; Jurka, J. Self-Synthesizing DNA Transposons in Eukaryotes. *Proc. Natl. Acad. Sci.* **2006,** *103*, 4540–4545.

Kazazian, H. H. Mobile Elements: Drivers of Genome Evolution. *Science* **2004,** *303* (5664), 1626–1632.

Knip, M.; de Pater, S.; Hooykaas, P. J. The *SLEEPER* Genes: A Transposase-Derived Angiosperm-Specific Gene Family. *BMC Plant Biol.* **2012,** *12*, 192.

Kobayashi, S.; Goto-Yamamoto, N.; Hirochika, H. Retrotransposon-Induced Mutations in Grape Skin Color. *Science* **2004,** *304*, 982.

Komatsu, M.; Shimamoto, K.; Kyozuka, J. Two-Step Regulation and Continuous Retrotransposition of the Rice LINE-Type Retrotransposon Karma. *Plant Cell.* **2003,** *15*, 1934–1944.

Kramerov, D.; Vassetzky, N. Origin and Evolution of SINEs in Eukaryotic Genomes. *Heredity.* **2011a,** *107* (6), 487–495.

Kramerov, D. A.; Vassetzky, N. S. SINEs. *Wiley Interdiscip. Rev. RNA.* **2011b,** *2* (6), 772–786.

Lazarow, K.; Doll, M. L.; Kunze, R. Molecular Biology of Maize Ac/Ds Elements: An Overview. *Plant Transp. Elem.* **2013,** 59–82.

Le, Q. H.; Wright, S.; Yu, Z.; Bureau, T. Transposon Diversity in *Arabidopsis thaliana. Proc. Natl. Acad. Sci.* **2000,** *97*, 7376–7381.

Leeton, P. R.; Smyth, D. R. An Abundant LINE-Like Element Amplified in the Genome of *Lilium speciosum. Mol. Gen. Genet.* **1993,** *237*, 97–104.

Li, X.; Chen, L.; Hong, M.; Zhang, Y.; Zu, F.; Wen, J.; Yi, B.; Ma, C.; Shen, J.; Tu, J.; Fu, T. A Large Insertion in *bHLH* Transcription Factor BrTT8 Resulting in Yellow Seed Coat in *Brassica rapa. PLoS One*. **2012,** *7*, e44145.

Li, Y.; Dooner, H. K. Excision of Helitron Transposons in Maize. *Genetics* **2009,** *182*, 399–402.

Lin, R.; Ding, L.; Casola, C.; Ripoll, D. R.; Feschotte, C.; Wang, H. Transposase-Derived Transcription Factors Regulate Light Signaling in *Arabidopsis*. *Science* **2007,** *318*, 1302–1305.

Lisch, D. How Important Are Transposons for Plant Evolution? *Nat. Rev. Genet.* **2013,** *14*, 49–61.

Liu, B.; Wendel, J. F. Retrotransposon Activation Followed by Rapid Repression in Introgressed Rice Plants. *Genome* **2000,** *43*, 874–880.

Lockton, S.; Gaut, B. S. The Evolution of Transposable Elements in Natural Populations of Self-Fertilizing Arabidopsis Thaliana and Its Outcrossing Relative *Arabidopsis lyrata. BMC Evol. Biol.* **2010,** *10* (10).

Mandel, J. R.; McAssey, E. V.; Nambeesan, S.; Garcia-Navarro, E.; Burke, J. M. Molecular Evolution of Candidate Genes for Crop-Related Traits in Sunflower (*Helianthus annuus* L.). *PLoS One*. **2014,** *9*, e99620.

Mandel, J. R.; Nambeesan, S.; Bowers, J. E.; Marek, L. F.; Ebert, D.; Rieseberg, L. H.; Knapp, S. J.; Burke, J. M. Association Mapping and the Genomic Consequences of Selection in Sunflower. *PLoS Genet*. **2013,** *9*, e1003378.

Martelotto, L. G.; Ortiz, J. P. A.; Stein, J.; Espinoza, F.; Quarin, C. L.; Pessino, S. C. Genome Rearrangements Derived from Autopolyploidization in *Paspalum* sp. *Plant Sci.* **2007,** *172* (5), 970–977.

Martienssen, R.; Barkan, A.; Taylor, W. C.; Freeling, M. Somatically Heritable Switches in the DNA Modification of Mu Transposable Elements Monitored with a Suppressible Mutant in Maize. *Genes Dev.* **1990,** *4*, 331–343.

Martin, S. L. Nucleic Acid Chaperone Properties of ORF1p from the Non-LTR Retrotransposon, LINE-1. *RNA Biol.* **2010,** *7* (6), 706–711.

Mattiello, L.; Kirst, M.; da Silva, F. R.; Jorge, R. A.; Menossi, M. Transcriptional Profile of Maize Roots Under Acid Soil Growth. *BMC Plant Biol.* **2010,** *10*, 196.

Mayer, K. F. X.; Waugh, R.; Brown, J. W. S.; Schulman, A.; Langridge, P.; Platzer, M.; Fincher, G. B.; Muehlbauer, G. J.; Sato, K.; Close, T. J.; Wise, R. P.; Stein, N. A Physical, Genetic and Functional Sequence Assembly of the Barley Genome. *Nature* **2012,** *491*, 711–716.

McClintock, B. Mutations in Maize and Chromosomal Aberrations in Neurospora. *Carnegie Inst. Wash. Yearb.* **1954,** *53*, 254–261.

McClintock, B. The Origin and Behavior of Mutable Loci in Maize. *Proc. Natl. Acad. Sci.* **1950,** *36*, 344–355.

Mirouze, M.; Paszkowski, J. Epigenetic Contribution to Stress Adaptation in Plants. *Curr. Opin. Plant Biol.* **2011,** *14* (3), 267–274.

Miura, K.; Agetsuma, M.; Kitano, H.; Yoshimura, A.; Matsuoka, M.; Jacobsen, S. E.; Ashikari, M. A Metastable DWARF1 Epigenetic Mutant Affecting Plant Stature in Rice. *Proc. Natl. Acad. Sci.* **2009,** *106* (27), 11218–11223.

Momose, M.; Abe, Y.; Ozeki, Y. Miniature Inverted-Repeat Transposable Elements of Stowaway Are Active in Potato. *Genetics* **2010,** *186*, 59–66.

Moran, J.; Gilbert, N. Mammalian LINE-1 Retrotransposons and Related Elements. In *Mobile DNA II*; Craig, N., Craigie, R., Gellert, M., Lambowitz, A. M., Eds.; ASM Press: Washington DC, 2002; pp 836–869.

Niu, X. M.; Xu, Y. C.; Li, Z. W.; Bian, Y. T.; Hou, X. H.; Chen, J. F.; Zou, Y. P.; Jiang, J.; Wu, Q.; Ge, S.; Balasubramanian, S.; Guo, Y. L. Transposable Elements Drive Rapid Phenotypic Variation in *Capsella rubella*. *PNAS* **2019,** *116* (14), 6908–6913.

Oliver, K. R.; McComb, J. A.; Greene, W. K. Transposable Elements: Powerful Contributors to Angiosperm Evolution and Diversity. *Genome Biol. Evol.* **2013,** *5* (10), 1886–1901.

Park, K. I.; Ishikawa, N.; Morita, Y.; Choi, J. D.; Hoshino, A.; Iida, S. A *bHLH* Regulatory Gene in the Common Morning Glory, *Ipomoea purpurea*, Controls Anthocyanin Biosynthesis in Flowers, Proan Thocyanidin and Phytomelanin Pigmentation in Seeds, and Seed Trichome Formation. *Plant J.* **2007,** 641–654.

Peterson, D. G.; Schulze, S. R.; Sciara, E. B.; Lee, S. A.; Bowers, J. E.; Nagel, A.; Jiang, N.; Tibbitts, D. C.; Wessler, S. R.; Paterson, A. H. Integration of Cot Analysis, DNA Cloning, and High Throughput Sequencing Facilitates Genome Characterization and Gene Discovery. *Genome Res.* **2002,** *12* (5), 795–807.

Peterson, P. A. A Mutable Pale Green Locus in Maize. *Genetics* **1953,** *38*, 682–683.

Piednoel, M.; Bonnivard, E. DIRS1-Like Retrotransposons Are Widely Distributed Among Decapoda and Are Particularly Present in Hydrothermal Vent Organisms. *BMC Evol. Biol.* **2009,** *9* (86).

Piegu, B.; Guyot, R.; Picault, N.; Roulin, A.; Saniya, A.; Kim, H.; Collura, K.; Brar, D. S.; Jackson, S.; Wing, R. A.; Panaud, O. Doubling Genome Size Without Polyploidization: Dynamics of Retrotransposition-Driven Genomic Expansions in *Oryza australiensis*, a Wild Relative of Rice. *Genome Res.* **2006,** *16* (10), 1262–1269.

Pritham, E. J.; Feschotte, C. Massive Amplification of Rolling-Circle Transposons in the Lineage of the Bat *Myotis lucifugus*. *Proc. Natl. Acad. Sci.* **2007,** *104*, 1895–1900.

Pritham, E. J.; Putliwala, T.; Feschotte, C. Mavericks, a Novel Class of Giant Transposable Elements Widespread in Eukaryotes and Related to DNA Viruses. *Gene* **2007,** *390* (102), 3–17.

Rhoades, M. M. Effect of the *Dt* Gene on the Mutability of the a1 Allele in Maize. *Genetics* **1938,** *23*, 377–395.

SanMiguel, P.; Gaut, B. S.; Tikhonov, A.; Nakajima, Y.; Bennetzen, J. L. The Paleontology of Intergene Retrotransposons of Maize. *Nat. Genet.* **1998,** *20* (1), 43–45.

Saze, H.; Kakutani, T. Heritable Epigenetic Mutation of a Transposon-Flanked *Arabidopsis* Gene Due to Lack of the Chromatin-Remodeling Factor DDM1. *EMBO J.* **2007,** *26*, 3641–3652.

Saze, H.; Shiraishi, A.; Miura, A.; Kakutani, T. Control of Genic DNA Methylation by a jmjC Domain-Containing Protein in *Arabidopsis thaliana. Science* **2008,** *319*, 462–465.

Schmidt, T. LINEs, SINEs and Repetitive DNA: Non-LTR Retrotransposons in Plant Genomes. *Plant Mol. Biol.* **1999,** *40*, 903–910.

Schnable, P. S.; Ware, D.; Fulton, R. S.; Stein, J. C.; Wei, F.; Pasternak, S.; Liang, C.; Zhang, J.; Fulton, L.; Graves, T. A.; Minx, P.; Reily, A. D.; Courtney, L.; Kruchowski, S. S.; Tomlinson, C.; Strong, C.; Delehaunty, K.; Fronick, C.; Courtney, B.; Rock, S. M.; Belter, E.; Du, F.; Kim, K.; Abbott, R. M.; Cotton, M.; Levy, A.; Marchetto, P.; Ochoa, K.; Jackson, S. M.; Gillam, B.; Chen, W.; Yan, L.; Higginbotham, J.; Cardenas, M.; Waligorski, J.; Applebaum, E.; Phelps, L.; Falcone, J.; Kanchi, K.; Thane, T.; Scimone, A.; Thane, N.; Henke, J.; Wang, T.; Ruppert, J.; Shah, N.; Rotter, K.; Hodges, J.; Ingenthron, E.; Cordes, M.; Kohlberg, S.; Sgro, J.; Delgado, B.; Mead, K.; Chinwalla, A.; Leonard, S.; Crouse, K.; Collura, K.; Kudrna, D.; Currie, J.; He, R.; Angelova, A.; Rajasekar, S.; Mueller, T.; Lomeli, R.; Scara, G.; Ko, A.; Delaney, K.; Wissotski, M.; Lopez, G.; Campos, D.; Braidotti, M.; Ashley, E.; Golser, W.; Kim, H.; Lee, S.; Lin, J.; Dujmic, Z.; Kim, W.; Talag, J.; Zuccolo, A.; Fan, C.; Sebastian, A.; Kramer, M.; Spiegel, L.; Nascimento, L.; Zutavern, T.; Miller, B.; Ambroise, C.; Muller, S.; Spooner, W.; Narechania, A.; Ren, L.; Wei, S.; Kumari, S.; Faga, B.; Levy, M. J.; McMahan, L.; Van Buren, P.; Vaughn, M. W.; Ying, K.; Yeh, C. T.; Emrich, S. J.; Jia, Y.; Kalyanaraman, A.; Hsia, A. P.; Barbazuk, W. B.; Baucom, R. S.; Brutnell, T. P.; Carpita, N. C.; Chaparro, C.; Chia, J. M.; Deragon, J. M.; Estill, J. C.; Fu, Y.; Jeddeloh, J. A.; Han, Y.; Lee, H.; Li, P.; Lisch, D. R.; Liu, S.; Liu, Z.; Nagel, D. H.; McCann, M. C.; SanMiguel, P.; Myers, A. M.; Nettleton, D.; Nguyen, J.; Penning, B. W.; Ponnala, L.; Schneider, K. L.; Schwartz, D. C.; Sharma, A.; Soderlund, C.; Springer, N. M.; Sun, Q.; Wang, H.; Waterman, M.; Westerman, R.; Wolfgruber, T. K.; Yang, L.; Yu, Y.; Zhang, L.; Zhou, S.; Zhu, Q.; Bennetzen, J. L.; Dawe, R. K.; Jiang, J.; Jiang, N.; Presting, G. G.; Wessler, S. R.; Aluru, S.; Martienssen, R. A.; Clifton, S. W.; McCombie, W. R.; Wing, R. A.; Wilson, R. K. The B73 Maize Genome: Complexity, Diversity, and Dynamics. *Science* **2009,** *326*, 1112–1115.

Schwarz-Sommer, Z.; Hue, I.; Huijser, P.; Flor, P. J.; Hansen, R.; Tetens, F.; Lönnig, W. E.; Saedler, H.; Sommer, H. Characterization of the *Antirrhinum* Floral Homeotic MADS-Box Gene Deficiens: Evidence for DNA Binding and Autoregulation of Its Persistent Expression Throughout Flower Development. *EMBO J.* **1992,** *11* (1), 251–263.

Schwarz-Sommer, Z.; Leclercq, L.; Gobel, E.; Saedler, H. Cin4, an Insert Altering the Structure of the A1 Gene in Zea mays, Exhibits Properties of Nonviral Retrotransposons. *EMBO J.* **1987a,** *6*, 3873–3880.

Schwarz-Sommer, Z.; Shepherd, N.; Tacke, E.; Gierl, A.; Rohde, W.; Leclercq, L.; Mattes, M.; Berndtgen, R.; Peterson, P. A.; Saedler, H. Influence of Transposable Elements on the Structure and Function of the *A7* Gene of *Zea mays*. *EMBO J.* **1987b,** *6*, 287–294.

Sharma, A.; Wolfgruber, T. K.; Presting, G. G. Tandem Repeats Derived from Centromeric Retrotransposons. *BMC Genom.* **2013,** *14*, 142.

Shen, J.; Liu, J.; Xie, K.; Xing, F.; Xiong, F.; Xiao, J.; Li, X.; Xiong, L. Translational Repression by a Miniature Inverted-Repeat Transposable Element in the 3′ Untranslated Region. *Nat. Commun.* **2017,** *8*, 14651.

Sommer, H.; Beltran, J. P.; Huijser, P.; Pape, H.; Lonnig, W. E.; Saedler, H.; Schwarz-Sommer, Z. Deficiens, a Homeotic Gene Involved in the Control of Flower Morphogenesis in *Antirrhinum majus*: The Protein Shows Homology to Transcription Factors. *EMBO J.* **1990,** *9*, 605–613.

Strayer, C.; Oyama, T.; Schultz, T. F.; Raman, R.; Somers, D. E.; Más, P.; Panda, S.; Kreps, J. A.; Kay, S. A. Cloning of the *Arabidopsis* Clock Gene TOC1, an Autoregulatory Response Regulator Homolog. *Science*. **2000,** *289*, 768–771.

Studer, A.; Zhao, Q.; Ross-Ibarra, J.; Doebley, J. Identification of a Functional Transposon Insertion in the Maize Domestication Gene *tb1*. *Nat. Genet.* **2011,** *43*, 1160–1163.

Tähtiharju, S.; Rijpkema, A. S.; Vetterli, A.; Albert, V. A.; Teeri, T. H.; Elomaa, P. Evolution and Diversification of the *CYC/TB1* Gene Family in *Asteraceae*: A Comparative Study in Gerbera (*Mutisieae*) and Sunflower (*Heliantheae*). *Mol. Biol. Evol.* **2012,** *29*, 1155–1166.

Tamakia, S.; Tsujib, H.; Matsumotob, A.; Fujitab, A.; Shimatanib, Z.; Teradac, R.; Sakamotoa, T.; Kurataa, T.; Shimamoto, K. FT-Like Proteins Induce Transposon Silencing in the Shoot Apex During Floral Induction in Rice. *Proc. Natl. Acad. Sci.* **2015,** *112* (8), E901–E910.

Tan, F.; Zhou, C.; Zhou, Q.; Zhou, S.; Yang, W.; Zhao, Y.; Li, G.; Zhou, D. X. Analysis of Chromatin Regulators Reveals Specific Features of Rice DNA Methylation Pathways. *Plant Physiol.* **2016,** *171 (3), 2041–2054.*

Tsugane, K.; Maekawa, M.; Takagi, K.; Takahara, H.; Qian, Q.; Eun, C. H.; Iida, S. An Active DNA Transposon nDart Causing Leaf Variegation and Mutable Dwarfism and Its Related Elements in Rice. *Plant J.* **2006,** *45*, 46–57.

Turcotte, K.; Bureau, T. Phylogenetic Analysis Reveals Stow-Away Like Elements May Represent a Fourth Family of the IS630-Tc1-Mariner Superfamily. *Genome* **2002,** *45*, 82–90.

Uchiyama, T.; Hiura, S.; Ebinuma, I.; Senda, M.; Mikami, T.; Martin, C.; Kishima, Y. A Pair of Transposons Coordinately Suppresses Gene Expression, Independent of Pathways Mediated by siRNA in *Antirrhinum*. *New Phytol.* **2013,** *197*, 431–440.

Umeda, M.; Ohtsubo, H.; Ohtsubo, E. Diversification of the Rice Waxy Gene by Insertion of Mobile DNA Elements into Introns. *Jpn. J. Genet.* **1991,** *66*, 569–586.

Ungerer, M.; Strakosh, S.; Zhen, Y. Genome Expansion in Three Hybrid Sunflower Species Is Associated with Retrotransposon Proliferation. *Curr. Biol.* **2006,** *16*, 872–873.

Vicient, C. M.; Kalendar, R.; Anamthawat, J. K.; Schulman, A. H. Structure, Functionality, and Evolution of the BARE-1 Retrotransposon of Barley. *Genetica* **1999,** *107* (1–3), 53–63

Wang, C. G.; Hui, L.; Xue, Z. Y.; Chen, C. B.; Gu, Y.; Sun, D. L.; Song, W. Q. Marker-Based Analysis of Genome Structure and DNA Methylation in a Watermelon (*Citrullus lanatus*) Ploidy Series. *Bot. Study*. **2009,** *50* (4), 389–402.

Wei, L.; Cao, X. The Effect of Transposable Elements on Phenotypic Variation: Insights from Plants to Humans. *Sci. China Life Sci.* **2016,** *59* (1), 24–37.

Wei, L.; Gu, L.; Song, X.; Cui, X.; Lu, Z.; Zhou, M.; Wang, L.; Hu, F.; Zhai, J.; Meyers, B. C.; Cao, X. Dicer-Like 3 Produces Transposable Element-Associated 24-nt siRNAs that Control Agricultural Traits in Rice. *Proc. Natl. Acad. Sci.* **2014,** *111*, 3877–3882.

Wessler, S. R. Plant Retrotransposons: Turned on by Stress. *Curr. Biol.* **1996,** *6*, 959–961.

Wicker, T.; Stein. N.; Albar, L.; Feuillet, C.; Schlagenhauf, E.; Keller, B. Analysis of a Contiguous 211 kb Sequence in Diploid Wheat (*Triticum monococcum* L.) Reveals Multiple Mechanisms of Genome Evolution. *Plant J.* **2001,** *26* (3), 307–316.

Wicker, T.; Sabot, F.; Hua-Van, A.; Bennetzen, J. L.; Capy, P.; Chalhoub, B.; Flavell, A.; Leroy, P.; Morgante, M.; Panaud, O.; Paux, E.; SanMiguel, P.; Schulman, A. H. A Unified Classification System for Eukaryotic Transposable Elements. *Nat. Rev. Genet.* **2007,** *8* (12), 973–982.

Wolfgruber, T. K.; Sharma, A.; Schneider, K. L.; Albert, P. S.; Koo, D. H.; Shi, J.; Gao, Z.; Han Lee, H.; Xu, R.; Allison, J.; Birchler, J. A.; Jiang, J.; Dawe, R. K.; Presting, G. G. Maize Centromere Structure and Evolution: Sequence Analysis of Centromeres 2 and 5 Reveals Dynamic Loci Shaped Primarily by Retrotransposons. *PLOS Genet.* **2009,** *5*, e1000743.

Xiao, H.; Jiang, N.; Schaffner, E.; Stockinger, E. J.; van der Knaap, E. A Retrotransposon-Mediated Gene Duplication Underlies Morphological Variation of Tomato Fruit. *Science* **2008,** *319*, 1527–1530.

Xu, L.; Yuan, K.; Yuan, M.; Meng, X.; Chen, M.; Wu, J.; Li, J.; Qi, Y. Regulation of Rice Tillering by RNA-Directed DNA Methylation at Miniature Inverted-Repeat Transposable Elements. *Mol. Plant* **2020,** *13*, 851–863.

Xue, W.; Xing, Y.; Weng, X.; Zhao, Y.; Tang, W.; Wang, L.; Zhou, H.; Yu, S.; Xu, C.; Li, X.; Zhang, Q. Natural Variation in Ghd7 Is an Important Regulator of Heading Date and Yield Potential in Rice. *Nat. Genet.* **2008,** *40*, 761–767.

Yamashita, H.; Tahara, M. A LINE-Type Retrotransposon Active in Meristem Stem Cells Causes Heritable Transpositions in the Sweet Potato Genome. *Plant Mol. Biol.* **2006,** *61*, 79–94.

Yan, L.; Loukoianov, A.; Blechl, A.; Tranquilli, G.; Ramakrishna, W.; SanMiguel, P.; Bennetzen, J. L.; Echenique, V.; Dubcovsky, J. The Wheat *VRN2* Gene Is a Flowering Repressor Down-Regulated by Vernalization. *Science* **2004,** *303*, 1640–1644.

Yu, C.; Zhang, J.; Peterson, T. Genome Rearrangements in Maize Induced by Alternative Transposition of Reversed Ac/ Ds Termini. *Genetics* **2011,** *188*, 59–67.

Zabala, G.; Vodkin, L. Novel Exon Combinations Generated by Alternative Splicing of Gene Fragments Mobilized by a CACTA Transposon in *Glycine max*. *BMC Plant Biol.* **2007,** *7*, 38.

Zabala, G.; Vodkin, L. O. The *wp* Mutation of *Glycine max* Carries a Gene-Fragment-Rich Transposon of the CACTA Superfamily. *Plant Cell.* **2005,** *17*, 2619–2632.

Zemach, A.; Kim, M. Y.; Silva, P.; Rodrigues, J. A.; Dotson, B.; Brooks, M. D.; Zilberman, D. Local DNA Hypomethylation Activates Genes in Rice Endosperm. *Proc. Natl. Acad. Sci.* **2010,** *107*, 18729–18734.

Zhang, J.; Liu, Y.; Xiaa, E.; Yao, Q.; Liu, X.; Gao, L. Autotetraploid Rice Methylome Analysis Reveals Methylation Variation of Transposable Elements and Their Effects on Gene Expression. *Proc. Nat. Acad. Sci.* **2015a,** *112* (50), 7022–7029.

Zhang, L.; Cheng, Z.; Qin, R.; Qiu, Y.; Wang, J.; Cui, X.; Gu, L.; Zhang, X.; Guo, X.; Wang, D.; Jiang, L.; Wu, C.; Wang, H.; Cao, X.; Wan, J. Identification and Characterization of an Epi-Allele of FIE1 Reveals—A Regulatory Linkage Between Two Epigenetic Marks in Rice. *Plant Cell.* **2012,** *24* (11), 4407–4421.

Zhang, X.; Sun, J.; Cao, X.; Song, X. Epigenetic Mutation of RAV6 Affects Leaf Angle and Seed Size in Rice. *Plant Physiol.* **2015b,** *169*, 2118–212.

Zhou, M. B.; Tao, G. Y.; Pi, P. Y.; Zhu, Y. H.; Bai, Y. H.; Meng, X. W. Genome-Wide Characterization and Evolution Analysis of Miniature Inverted-Repeat Transposable Elements (MITEs) in Moso Bamboo (*Phyllostachys heterocycla*). *Planta* **2016,** *244* (4), 775–787.

Zuker, C.; Lodish, H. F. Repetitive DNA Sequences Co-Transcribed with Developmentally Regulated *Dictyostelium discoideum* mRNAs. *Proc. Natl. Acad. Sci.* **1981,** *78* (9), 5386–5390.

CHAPTER 11

Regulation of Transposable Elements in Plant Genomes

SHANKAR BHUJBAL, SUMAN BAKSHI, ARCHANA RAI, and SANJAY J. JAMBHULKAR*

Nuclear Agriculture and Biotechnology Division,
Bhabha Atomic Research Centre, Trombay, Mumbai 400085, India

**Corresponding author. E-mail: sjj@barc.gov.in*

ABSTRACT

Transposable elements (TEs) are repetitive deoxyribonucleic acids (DNA) sequences with an ability to change its position within the genome. These TEs make up substantial share ranging from 20% to 85% of plant genome. They are highly mutagenic and most of them are largely transcriptionally and transpositionally inactive under natural conditions. Environmental stress causes epigenetic de-repression of TEs, which influence on their transcription level resulting in increasing transpositional activity. Plant host genomes have evolved various TE suppression mechanisms, which operate before and after transcription. Transcriptional silencing is achieved through methylation of DNA bases, modifications of histone proteins, and alterations in nucleosome positions. DNA methylation depends largely on RNA-directed DNA methylation (RdDM) pathway. These methylated TEs are preferentially targeted by host genome to recruit the RNA polymerases (PolIV and Pol V) through RdDM for silencing and maintain the methylation marks during DNA replication and cell division. Further, chromatin modifications for suppressing TE transcription cause alteration of histone tails, enhancement of histone methylation especially of nucleosome associated with TEs and modulation

Plant Transposable Elements: Biology and Biotechnology, Deepu Pandita, Anu Pandita, (Eds.)

of chromatin packing and condensation. TEs that dodge the silencing or other newly formed transposon in the genome are selected by the RNA-dependent RNA Polymerase 6-RNA directed DNA methylation process that catalyze post-transcriptional degradation of mRNAs of transposon elements. TEs repetitive sequences can also form RNA hairpin structures and later processed by miRNA biogenesis pathways. Long noncoding RNAs found in plant genome predominately arose from TEs, which suggest that transposons play part in plant evolution by co-opting RNA route.

11.1 INTRODUCTION

Transposable elements (TEs) occupy substantial portion of plant genomes hence, determine size of genome, and heavily influence plant evolution. Angiosperm has variable size of genomes and is heavily associated with TEs content. Self-amplifying TEs are major contributor to the expansion of the plant genome (Naville et al., 2019; Lisch, 2013). TE content varies with size of plant genome. The plants like *Arabidopsis thaliana* or *Brachypodium distachyon* having small genome size consisted of 20–30% TEs of the total deoxyribonucleic acids (DNA) content whereas larger genomes have proportionately higher TE-derived DNA. Barley and Maize have more than 85% TE-derived DNA (Schnable et al., 2009). The advancement of sequencing has significantly contributed in understanding role of TEs in plant genome evolution and gene regulation. TEs action may amplify or assemble many genes in genome and can also modify individual gene structure and its regulation (Bennetzen, 2000). However, there are instances where these self-propagating elements cause genomic instability through chromosome breakage, illegitimate recombination, and genome rearrangement (Kim, 2017). Therefore, it is essential to control the TEs activity to evade its interference on the genome structure, regulatory and developmental processes of an organism. Plants control transposon expression and mobility through transposition repression, which involves mechanisms like DNA methylation, covalent modifications of histones, histone variants incorporation, chromatin-remodeling, and RNAi using small RNAs (Quesneville, 2020).

The most common transposon defense system constitutes small noncoding RNAs, which are accountable for causing silencing and prevents mobility of these of TEs. These small ncRNAs are one of the major factors that cause epigenetic modifications involving methylation of DNA and culminates into silencing which is heritable (Lisch, 2013; Deniz et al., 2019). DNA methylation is followed by silencing of TEs at transcriptional level and relies heavily on RNA-directed DNA methylation (RdDM) pathway. Addition of DNA methylation is trailed by maintenance phase to retain TEs in quiescence phase (Roessler et al., 2018). Another epigenetic regulation of TEs is through histone modification and its interaction with DNA methylation further assure the stringent control on the activity of the TE (Bannister and Kouzarides, 2011). Histone methylation and deacetylation functions as epigenetic twists generally associated with transcriptionally inactive chromatin and control TEs transcription (Zhang, 2018). Post-transcriptional silencing act as alternative mechanism and initiates its action under situations of transcriptional silencing failure under environmental stresses or developmental relaxation of TE silencing (Bousios and Gaut, 2016). Besides small interfering RNA (siRNAs), a variety of long noncoding RNAs (lncRNAs) are synthesized from plant TEs and these lncRNAs play role to epigenetically modify the gene expression by modulating histones and chromatin structural organizations (Ariel and Manavella, 2021).

11.2 DNA METHYLATION AND DEMETHYLATION

Methylated DNA is among the major TEs repressor limiting their genotoxic potential (Zhou et al., 2020). Regulation through DNA methylation restricts transposon activation, their propagation, and survival within the host genome. Using this mechanism, host is able to access the regulatory sequences encoded by the TE (Chuong et al., 2017). However, it happens that DNA methylation marks spreads to the gene in euchromatin region. Thus, to avoid the blocking of the important DNA sequences upon methylation, demethylation plays a significant role. DNA glycosylases and DNA cytosine deaminases are involved in DNA demethylation, which can excise bases and establish genomic mismatches respectively along with other DNA repair factors (Kohli and Zhang, 2013). Transposons and other repetitive DNA elements are the preferential targets of the DNA methylation. It commonly occurs at cytosine bases and classified as the symmetric methylation (methylation on both DNA strands) of CG and CHG sequence (where H = A, T, or C) and the asymmetric methylation (methylation on only one DNA strand) of CHH. Genome-wide ~24%, 6.7%, and 1.7% of DNA methylation levels are observed for CG, CHG, and CHH contexts, respectively (Law and Jacobsen, 2010). At each successive generation, asymmetric DNA methylation is regained by the mechanism involving small interfering RNAs (siRNAs) (Jin et al., 2011).

11.2.1 TRANSPOSON DNA METHYLATION THROUGH RNA DIRECTED DNA METHYLATION (RDDM)

RdDM is a method through which methylation of specific DNA takes place using noncoding RNA molecules. RdDM mechanism is well characterized in *A. thaliana* (Erdmann and Picard, 2020). It was first revealed in tobacco where infectious RNA viroids homologous to plant genomic sequences were found to become cytosine methylated. Consequently, it was shown that double-stranded RNA (dsRNA) generates siRNAs, which cause dense homologous DNA cytosine methylation in all related sequences (Wassenegger et al., 1994; Mette et al., 2000). In the plant, RdDM initiation process was first observed in Arabidopsis *FWA* gene expression analysis and *FWA* gene was found to be imprinted and expressed only in seeds (Chan, 2004; Lippman et al., 2004). RdDM mainly influence the short TEs methylation as revealed by RdDM mutant analysis (Herr et al., 2005), which was further confirmed by genome-wide DNA methylation analysis and small RNA surveys (Tran et al., 2005; Lee et al., 2012).

RdDM starts with RNA transcribed by RNA POLYMERASE IV (POL IV). The ssRNA is processed into channeled synthesis of dsRNA precursors for siRNA biogenesis by RNA-DEPENDENT RNA POLYMERASE 2 (*RDRP2* or *RDR2*), which is coupled with POL IV (Haag et al., 2012). The dsRNA is fragmented into 24nt siRNAs by dicer-like protein 3 (*DCL3*). The siRNAs are carried by Argonaute (AGO) AGO4 and AGO6 family proteins, which recognizes homologous nascent nuclear transcripts produced by POLV. This recognition leads interaction of AGO4 with Domains Rearranged Methyl Transferases (DRM2), which catalyzes to cytosine methylation of the DNA (Gao et al., 2010). Along with sequence-specific pairing, protein interactions between AGO4 and the NRPE1 (largest subunit of POL V) and RDM3 (POLV-associated putative transcription elongation factor) are also important for RdDM (Bies-Etheve et al., 2009; He et al., 2009). ncRNAs transcribed by POL V remains on the chromatin with the help of RRP6-LIKE 1 (*RRP6L1*) protein (to act as scaffold RNA). The stabilization of the pairing between siRNA and scaffold RNA is mediated by INVOLVED IN DE NOVO 2 (IDN2)-IDN2 PARALOGUE (IDP) system. This complex further binds RNA to interact with SWI/SNF chromatin remodeling complex composed of SUBUNIT SW13B and participates in POL V mediated transcriptional silencing by changing nucleosome positioning (Zhang et al., 2012; Ausin et al., 2012; Zhu et al., 2013). There are other pathways for DNA methylation along with RdDM pathway. At transcriptionally active transposons and *trans*-acting siRNA genes, POLII enzyme synthesize aberrant transcript, which is changed into dsRNA by RDR6 and further cleaved by dicer-like nucleases (DCL2 and DCL4) into 21 and 22nt small RNA (sRNA) molecules. These sRNAs are loaded into AGO4 and AGO6, which identify homologous nascent nuclear transcripts and this detection leads to encoding DNA methylation by DRM (Erdmann and Picard, 2020). The elements of RdDM machinery and their functions are mentioned in Table 11.1.

11.2.2 MAINTENANCE OF DNA METHYLATION

RdDM maintains short transposons asymmetric (CHH) methylation, the long transposons edges methylation (Stroud et al., 2014) and targets TE dispersed at euchromatic territory (Cokus et al., 2008; Zemach et al., 2013). In the maize genome, active genes from the main genome are frequently fragmented by RdDM mediated CHH methylation Islands. Absence of which often enables transcription in nearby transposons suggesting the role of RdDM in prohibiting reactivation of silenced transposon by active genes in the vicinity. (Li et al., 2015). In sugar beets, asymmetric methylation using RdDM is involved in silencing DNA transposons than retrotransposons (Zakrzewski et al., 2017). RdDM is showed negative correlation with GC content and mostly histone modifications present in heterochromatic region including *H3K9me2* and *H3K27me1* (Cokus et al., 2008; Zemach et al., 2013). This was confirmed in another study where induction of RdDM was observed during the extensive decondensation of pericentromeric heterochromatin of male gametophytes vegetative cell nuclei (Schoft et al., 2009). RdDM mediated DNA methylation maintenance through RNA polV require the *SUVH2* and *SUVH9,* which is SU (VAR)3-9 homologs for the occupancy on the chromatin. SRA domain of these proteins predominantly bind to CHG and CG context-dependent methylated DNA. SET domain of these proteins lack the catalytic activity essential for the methylation of the histone H3 at K9 position. The transcripts produced by RNA polV ultimately leads to methylation through RdDM with the help of the *DDR1* complex (Malagnac, 2002; Kuhlmann et al., 2012; Law, 2010; Johnson et al., 2014).

DNA replication generates semimethylated symmetrical CG dinucleotides. These sites are recognized by VARIANT IN METHYLATION proteins (VIM) and appoint DNA METHYLTRANSFERASE 1 (*MET1*) to cytosine-guanine (CG) methylation of daughter strand (Law and Jacobsen,

TABLE 11.1 Factors Essential for RNA-Directed DNA Methylation (RdDM).

Factors involved in RdDM	Factor types	Function
RNA polymerase	NRPD1 and the Pol IV complex	Pol IV attaches to heterochromatic regions and synthesis precursors of the sRNAs, which play role in canonical RNA-directed DNA methylation
	NRPE1 and the Pol V complex	DNA methylation of target loci
RNA-dependent RNA polymerase	RDR2	It forms a complex with Pol IV and a component of canonical RdDM.
	RDR6	Converts sRNAs to dsRNAs and processed into 21–22 nucleotides RNAs by dicer proteins DCL2 & DCL4. Involved in PTGS and noncanonical RNA-directed DNA methylation
Dicer like proteins (Endoribonuclease)	DCL1	miRNA and sRNA production
	DCL2	sRNA production
	DCL3	sRNA production
	DCL4	sRNA production
Argonaute protein	AGO1	sRNA production and involved in PTGS pathway
	AGO4	Function in canonical pathway of RNA-directed DNA methylation
	AGO6	Functions in RNA-directed DNA methylation (either canonical or noncanonical)
	AGO9	Specialized AGO express in germline to silence TEs
RNA methylase	HEN1	sRNA production
Proteins bind to Methyl-DNA	SUVH2, SUVH9	Methylation of DNA in target loci
DNA methyltransferase	DRM1	DNA methylation of target sites
	DRM2	DNA methylation of target sites
Proteins bind to DNA and chromatin	SHH1/DTF1	Synthesis of sRNA

TABLE 11.1 *(Continued)*

Factors involved in RdDM	**Factor types**	**Function**
Putative chromatin remodelers	CLSY1, CLSY2	Regulate sRNAs derived by Pol IV action, which are located in chromosome arms.
	CLSY3, CLSY4	Regulate Pol IV-derived sRNAs, which are located in the pericentromere.
Putative chromatin remodeling complex	DDR complex (RDM1, DMS3, DRD1)	DNA methylation of target loci
Transcription factor	SPT5L/RDM3/KTF1	Methylation of DNA in target loci
Histone deacetylase	HDA6	Synthesis of sRNA
dsRNA-binding protein	IDN2-IDP complex	Methylation of DNA in target loci
GHKL ATPases	MORC1, MORC6	Methylation of DNA in target loci
GW repeat- and PHD finger-containing protein	NERD	sRNA synthesis, DNA methylation of target sites
Chromatin remodeling complex	SWI/SNF complex	SWI/SNF complex influences the accessibility of chromatin and simplify the access of DNA methyltransferase (DRM2) to DNA and thereby promote RNA-directed DNA methylation

Source: Information sourced Erdmann and Picard (2020).

2010). Methylation by *MET1* is highly efficient and can methylate replicating CG semiconservatively on its own. Whereas, DNA methyltransferase chromomethylase 3 (*CMT3*) catalyzed the maintenance of CHG methylation. Chromo domains of *CMT3* recognize and bind to *H3K9me2* and its bromo-adjacent homology domain methylate cytosine in the CHG context (where H is any nucleotide but not G) thereby suggesting the self-regulatory reinforced feedback silencing mechanism (Du et al., 2012; Matzke et al., 2014). DNA sequence methylated at CHH is maintained by both RdDM through *DRM2* and chromomethyltransferase 2 and *CMT2* dependent DNA methylation process (Li et al., 2020). *CMT2* methylates mostly the TEs at the heterochromatic region rich in histone H1 with the help of chromatin remodeler DDM1 that make chromatin accessible to the *CMT2*. The *snf2* family member DDM1 remodel the chromatin and alter the nucleosome composition and placement thereby modulating the access of the DNA to different proteins (Jeddeloh et al.,1999; Zemach, 2013; Halibart-Puzio et al., 2015).

11.2.3 MECHANISMS OF DEMETHYLATION IN PLANTS

Numerous short TEs or TE-derived DNA sequences prevail in gene rich euchromatic regions, which are methylated by RdDM (Tran et al., 2005; Stroud et al., 2014). Demethylation of DNA is counteracting RdDM and prevent neighboring genes from methylation. DNA demethylation targets TE sequences positioned at the 5′ regions of biotic or abiotic responsive genes and activate those genes (Le et al., 2014). During DNA replication, DNA demethylation can occur passively when DNA is not methylated in daughter strand; however, DNA demethylases family (DNA glycosylase) is involved in active DNA demethylation. These DNA glycosylase enzymes cleave 5-methylcytosine from the target DNA sequence creating a gap in the DNA strand. Further base excision repair system fills the gap with unmethylated cytosine (Zhu et al., 2009). Demeter (DME), REPRESSOR OF SILENCING 1 (ROS1)/demeter-like1 (*DML1*), DML2, and DML3 are the demethylases studied in *A. thaliana* (Penterman et al., 2007). DNA demethylases work irrespective of the sequence context. Only DME is primarily synthesized in the gametic cells, whereas rest other enzymes are synthesized in all vegetative cells. In vegetative tissues, *ROS1* is the main 5-meC DNA glycosylase and ROS1 usually targetes the transposons, which is present near the protein-coding genes. Hypermethylation has been observed from TE edges to neighboring sequences in ROS1 mutants, which indicate that transcriptional inactivation of genes which present nearby the TE, is being protected by DNA demethylation by *ROS1* (Gong et al., 2002; Zhu et al., 2007). *ROS1* activity also reactivates the TEs expression, which performs regulatory functions during vegetative development for instance leaf epidermal cells pattern (Yamamuro et al., 2014). ROS1 targeted region of the genome is marked by acetylated *H2K18* and depletion of *H3K27me* and *H3K9me2*. ROS1 performs the demethylation of the DNA preventing transcriptional gene silencing (TGS) thereby maintaining the active state of the transgene and TE sequences (Zhang, 2018).

11.2.4 ALTERATION IN TRANSPOSON METHYLATION STATUS

DNA methylation patterns also change due to plant hybridization as suggested by Liu et al. (2004). Remarkable changes in methylated cytosine profile and transcription of genes and TE were observed in two stable rice lines with introgression from *Zizanialatifolia* wild rice (Liu et al., 2004). Over a long period of interaction between transposons with their host plant, transposons have developed strategies to evade epigenetic silencing. Some transposon genes may play a dynamic role by reversing DNA methylation and prohibiting their own transcriptional silencing For example, in maize *MuDR* element reverse the methylation of nonautonomous *Mu* elements, which is otherwise consistently methylated (Lisch et al., 1995; Benito and Walbot, 1997). A similar effect on methylated *Spm* elements has also been observed due to *Spm* transposase and it involves active demethylation of the methylated *Spm* (Cui et al., 2002).

11.3 ROLE OF HISTONE MODIFICATIONS IN TE SILENCING

Plant genomic DNA remained packed around the histone octamer called as nucleosomes. Histone protein tail at their N-terminal region is loaded with lysine residue contributes to different histone modifications. These modifications are used as identification marks for the particular region of the genome and transferred over generations as epigenetic memory marks. TEs rich region of the genome are kept under control by tightly packed DNA with histone proteins called as heterochromatin (Wood et al., 2013). Depending upon the expression status of the particular DNA sequence the chromatin is categorized as euchromatin, facultative euchromatin, and heterochromatin (Trojer and Danny Reinberg, 2007). Active and inactive sequences are marked with different chemical modifications to both DNA and histones. Combined modifications in

DNA and histone depict the status and the position of the corresponding DNA in the chromatin, which largely determines whether TE will be transcriptionally active or inactive. Heterochromatization represses the expression of the TE by preventing the entry of the RNA polymerases and other DNA binding transcription factors to the DNA. Epigenetic marks of the chromatin help histone reader and writers to access the DNA sequence for necessary changes such as DNA methylation or demethylation (Cui et al., 2013).

11.3.1 HISTONE MARKS ASSOCIATED WITH THE TRANSCRIPTIONAL SILENCING OF TE

Differential combinations of the histone marks have been found to be associated with the particular state of the expression of the genomic segment (Bernatavichute et al., 2008; Roudier et al., 2011). Intronic heterochromatin repeats are characterized by the *H3K9me2* and *H3K4me3* and identified by EDM2 (ENHANCED DOWNY MILDEW 2) with composite homeodomain (PHD) (Zhang, 2018; Pontvianne et al., 2010). *H3K4me3* is associated with transcriptional activeness of the chromatin whereas the genes that are present in the actively transcribed region and need to be repressed for the developmental purposes are marked by *H3K27me3* (Li et al., 2020). The modification of the *H3K27me3* is largely responsible for silencing of the genes, which were majorly coding the different transcription factors (Lafos et al., 2011). *H3K4* di and trimethylation are observed in case of the active genes. The monomethylated *H3K4* is rich in the gene body and scarce in the boundary regions of the gene whereas di and trimethylated *H3K4* is found to be present at the upstream region of the gene. The upstream region of the promoter is significantly modified by *H3K4* methylation (Roudier et al., 2011; Pontvianne et al., 2010; Feng and Jacobsen, 2011). In higher plants like Arabidopsis, the major heterochromatin mark is the dimethylation of the *H3K9*. It is also observed specifically in regions rich in TEs and the repetitive DNA elements from the euchromatin where DNA is prominently methylated at CHG sequence context (Feng and Jacobsen, 2011; Underwood et al., 2017).

11.3.2 HISTONE MODIFICATIONS IN TE SILENCING

Like methylation, histones are also modified by acetylation, ubiquitination, etc. Local composition of the histone variants in the chromatin changes depending upon the type of the chromatin and gene expression status. The process involves the modulation of the chromatin architecture by modifying histones according to the active or inactive state of the chromatin.

11.3.2.1 HISTONE H1

Histone H1 is an essential component of the chromatin organization of the eukaryotic genome including plants and animals. Histone H1 is associated with DNA methylation as it is found to be abundantly present in the heterochromatic regions than euchromatic regions. Methylated TEs show special enrichment of the histone H1.1 and H1.2 variants (Rutowicz et al., 2015). Loss of histone H1 has been recently shown to cause global reorganization of the nucleosomes, dispersion of the heterochromatin, and activation of the H1 bound genes but it could weakly activate the TEs alone. However, loss of H1 histone with MET1 mutation showed strong activation of the TEs suggesting the requirement of both the components for effective TEs silencing (Choi et al., 2020). It is evident from in vitro analysis that H1 was preferentially inhibiting the transcription initiation process. H1 is required by DDM1 for the access to the inaccessible heterochromatin in order to stably maintain the methylation in the heterochromatic DNA in cooperation with the RdDM pathway (Zemach et al., 2013).

11.3.2.2 HISTONE H2

Variant of histone H2A, that is, H2A.Z is negatively related with the DNA methylation and it protects the DNA from methylation in the actively transcribed genes bodies. Exclusion of H2A.Z can influence chromatin structure and maintain gene silencing. Deposition of the histone H2A.Z and H3.3 at 5′ and 3′ ends of the transcriptionally active genes, respectively, defines the boundary of the euchromatin in Arabidopsis. Likewise independent of *H3K9me2* or DNA methylation, the histone H2A.W variant specifically located to heterochromatin in cooperation with DNA methylation silence the TEs through independent pathways. The chromatin remodeling complex SWR1-C is required for the deposition of the H2A.Z to chromatin (Zilberman et al., 2008; Yelagandula et al., 2014; Zambrano et al., 2018). Histone H2B deubiquitination by SUP32 shows the importance of histone modification that relates heterochromatic histone modifications with DNA methylation (Sridhar et al., 2007).

11.3.3 HISTONE DEMETHYLATION AND DEACETYLATION

To prevent the establishment of the *H3K9me2* mark of heterochromatization, histone demethylases are also work. *IBM1* of the family *JHDM2* contains the domain jmjC having the activity of the histone demethylation. *IBM1* is found to impair the DNA methylation by demethylating the *H3K9* in CHG sequence context and IDM1 in CG context. One more jmjC protein *JMJ14* acts downstream of the RdDM component RDR2 and AGO4 that demethylate histone *H3K4me3*marks and leads to consequent DNA methylation. TE activation is suppressed by *H3K4* demethylase proteins *JMJ14*, *LDL1*, and *LDL2,* which remove the mark of genic chromatin. There is an involvement of the chaperons in the process of the DNA demethylation (He et al., 2011; Zhang, 2018; Roudier et al., 2011).

Mutation of *HDA6* (a histone deacetylase that acts upon histone H4) results in loss of TGS at some RdDM targets indicating a function of*HDA6* in RdDM. Its mutant causes reduced methylation of the symmetrical genomic regions suggesting its role in the maintenance indicating the role of histone deacetylation by *HDA6*in symmetric DNA methylation at particular chromatin regions while suppressing asymmetric DNA methylation (Liu et al., 2012).

11.3.4 INTERPLAY BETWEEN HISTONE MODIFICATIONS AND TRANSCRIPTIONAL SILENCING

Histone methyltransferases KRYPTONITE (*KYP*)/*SUVH4* and *SUVH6* performs methylation of *H3K9.* The SRA domain of the *SUVH4* binds the methylated cytosine owing to the requirement of the methylated DNA for histone methylation by *SUVH4/KYP*. After binding SET domains performs the dimethylation of histone at *H3K9* (Johnson et al., 2007; Zhang, 2018). Dimethylation of the *H3K9* acts as identification mark for the binding of the multiple DNA methylation enzymes.

TEs were transcriptionally and transpositionally activated in the hypomethylated condition as a result of decreased DNA methylation in *DDM1* mutant (Hirochika et al., 2000). *DDM1* acts through rearrangement of the nucleosome structure utilizing the ATP and walking over the DNA. The behavior of the *DDM1* with RdDM and sequence preference is largely unknown. Short TEs residing in the vicinity of the genes from the euchromatin are silenced by the DNA methylation carried out by the DNA methyl transferase *DRM2* through RdDM pathway of silencing (Matzke et al., 2014).

11.4 POST-TRANSCRIPTIONAL REGULATION OF TES

TEs activity is restrained by host plant both before and after transcription. When RdDM pathway involving 24nt siRNAs and homology-based silencing failed to silence TEs then post-transcriptional silencing pathway to silence the active TEs has to be initiated by RNAi. Post-transcriptional silencing chiefly involves RNAi that identifies and causes degradation of mRNAs of transposon elements, which is produced through action of RNA polymerase II. The components essential for causing degradation of TE mRNAs are RNA-dependent RNA POL VI that predominantly converts a ssRNA to dsRNA and two dicer proteins namely *DCL2* and *DCL4* that assists in cleavage of dsRNAs to produce 21–22 nucleotide siRNAs and AGO1 protein, which guides siRNAs to mRNAs for degradation (Fultz et al., 2015).

11.4.1 RNAI PATHWAY AND RECOGNITION OF TES MRNA TRANSCRIPTS

The commencement of post-transcriptional silencing demands on recognition of Pol II-derived TE mRNAs. Processing of ssRNA into dsRNA by RNA-dependent RNA pol VI facilitates the identification of mRNA of actively transcribing TEs (Matzke et al., 2014). As TEs are epigenetically activated, TE mRNA transcripts are favored to be targeted by miRNAs (protein DCL1-dependent) bound by Argonaute1 protein. Successful breaking of TE mRNA attracts RNA-dependent RNA pol VI and protein DCL4, which produce 21nucleotide long epigenetically-activated siRNAs from transposon open reading frames. However, in a post-transcriptional silencing that are loaded onto AGO1 protein and deters engagement of RNA-dependent RNA polymerase 2 (RDR2) and RdDM (Creasey et al., 2014). The miRNAs are exhaustively involved in the cleavage of mRNA and the specificity to pick TEs mRNA over genic mRNAs still remain enigmatic. The miRNA can target mRNAs of selective transposon element families in Arabidopsis and resulted into silencing and this indicates that other mechanisms must be involved in nonhomology-based silencing. Hairpin-derived small RNAs also involved in TE mRNA recognition for example; rearrangement of inverted repeats in a TE known as MU Killer known to initiate RNA interference and followed by other MuDR members silencing (Slotkin et al., 2005). Du-noyer et al. (2010) also reported direct formation of siRNAs from hairpins and found no role of dsRNA, RNA-dependent RNA polymerase 6, and dicer-like proteins in Arabidopsis. The hairpin structures are converted to hpRNA by DCL proteins that results into RDR VI generation of dsRNA either as primers or through their loading onto an AGO protein and further initiates the RNAi cycle. On the other hand, the hairpin structure could result into RDRVI synthesis independent of the hpRNAs (Devert et al., 2015). These hypothesized models imply that palindromic sequences might be pivotal indicators for the TE suppression. These explanations indicate the pathway of formation of the hairpin structures from palindromic sequences in Sire virus mRNA involved in silencing. The small RNAs (hpRNAs) are generated from the hairpin structures with the help of dicer-like proteins and the hpRNAs further facilitate RNA-dependent RNA pol VI (RDR6) synthesis of dsRNA either. These studies showed that palindromic regions may be key indicators for the commencement of transposon element silencing.

11.4.2 SWITCHING BETWEEN RDDM AND RNAI

The post-transcriptional regulation involving 21–22nt siRNAs can degrade large number of TE mRNAs and also initiate pretranscriptional TE silencing (Nuthikattu et al., 2013; McCue et al., 2015). Thus, it can be inferred that the 21–22nt siRNAs serve twin purpose through their involvement in post-transcriptional silencing and beginning pretranscriptional TS suppression through RdDM. The shift from post-transcriptional RNAi to RdDM was hypothesized by McCue et al. (2015) according to which involvement of *AGO6* to carry 21–22nt siRNAs and then that directed chromatin modifications rather than AGO1 needed for PTGS. Whereas, Marí-Ordóñezet al. (2013) proposed that DCL2/DCL4 fails to process large amount of TE mRNA and therefore activates the biogenesis of 24nt siRNA by specific dicer like protein DCL3 and resulted in initiation of pretranscriptional RNA-directed DNA methylation. On initiation of RdDM pathway, TE got inactivated and the chromatin modifications are accurately passed during DNA replication resulted in inheritance of silenced chromatin in daughter cells (Panda and Slotkin, 2013).

The post-transcriptional regulation involving 21–22nt siRNAs can degrade huge number of transposon element mRNAs and also initiate pretranscriptional silencing of transposon elements (Nuthikattu et al., 2013). Thus, it can be inferred that the 21–22nt siRNAs serve twin purpose through their involvement in post-transcriptional silencing and start of pretranscriptional silencing of transposon elements through RdDM. The shift from post-transcriptional RNA interference to RNA-directed DNA methylation was hypothesized by McCue et al. (2015) according to which involvement of *AGO6* to carry 21–22nt siRNAs and then that direct chromatin modifications rather than AGO1 needed for PTGS. Whereas, Marí-Ordóñezet al. (2013) proposed that DCL2/DCL4 fails to process large amount of TE mRNA and therefore activates the biogenesis of 24 nucleotide long siRNA by specific protein DCL3 and resulted in initiation of pretranscriptional RdDM. After the onset of the RdDM pathway TEs got inactivated and the chromatin modifications of silenced chromatin were inherited to the daughter cells (Panda and Slotkin, 2013).

Many TEs resume ability to transpose and get activated due to epigenetic loss or remodeling, which was observed in *mop1* mutants of maize (Jia et al., 2009), pollen vegetative nucleus in Arabidopsis (Slotkin et al., 2009), transition of vegetative to reproductive phase in Maize (Li et al., 2010) and copia type retrotransposon in Arabidopsis under temperature stress (Ito et al., 2011). It was observed in Arabidopsis and maize that silencing of reactivated TEs occurred in trans by the Pol IV-RdDM pathway (Slotkin et al., 2005). However, the proliferation of horizontally invaded TEs in new host is regulated by presence or absence transcripts of silenced endogenous TEs endowed with sequence similarity to the invading TEs. Instances have been found where the 24 siRNAs-based silencing is not effective to silence the active TEs. In Arabidopsis, the "Evade" element existed in active state regardless of silenced TE families in the host genome (Mirouze et al., 2009). In maize, LTR transposon showed homology to large number of 21–22nucleotide siRNAs indicated that members of LTR transposons have been silenced through post-transcriptional RNAi (Bousios and Gaut, 2016).

11.4.3 TRANSPOSON ELEMENTS REMAIN ACTIVE REGARDLESS OF DIFFERENT METHODS OF SILENCING

Plant host can commence silencing by several methods, which included RdDM pathway, post-transcriptional silencing by 22nt siRNAs and microRNA however, some TEs can remain active. Cavrak et al. (2014) reported that LTR copia type retrotransposon named ONSEN in Arabidopsis is activated by heat shock transcription factor during heat stress period. Other copia-type retrotransposon evades in Arabidopsis remain active and transposes into genes and silenced loosely related members in the genome (Mirouze et al., 2009). The transposon element escaped from silencing due to deficiency of TE siRNAs that is produced by retrotransposon itself or closely related members of the family and resulting into high copy number.

In many TEs, the lack of silencing is perplexing because silenced elements of the same transposon element family in the same genome are generating ample Pol IV-derived 24nucleotide siRNAs that can trans-silence the active TE copies. For instance, in mutator family, TE copies stay active in spite of large amount of 24nt siRNAs generated from silenced and fragmented homologous TE copies (Rudenko et al., 2003). There are hypothesized theories proposed in different studies about the maintenance of active state of certain TEs irrespective of other family members are silenced in the same genome. The transposon elements MuDR in maize and Mu killer brought together genetically. MuDR transposon elements become epigenetically silenced after losing the Muk in succeeding generations. However, in certain positions in the genome, these MuDR elements in the maize are capable of erasing epigenetic silencing (Singh et al., 2008). It has been reported that insertions in genic 3′ untranslated regions to dodge silencing of active Mutator TEs in Arabidopsis. The transposon *AtMu1c* expression found to be highest for insertions within 3′ untranslated regions signifying that by inserting into specific locations in genome is one of mechanism adopted by transposon to circumvent epigenetic silencing (Kabelitz et al., 2014). The general rule for TE copies escaped from silencing due to the selective requirement to preserve the genes in an active chromatin region of genome. The transposon element families have preference for integration in the genome and the competence to escape transposon element silencing is anticipated to be higher for DNA TEs and smaller for LTR retrotransposons that mainly incorporates into heterochromatin.

11.4.4 TE ROLE THROUGH LNCRNA

The long noncoding RNAs are transcript longer that 200 bp in size that are not translated into functional proteins and found in many plant genomes (Liu et al., 2015; Budak et al., 2020). Transposon elements have been emerging as source of large number of long coding RNAs and perform many biological functions by participating in epigenetic, transcriptional, post-transcriptional, or translational regulation of target genes (Chen et al., 2020; Ariel and Manavella, 2021). These lncRNAs contribute to varied molecular process through interaction with DNA, RNA and proteins hence, regulate expression of genes and activity of proteins and suggested dynamic evolution of plant genomes through co-option of TEs in the form of RNA (Urquiaga et al., 2020).

11.5 CONCLUSION

TEs transposition signifies an effective mechanism of genome expansion on evolutionary scale. For the survival and maintenance in host genome, TEs attempts to maintain a balance to promote its amplification to the level that it should not cause disadvantage to the fitness of host. Host plant is endowed with variety of mechanisms for controlling TE activity and these include DNA methylation and RNAi using small RNAs, histone modifications, addition of histone variants, and additionally chromatin-remodeling enzymes. The silencing mechanism is balanced by regulated activation of TE by the processes like demethylation of DNA as well as histones molecules for plant survival and growth. TE silencing mechanisms recognized to be relaxed to permit developmental regulation of host gene expression during early embryonic development. The TEs acquired strategies to escape from host regulation for instance; some TEs could acquire stress-responsive motifs that are recognized by host regulatory proteins to self-regulate their activity in response to stress. There are several dimensions to understand the behavior of TEs particularly in nonmodel plant genomes about transposon bursts, preferential insertion sites, and their activation, which is influenced by environmental conditions. Further investigations and analysis of transposon elements in sequenced genomes could reveal additional information about the hidden genetic treasure and their regulation by host genomes.

KEYWORDS

- **transposable elements**
- **TEs regulation**
- **plant genome**
- **methylation**
- **RNA directed DNA methylation (RdDM)**
- **histone modification**
- **epigenetic transcriptional silencing**
- **genome evolution**
- **long noncoding RNA**

REFERENCES

Ariel, F. D.; Manavella, P. A. When Junk DNA Turns Functional: Transposon-Derived Non Coding RNAs in Plants. *J. Exp. Bot.* **2021,** *72* (11), 4132–4143. DOI: 10.1093/jxb/erab073.

Ausin, I.; Greenberg, M. V. C.; Simanshu, D. K.; Hale, C. J.; Vashisht, A. A.; Simon, S. A.; Lee, T.; Feng, S.; Española, S. D.; Meyers, B. C.; Wohlschlegel, J. A.; Patel, D. J.; Jacobsen, S. E. *INVOLVED IN DE NOVO 2*-Containing Complex Involved in RNA-Directed DNA Methylation in Arabidopsis. *Proc. Natl. Acad. Sci. U.S.A.* **2012,** *109*, 8374–8381.

Bannister, A. J.; Kouzarides, T. Regulation of Chromatin by Histone Modifications. *Cell Res.* **2011,** *21* (3), 381–395.

Benito, M. I.; Walbot, V. Characterization of the Maize Mutator Transposable Element MURA Transposase as a DNA-Binding Protein. *Mol. Cell. Biol.* **1997**, *17*, 5165–5175.

Bennetzen, J. L. Transposable Element Contributions to Plant Gene and Genome Evolution. *Plant Mol. Biol.* **2000,** *42*, 251–269.

Bernatavichute, Y. V.; Zhang, X.; Cokus, S.; Pellegrini, M.; Jacobsen, S. E. Genome-Wide Association of Histone H3 Lysine Nine Methylation with CHG DNA Methylation in Arabidopsis thaliana. *PLoS One* **2008**, *3*, 1–11.

Bies-Etheve, N.; Pontier, D.; Lahmy, S.; Picart, C.; Vega, D.; Cooke R.; Lagrange, T. RNA-Directed DNA Methylation Requires an AGO4-Interacting Member of the SPT5 Elongation Factor Family. *EMBO Rep.* **2009**, *10*, 649–654.

Bousios, A.; Gaut, B. S. Mechanistic and Evolutionary Questions About Epigenetic Conflicts Between Transposable Elements and Their Plant Hosts. *Curr. Opin. Plant Biol.* **2016,** 123–133.

Budak, H.; Kaya, S. B.; Cagirici, H. B. L. Long Non-Coding RNA in Plants in the Era of Reference Sequences. *Front. Plant Sci.* **2020,** *11*, 276. DOI: 10.3389/fpls.2020.00276.

Cavrak, V. V.; Lettner, N.; Jamge, S.; Kosarewicz, A.; Bayer, L. M.; Mittelsten Scheid, O. How a Retrotransposon Exploits the Plant's Heat Stress Response for Its Activation. *PLoS Genet.* **2014,** 10, e1004115. DOI: https://doi.org/10.1371/journal.pgen.1004115.

Chan, S. W.; Zilberman, D.; Xie, Z.; Johansen, L. K.; Carrington, J. C.; Jacobsen, S. E. RNA Silencing Genes Control De Novo DNA Methylation. *Science* **2004,** *303* (5662), 1336. DOI: 10.1126/science.1095989.

Chen, L.; Zhu, Q. H.; Kaufmann, K. Long non-coding RNAs in plants: emerging modulators of gene activity in development and stress responses. *Planta* **2020,** 252, 92 doi:10.1007/s00425-020-03480-5

Cho, J.; Paszkowski, J. Regulation of Rice Root Development by a Retrotransposon Acting as a Micro RNA Sponge. *Elife* **2017,** *6*, e30038. DOI: 10.7554/eLife.30038.

Choi, J.; Lyons, D. B.; Kim, M. Y.; Moore, J. D.; Zilberman, D. DNA Methylation and Histone H1 Jointly Repress Transposable Elements and Aberrant Intragenic Transcripts. *Mol. Cell.* **2020,** *77*, 310–323.

Chuong, E. B.; Elde, N. C.; Feschotte, C. Regulatory Activities of Transposable Elements: From Conflicts to Benefits. *Nat. Rev. Genet.* **2017,** *18*, 71–86.

Cokus, S. J.; Feng, S.; Zhang, X.; Chen, Z.; Merriman, B.; Haudenschild, C. D.; Pradhan, S.; Nelson, S. F.; Pellegrini, M.; Jacobsen, S. E. Shotgun Bisulphite Sequencing of the Arabidopsis Genome Reveals DNA Methylation Patterning. *Nature* **2008,** *452*, 215–219.

Creasey, K. M.; Zhai, J.; Borges, F.; Van, E. F.; Regulski, M.; Meyers, B. C.; Martienssen, R. A. miRNAs Trigger Widespread Epigenetically Activated siRNAs from Transposons in Arabidopsis. *Nature* **2014,** *508*, 411–415.

Cui, H.; Fedoroff, N. V. Inducible DNA Demethylation Mediated by the Maize Suppressor-Mutator Transposon-Encoded TnpA Protein. *Plant Cell.* **2002**, *14*, 2883–2899.

Cui, X.; Jin, P.; Cui, S. G. L.; Lu, Z.; Xue, Y.; Wei, L.; Qi, J.; Song, X.; Luo, M.; An, G.; Cao, X. Control of Transposon Activity by a Histone H3K4 Demethylase in Rice. *PNAS* **2013,** *110* (5), 1953–1958.

Deniz, O.; Frost, J. M.; Branco, M. R. Regulation of Transposable Elements by DNA Modification. *Nat. Rev. Gen.* **2019,** *20* (7):417–431.

Devert, A.; Fabre, N.; Floris, M.; Canard, B.; Robaglia, C.; Crete, P. Primer-Dependent and Primer-Independent Initiation of Double Stranded RNA Synthesis by Purified Arabidopsis RNA-Dependent RNA Polymerases RDR2 and RDR6. *PLoS One* **2015,** *10*, e0120100. DOI: 10.1371/journal.pone.0120100.

Du, J.; Zhong, X.; Bernatavichute, Y. V.; Stroud, H.; Feng, S.; Caro, E.; Vashisht, A. A.; Terragni, J.; Chin, H. G.; Tu, A.; Hetzel, J.; Wohlschlegel, J. A.; Pradhan, S.; Patel, D. J.; Jacobsen, S. E. Dual Binding of Chromomethylase Domains to H3K9me2-Containing Nucleosomes Directs DNA Methylation in Plants. *Cell* **2012,** *151*, 167–180

Du-noyer, P.; Brosnan, C. A.; Schott, G.; Wang, Y.; Jay, F.; Alioua, A.; Himber, C.; Voinnet,O. An Endogenous, Systemic RNAi Pathway in Plants. *EMBO J.* **2010,** *29*, 1699–1712.

Erdmann, R. M.; Picard, C. L. RNA-Directed DNA Methylation. PLoS GENET. **2020,** *16*, e1009034. DOI: https://doi.org/10.1371/journal.pgen.1009034.

Feng, S.; Jacobsen, S. E.; Epigenetic Modifications in Plants: An Evolutionary Perspective. *Curr. Opin. Plant Biol.* **2011,** *14*, 1179–1186.

Fultz, D.; Choudhary, S. G.; Slotkin, R. K. Silencing of Active Transposable Elements in Plants. *Curr. Opin. Plant Biol.* **2015,** *27*, 67–76.

Gao, Z.; Liu, H. L.; Daxinger, L. Pontes, O.; He, X.; Qian, Weiqiang.; Lin, H.; Xie, M.; Lorkovic, Z. J.; Zhang, S.; Miki, D.; Zhan, X.; Pontier, D.; Lagrange, T.; Jin, H.; Matzke, A. J. M.; Matzke, M.; Pikaard, C. S.; Zhu, J. K. An RNA Polymerase II- and AGO4-Associated Protein Acts in RNA-Directed DNA Methylation. *Nature* **2010,** *465*, 106–109.

Gong, Z.; Teresa, M. R. T.; Ariza, R. R.; Roldán-Arjona, T.; David, L.; Zhu, J. K. ROS1, a Repressor of Transcriptional Gene Silencing in Arabidopsis, Encodes a DNA Glycosylase/Lyase. *Cell* **2002,** *111*, 803–814.

Haag, J. R.; Ream, T. S.; Marasco, M. Nicora, C. D.; Norbeck, A. D.; Pasa-Tolic, L.; Pikaard, C. S. In Vitro Transcription Activities of Pol IV, Pol V, and RDR2 Reveal Coupling of Pol IV and RDR2 for dsRNA Synthesis in Plant RNA Silencing. *Mol. Cell.* **2012,** *48*, 811–818.

Halibart-Puzio, J.; Lirski, M.; Kotliński, M.; Kroteń, M. A.; Knizewski, L.; Lange, B.; Muszewska, A.; Śniegowska-Świerk, K.; Kościelniak, J.; Iwanicka-Nowicka, R.; Buza, K.; Janowiak, F.; Żmuda, K.; Jõesaar, I.; Laskowska-Kaszub, K.; Fogtman, A.; Kollist, H.; Zielenkiewicz, P.; Tiuryn, J.; Siedlecki, P.; Swiezewski, S.; Ginalski, K.; Koblowska, M.; Archacki, R.; Wilczynski, B.; Rapacz, M.; Jerzmanowski, A. A. Specialized Histone H1 Variant Is Required for Adaptive Responses to Complex Abiotic Stress and Related DNA Methylation in Arabidopsis. *Plant Physiol.* **2015,** *169* (3), 2080–2101.

He, X. J.; Chen, T.; Zhu, J. K. Regulation and Function of DNA Methylation in Plants and Animals. *Cell Res.* **2011,** *21* (3), 442–465.

He, X. J.; Hsu, Y. F.; Zhu, S.; Wierzbicki, A. T.; Pontes, O.; Pikaard, C. S.; Liu, H. L.; Wang, C. S.; Jin, H.; Zhu, J. K. An Effector of RNA-Directed DNA Methylation in Arabidopsis is an ARGONAUTE 4-and RNA-Binding Protein. *Cell* **2009,** *137*, 498–508.

Herr, A. J.; Jensen, M. B.; Dalmay, T.; Baulcombe, D. C. RNA Polymerase IV Directs Silencing of Endogenous DNA. *Science* **2005,** *308*, 118–120.

Hirochika, H.; Okamoto, H.; Kakutani, T. Silencing of Retrotransposons in Arabidopsis and Reactivation by the ddm1 Mutation. *Plant Cell* **2000,** *12*, 357–368.

Ito, H.; Gaubert, H.; Bucher, E.; Mirouze, M.; Vaillant, I.; Paszkowski, J. An siRNA Pathway Prevents Transgenerational Retrotransposition in Plants Subjected to Stress. *Nature* **2011,** *472*, 115–119.

Jeddeloh, J. A.; Stokes, T. L. Richards, E. J. Maintenance of Genomic Methylation Requires a SWI2/SNF2-Like Protein. *Nat. Genet.* **1999,** *22*, 94–97.

Jia, Y.; Lisch, D. R.; Ohtsu, K.; Scanlon, M. J.; Nettleton, D.; Schnable, P. S. Loss of RNA–Dependent RNA Polymerase 2 (RDR2) Function Causes Widespread and Unexpected Changes in the Expression of Transposons, Genes, and 24-nt Small RNAs. *PLoS Genet.* **2009,** 5 (11), e1000737. DOI: https://doi.org/10.1371/journal.pgen.1000737.

Jin, B.; Li, Y.; Robertson, K. D. DNA Methylation: Superior or Subordinate in the Epigenetic Hierarchy? *Genes Cancer* **2011,** *2* (6), 607–617.

Johnson, L. M.; Bostic, M.; Zhang, X.; Kraft, E.; Henderson, I.; Callis, J.; Jacobsen, S. E. The SRA Methyl-Cytosine-Binding Domain Links DNA and Histone Methylation. *Curr. Biol.* **2007,** *17* (4), 379–384.

Johnson, L. M.; Du, J.; Hale, C. J.; Bischof, S.; Feng, S.; Chodavarapu, R. K.; Zhong, X.; Marson, G.; Pellegrini, M.; Segal, D. J.; Patel, D. J.; Jacobsen, S. E. SRA/SET Domain-Containing Proteins Link RNA Polymerase V Occupancy to DNA Methylation. *Nature* **2014,** *507* (7490), 124–128.

Kabelitz, T.; Kappel, C.; Henneberger, K.; Benke, E.; Nöh, C.; Bäurle, I. eQTL Mapping of Transposon Silencing Reveals a Position-Dependent Stable Escape from Epigenetic Silencing and Transposition of AtMu1 in the Arabidopsis Lineage. *Plant cell* **2014,** *26* (8), 3261–3271. DOI: https://doi.org/10.1105/tpc.114.128512.

Kim, N. S. The genomes and transposable elements in plants: are they friends or foes? *Genes Genome* **2017,** *39*, 359–370.

Kohli, R. M.; Zhang, Y. TET Enzymes, TDG and the Dynamics of DNA Demethylation. *Nature* **2013,** *502* (7472), 472–479.

Kuhlmann, M.; Mette, M. F. Developmentally Non-Redundant SET Domain Proteins SUVH2 and SUVH9 are Required for Transcriptional Gene Slencing in Arabidopsis Thaliana. *Plant Mol. Biol.* **2012,** *79*, 623–633.

Lafos, M.; Kroll, P.; Hohenstatt, M. L.; Thorpe, F. L.; Clarenz, O.; Schubert, D. Dynamic Regulation of H3K27 Trimethylation During Arabidopsis Differentiation. *PLoS Genet.* **2011,** *7* (4), 1–20.

Law, J. A.; Du, J.; Hale, C. J.; Feng, S.; Krajewski, K.; Palanca, A. M. S.; Strahl, B. D.; Patel, D. J.; Jacobsen, S. E. Polymerase IV Occupancy at RNA-Directed DNA Methylation Sites Requires SHH1. *Nature* **2013,** *498*, 385–389.

Law, J. A.; Jacobsen, S. E. Establishing, Maintaining and Modifying DNA Methylation Patterns in Plants and Animals. *Nat. Rev. Genet.* **2010**, *11*, 204–220.

Le, T. N.; Schumann, U.; Smith, N. A.; Tiwari, S.; Au, P. C. K.; Zhu, Q. H.; Taylor, J. M.; Kazan, K.; Llewellyn, D. J.; Zhang, R.; Dennis, E. S.; Wang, M. B. DNA Demethylases Target Promoter Transposable Elements to Positively Regulate Stress Responsive Genes in Arabidopsis. *Genome Biol.* **2014,** *15* (9), 458. DOI: 10.1186/s13059-014-0458-3.

Lee, T. F.; Gurazada, S. G.; Zhai, J.; Li, S.; Simon, S. A.; Matzke, M. A.; Chen, X.; Meyers, B. C. RNA Polymerase V-Dependent Small RNAs in Arabidopsis Originate from Small, Intergenic Loci Including Most SINE Repeats. *Epigenetics* **2012,** *7*, 781–795.

Li, H.; Freeling, M.; Lisch, D. Epigenetic Reprogramming During Vegetative Phase Change in Maize. *Proc. Natl. Acad. Sci. U.S.A.* **2010,** *107*, 22184–22189.

Li, J.; Yang, D. L.; Huang, H.; Zhang, G.; He, L.; Pang, J.; Lozano-Durán, R.; Lang, Z.; Zhu, J. K. Epigenetic Memory Marks Determine Epiallele Stability at Loci Targeted by De Novo DNA Methylation. *Nat. Plant* **2020,** *6* (6), 661–674.

Li, Q.; Gent, J. I.; Zynda, G.; Song, J.; Makarevitch, I.; Hirsch, C. D.; Hirsch, C. N.; Dawe, R. K.; Madzima, T. F.; McGinnis, K. M.; Lisch, D.; Schmitz, R. J.; Vaughn, M. W.; Springera, N. M. RNA-Directed DNA Methylation Enforces Boundaries Between Heterochromatin and Euchromatin in the Maize Genome. *Proc. Natl. Acad. Sci. U.S.A.* **2015,** *112*, 14728–14733.

Lippman, Z.; Gendrel, A.; Black, M.; Vaughn, M. W.; Dedhia, N.; McCombie, W. R.; Lavine, K.; Mittal, V.; May, B.; Kasschau, K. D.; Carrington, J. C.; Doerge, R. W.; Colot, V.; Martienssen, R. Role of Transposable Elements in Heterochromatin and Epigenetic Control. *Nature* **2004,** *430*, 471–476.

Lisch, D. How Important Are Transposonsfor Plant Evolution? *Nature* **2013,** *14*, 49–61.

Lisch, D.; Chomet, P.; Freeling, M. Genetic Characterization of the Mutator System in Maize: Behaviour and Regulation of Mu Transposons in a Minimal Line. *Genetics* **1995,** *139*, 1777–1796.

Liu, J.; Wang, H.; Chua, N. H. Long Noncoding RNA Transcriptome of Plants. *Plant Biotechnol. J.* **2015,** *13*, 319–328.

Liu, X.; Yu, C. W.; Duan, J.; Luo, M.; Wang, K.; Tian, G.; Cui, Y.; Wu, K. HDA6 Directly Interacts with DNA Methyltransferase MET1 and Maintains Transposable Element Silencing in Arabidopsis. *Plant Physiol.* **2012,** *158*, 119–129.

Liu, J.; Jung, C.; Xu, J.; Wang, H.; Deng, S.; Bernad, L.; Arenas-Huertero, C.; Chua, N. H. Genome-Wide Analysis Uncovers Regulation of Long Intergenic Noncoding RNAs in Arabidopsis. *Plant Cell.* **2012,** *24*, 4333–4345.

Liu, Z.; Wang, Y.; Shen, Y. Guo, W.; Hao, S.; Liu, B. Extensive Alterations in DNA Methylation and Transcription in Rice Caused by Introgression from Zizania Latifolia. *Plant Mol. Biol.* **2004,** *54*, 571–582.

Malagnac, F.; Bartee, L.; Bender, J. An Arabidopsis SET Domain Protein Required for Maintenance but Not Establishment of DNA methylation. *EMBO J.* **2002,** *21* (24), 6842−6852.

Marí-Ordóñez, A.; Marchais, A.; Etcheverry, M.; Martin, A.; Colot, V.; Voinnet, O. Reconstructing De Novo Silencing of an Active Plant Retrotransposon. *Nat. Genet.* **2013,** *45* (9), 1029−1039. DOI: 10.1038/ ng.2703.

Matzke, M.; Aufsatz, W.; Kanno, T.; Daxinger, L.; Papp, I.; Mette, M. F.; Matzke, A. J. Genetic Analysis of RNA–Mediated Transcriptional Gene Silencing. *Biochem. Biophys. Acta* **2004,** *1677*, 129–141.

Matzke, M. A.; Mosher, R. A. RNA-Diirected DNA Methylation: An Epigenetic Pathway of Increasing Complexity. *Nat. Rev. Genet.* **2014,** *15* (6), 394–408. DOI: 10.1038/nrg3683.

McCue, A. D.; Nuthikattu, S.; Reeder, S. H.; Slotkin, R. K. Gene Expression and Stress Response Mediated by the Epigenetic Regulation of a Transposable Element Small RNA. *PLoS Genet.* **2012,** *8* (2). DOI: 10.1371/journal.pgen.1002474.

Mette, M. F.; Aufsatz, W.; van der Winden, J.; Matzke, M. A.; Matzke, A. J. Transcriptional Silencing and Promoter Methylation Triggered by Double-Stranded RNA. *EMBO J.* **2000,** *19* (19), 5194–5201.

Mirouze, M.; Reinders, J.; Bucher, E.; Nishimura, T.; Schneeberger, K.; Ossowski, S.; Cao, J.; Weigel, D.; Paszkowski, J.; Mathieu, O. Selective Epigenetic Control of Retrotransposition in Arabidopsis. *Nature* **2009,** *461*, 427–430.

Naville, M.; Henriet. S.; Warren, I.; Sumic, S.; Reeve, M.; Vloff, J.; Chourrout, D. Massive Changes of Genome Size Driven by Expansions of Non-Autonomous Transposable Elements. *Curr. Biol.* **2019,** *29*, 1161–1168.

Nuthikattu, S.; McCue, A. D.; Panda, K.; Fultz, D.; DeFraia, C.; Thomas, E. N.; Slotkin, R. K. The Initiation of Epigenetic Silencing of Active Transposable Elements Is Triggered by RDR6 and 21–22 Nucleotide Small Interfering RNAs. *Plant Physiol.* **2013,** *162*, 116–131. DOI: 10.1104/pp.113.216481.

Panda, K.; Slotkin, R. K. Proposed Mechanism for the Initiation of Transposable Element Silencing by the RDR6-Directed DNA Methylation Pathway. *Plant Signal. Behav.* **2013,** *8* (8), e25206. DOI: 10.4161/psb.25206.

Penterman, J.; Zilberman, D.; Huh, J. H.; Ballinger, T.; Henikoff, S.; Fischer, R. L. DNA Demethylation in the Arabidopsis Genome. *Proc. Natl. Acad. Sci. U.S.A.* **2007,** *104*, 6752–6757.

Pontvianne, F.; Blevins, T.; Pikaard, C. S. Arabidopsis Histone Lysine Methyltransferases. *Adv. Bot. Res.* **2010,** *53*, 1–22.

Quesneville, H. Twenty Years of Transposable Element Analysis in the Arabidopsis Thaliana Genome. *Mob. DNA* **2020,** *11*, 28. DOI: https://doi.org/10.1186/s13100-020-00223-x.

Roessler, K.; Bousios, A.; Meca, E.; Gaut, B. S. Modeling Interactions Between Transposable Elements and the Plant Epigenetic Response: A Surprising Reliance on Element Retention. *Genome Biol. Evol.* **2018,** *10* (3), 803–815.

Roudier, F.; Ahmed, I.; Berard, C.; Sarazin, A.; Mary-Huard, T.; Cortijo, S.; Bouyer, D.; Caillieux, E.; Duvernois-Berthet, E.; Al-Shikhley, L.; Giraut, L.; Després, B.; Drevensek, S.; Barneche, F.; Dèrozier, S.; Brunaud, V.; Aubourg, S.; Schnittger, A.; Bowler, C.; Martin-Magniette, M. L.; Robin, S.; Caboche, M.; Colot, V. Integrative Epigenomic Mapping Defines Four Main Chromatin States in Arabidopsis. *EMBO J.* **2011,** *30*, 1928–1938.

Rudenko, G. N.; Ono, A.; Walbot, V. Initiation of Silencing of Maize MuDR/Mu Transposable Elements. *Plant J.* **2003,** *33*, 1013–1025.

Rutowicz, K.; Puzio, M.; Puzio, J. H.; et al. A Specialized Histone H1 Variant Is Required for Adaptive Responses to Complex Abiotic Stress and Related DNA Methylation in Arabidopsis. *Plant Physiol.* **2015,** *169*, 2080–2101.

Schnable, P. S.; Ware, D.; Fulton, R. S.; Stein, J. C.; Wei, F.; Pasternak, S.; Liang, C.; Zhang, J.; Fulton, L.; Graves, T. A.; et al. The B73 Maize Genome: Complexity, Diversity, and Dynamics. *Science* **2014,** *1112*, 1112–1115.

Schoft, V. K.; Chumak, N.; Mosiolek, M.; Slusarz, L.; Komnenovic, V.; Brownfield, L.; Twell, D.; Kakutani, T.; Tamaru, H. Induction of RNA-Directed DNA Methylation upon Decondensation of Constitutive Heterochromatin. *EMBO Rep.* **2009,** *10*, 1015–1021.

Singh, J.; Freelin, M.; Lisch, D. A Position Effect on the Heritability of Epigenetic Silencing. *PLoS Genet.* **2008,** *4* (10), e1000216. DOI: 10.1371/journal.pgen.1000216.

Slotkin, R. K.; Vaughn, M.; Borges, F.; Tanurdzic, M.; Becker, J. D.; Feijo, J. A.; Martienssen, R. A. Epigenetic Reprogramming and Small RNA Silencing of Transposable Elements in Pollen. *Cell* **2009,** *136*, 461–472.

Slotkin, R. K.; Freeling, M.; Lisch, D. Heritable Transposon Silencing Initiated by a Naturally Occurring Transposon Inverted Duplication. *Nat. Genet.* **2005,** *37*, 641–644.

Sridhar, V.; Kapoor, A.; Zhang, K.; Zhu, J.; Zhou, T.; Hasegawa, P. M.; Bressan, R. A.; Zhu, J. K. Control of DNA Methylation and Heterochromatic Silencing by Histone H2B Deubiquitination. *Nature* **2007,** *447* (7145), 735–738.

Stroud, H.; Do, T.; Du, J.; Zhong, X.; Feng, S.; Johnson, L.; Patel, D. J.; Jacobsen, S. E. Non-CG Methylation Patterns Shape the Epigenetic Landscape in Arabidopsis. *Nat. Struct. Mol. Biol.* **2014,** *21* (1), 64–72.

Tran, R. K.; Zilberman, D.; de Bustos, C.; Ditt, R. F.; Henikoff, J. G.; Lindroth, A. M.; Delrow, J.; Boyle, T.; Kwong, S.; Bryson, T. D.; Jacobsen, S. E.; Henikoff, S. Chromatin and siRNA Pathways Cooperate to Maintain DNA Methylation of Small Transposable Elements in Arabidopsis. *Genome Biol.* **2005**, *6* (11), R90. DOI: 10.1186/gb-2005-6-11-r90.

Trojer, P.; Reinberg, D. Facultative Heterochromatin: Is There a Distinctive Molecular Signature? *Mol. Cell* **2007,** *28* (1), 1–13.

Urquiaga, M. C. O.; Thiebaut, F.; Hemerly, A. S.; Ferreira, P. C. G. From Trash to Luxury: The Potential Role of Plant LncRNA in DNA Methylation During Abiotic Stress. *Front. Plant Sci.* **2020,** *11*, 603246. DOI: 10.3389/fpls.2020.603246.

Underwood, C. J.; Henderson, I. R. Martienssen, R. A. Genetic and Epigenetic Variation of Transposable Elements in Arabidopsis. *Curr. Opin. Plant Biol.* **2017,** *36*, 135–141.

Wassenegger, M.; Heimes, S.; Riedel, L.; Sänger, H. L. RNA-Directed De Novo Methylation of Genomic Sequences in Plants. *Cell* **1994,** *76* (3), 567–576. DOI: 10.1016/0092-8674(94)90119-8.

Wood, J. G.; Helfand, H. L. Chromatin Structure and Transposable Elements in Organismal Aging. *Front. Genet.* **2013,** *4* (274), 1–12.

Yamamuro, C.; Miki, D.; Zheng, Z. Yamamuro, C.; Miki, D.; Zheng, Z.; Ma, J.; Wang, J.; Yang, Z.; Dong, J.; Zhu, J. K. Over Production of Stomatal Lineage Cells in Arabidopsis Mutants Defective in Active DNA Demethylation. *Nat. Commun.* **2014,** *5*, 4062. DOI: doi. org/ 10.1038/ncomms5062.

Yelagandula, R.; Stroud, H.; Holec, S.; Zhou, K.; Feng, S.; Zhong, X.; Muthurajan, U. M.; Nie, X.; Kawashima, T.; Groth, M.; Luger, M.; Jacobsen, S. E.; Berger, F. The Histone Variant H2A.W Defines Heterochromatin and Promotes Chromatin Condensation in Arabidopsis. *Cell* **2014,** *158*, 98–109.

Zakrzewski, F.; Schmidt, M.; Van Lijsebettens, M.; Schmidt, T. DNA Methylation of Retrotransposons, DNA Transposons and Genes in Sugar Beet (*Beta vulgaris* L.). *Plant J.* **2017,** *90* (6), 1156–1175.

Zambrano, A. G.; Crevillen, P.; Zorilla, J. M..; López, J. A.; Romero, J. A.; Roszak, P.; González, J. S.; Silvia Jurado, S.; Vázquez, J.; Köhler, C.; Solano, R.; Manuel, P.; Jarillo, J. A. Arabidopsis SWC4 Binds DNA and Recruits the SWR1 Complex to Modulate Histone H2A.Z Deposition at Key Regulatory Genes. *Mol. Plant.* **2018,** *11*, 815–832.

Zemach, A.; Kim, M. Y.; Hsieh, P. H.; Derr, C. D.; Williams, L. E.; Thao, K.; Harmer, S. L.; Zilberman, D. The Arabidopsis Nucleosome Remodeler DDM1 Allows DNA Methyltransferases to Access H1-Containing Heterochromatin. *Cell* **2013,** *153*, 193–205.

Zhang, H.; Zhu, J. K. Active DNA Demethylation in Plants and Animals. *Cold Spring Harb Symp. Quant Biol.* **2012,** *77*, 161–173.

Zhang, H.; Zhaoba, L.; Ian-Kang, Z. Dynamics and Function of DNA Methylation in Plants. *Nat. Rev.* **2018,** *19*, 489–506.

Zhou, W.; Liang, G.; Molloy, PL.; Jones, P. A. DNA Methylation Enables Transposable Element-Driven Genome Expansion. *Proc. Natl. Acad. Sci. U.S.A.* **2020,** *117* (32), 19359–19366. DOI: 10.1073/pnas.1921719117.

Zhu, J. H.; Kapoor, A.; Sridhar, V. V.; Agius, F.; Zhu, J. K. The DNA Glycosylase/Lyase ROS1 Functions in Pruning DNA Methylation Patterns in Arabidopsis. *Curr. Biol.* **2007,** *17*, 54–59.

Zhu, J. K. Active DNA Demethylation Mediated by DNA Glycosylases. *Annu. Rev. Genet.* **2009,** *43*, 143–166.

Zhu, Y. Y.; Rowley, M. J.; Bohmdorfer, G.; Wierzbicki, A. T. SWI/SNF Chromatin-Remodeling Complex Acts in Noncoding RNA-Mediated Transcriptional Silencing. *Mol. Cell.* **2013,** *49*, 298–309.

Zilberman, D.; Coleman-Derr, D.; Ballinger, T.; Henikoff, S. Histone H2A.Z and DNA Methylation Are Mutually Antagonistic Chromatin Marks. *Nature* **2008,** *456*, 125–129.

CHAPTER 12

Transposable Elements and DNA Repair at Cellular Level

ROUF MAQBOOL,[1] HUMAIRA SHAH,[1] TABASUM ASHRAF,[1]
MASARAT BASHIR,[2] and ASHRAF DAR[1*]

[1]*Department of Biochemistry, University of Kashmir, Srinagar, Jammu & Kashmir 190006, India*

[2]*CoTS, Sheri-Kashmir University of Agricultural Sciences and Technology, Shalimar, Jammu & Kashmir, India*

[*]*Corresponding author. E-mail: ashrafdar@kashmiruniversity.ac.in*

ABSTRACT

Transposable elements (TEs), also called transposons, are fragments of DNA present in both prokaryotic and eukaryotic genomes that have ability to mobilize. The mobility of these TEs often results in the multiplication of their sequences across the host genomes. During the process of mobilization, class II TEs (cut and paste transposons) inherently cause damage in DNA both at doner excision and at integration sites. The degree of DNA damage caused by transposition varies from single-nucleotide mismatch to the severe double-stranded breaks. Most, but not all, class II TEs encode transposase enzymes. These enzymes cleave the DNA at two ends of TE for its excision from the parent genomic locus. The transposase also cuts the DNA at the target site where TE is to be integrated. Class III TEs (or miniature inverted-repeat TEs) amplify in parent genome at the pattern of class II TEs. The class I TEs or retrotransposons (copy and paste transposons) are copied via an RNA intermediate and pasted at target site. The class I TEs with the help of retrotransposon

Plant Transposable Elements: Biology and Biotechnology, Deepu Pandita, Anu Pandita, (Eds.)

encoded integrase enzyme are integrated at the target site. The DNA damage at the excision or integration sites caused by the transposition enzymes is repaired by various cellular DNA repair machineries. In this chapter, we will sum up the role played by several DNA repair pathways in repairing DNA damage at excision and integration sites caused during the transposition process.

12.1 INTRODUCTION

Transposable elements (TEs) are the mobile DNA sequences or jumping genes that migrate from one position to the other in the same genome. TEs were first discovered in maize (corn) in 1950s by McClintock (1950). Since then, they have been found in almost all organisms from simple bacteria to the more complex humans. TEs make a large portion of eukaryotic genomes. In maize, ~90% of genome is comprised of TEs, whereas ~50% of human genome is believed to have evolved from TEs (SanMiguel et al., 1996). The whole phenomenon of transposition can have a significant impact on the genetic information of the host. The deletion (excision), inversion, or integration of TEs can alter the host DNA in such a manner that can further promote changes in gene expression. Some of the TEs possess outwardly firing promoters and their integration at the new locus can stimulate the expression of downstream genes (Trizzino et al., 2017). There are TEs that can also play a role in DNA rearrangements by acting as substrates for different recombination reactions. Homologous recombination (HR) between TEs at different genomic loci can lead to deletions, inversions, and replicon fusions. Therefore, TEs serve as devices of genomic alteration and genetic diversity in host organisms (Payer and Burns, 2019).

At present three categories of TEs are known. These include class II TEs (DNA transposons), class III TEs (miniature inverted-repeat transposable elements or MITEs), and class I TEs (retrotransposons) (Szuplewska et al., 2015). Class II TEs or cut and paste transposons excise at one genomic locus and integrate at a different locus. These TEs are known to be flanked by the terminal inverted repeats (TIRs) followed by target site duplications (TSDs). The class II TEs in most cases encode transposase enzyme that cleaves the ends of TE at parental location to release it and integrate it at a new genomic locus. Before integration of released TE, the transposase also cuts the DNA at target site, so that released TE is ligated there (Han et al., 2013) (Fig. 12.1). The TEs that encode their own transposase required for their transposition reaction are known as autonomous transposons. In case of nonautonomous transposons, the TEs themselves do not encode a transposase enzyme but depend on the transposase encoded elsewhere in the genome for their transposition (Pray, 2008).

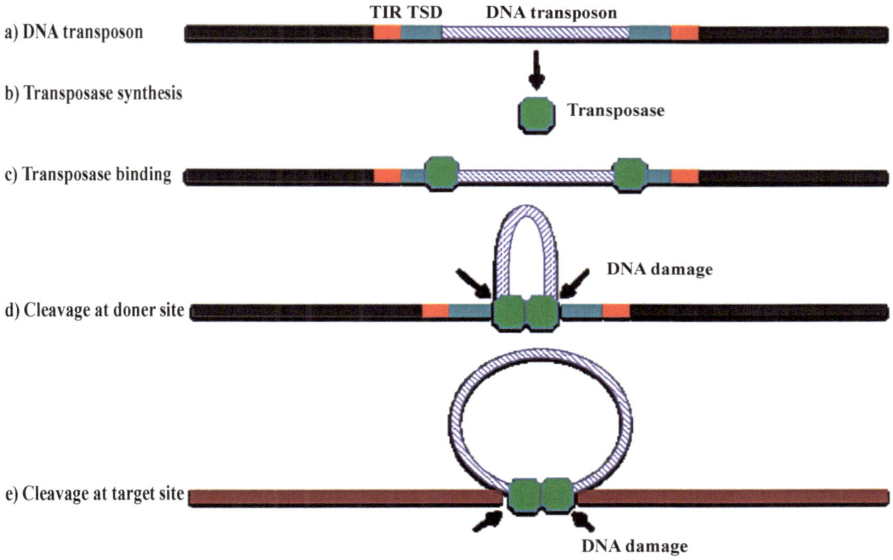

FIGURE 12.1 Schematic representation and transposition of class II TEs.

Class III TEs (MITEs) are very small mobile DNA elements of 50–500 bp length found in plants, bacteria, fungi, and animals. Like class II TEs, MITEs are characterized by 10–15 bp TIRs and two flanking target site duplications (Fig. 12.2) (Hu et al., 2018; Fattash et al., 2013). MITEs are nonautonomous transposons and do not encode their own transposase. They use a transposase enzyme synthesized elsewhere for transposition reaction. MITEs are very abundant and have usually been found in gene-rich areas of the chromosomes and as a result they affect expression of adjacent genes (Yang et al., 2005; Santiago et al., 2002).

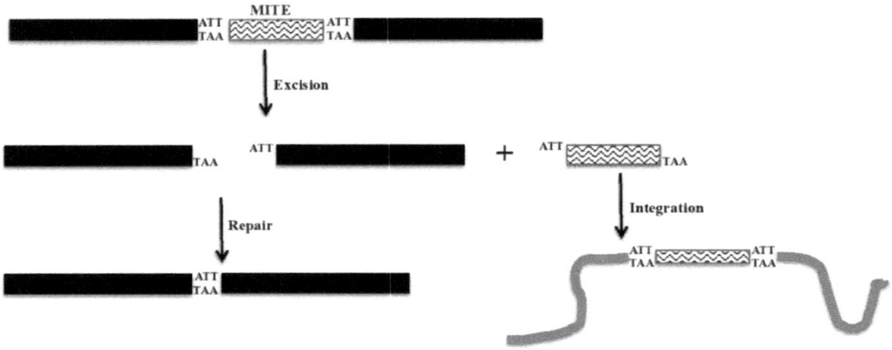

FIGURE 12.2 Schematic representation and transposition of class III TEs or MITEs.

Unlike class II and class III, class I TEs (or retrotransposons) propagate via an RNA intermediate. These copy and paste TEs reverse transcribe the intermediate RNA into DNA by reverse transcription. The DNA produced is then integrated elsewhere in the genome (Fig. 12.3). Retrotransposons are only found in eukaryotic organisms and share significant features with retrovirus like HIV (Drost and Sanchez, 2019; Sanchez et al., 2017). Two sub types of retrotransposons are long terminal repeat (LTR) retrotransposons and non-LTR retrotransposons. The LTR retrotransposons are ~5 kb long and contain LTRs of few hundred bases at ends. Between LTRs in LTR retrotransposons are genes that encode for retrovirus-like proteins gag and pol. Pol gene product possesses reverse transcriptase, integrase, and ribonuclease H activities. These enzyme activities are required for reverse transcription, retrotransposon integration, and phosphodiester bond cleavage between ribonucleotides (Orozco-Arias et al., 2019; Wicker et al., 2007). Lacking terminal repeats, non-LTR retrotransposons house various genes that code for nuclease, reverse transcriptase, RNA binding protein, and ribonuclease H enzymes. These non-LTR retrotransposon encoded enzymatic activities are required for transposition of non-LTR retrotransposons.

The mobilization of TEs generates DNA damage both at the excision and integration sites. None of the enzyme activities encoded by the transposons have DNA damage repair function. The DNA repair machineries in the host cells essentially plays a key role in repairing the broken DNA ends at excision and integration sites of TEs. This highlights the significance of cellular repair machinery in transposition process and therefore maintaining the integrity of host genome.

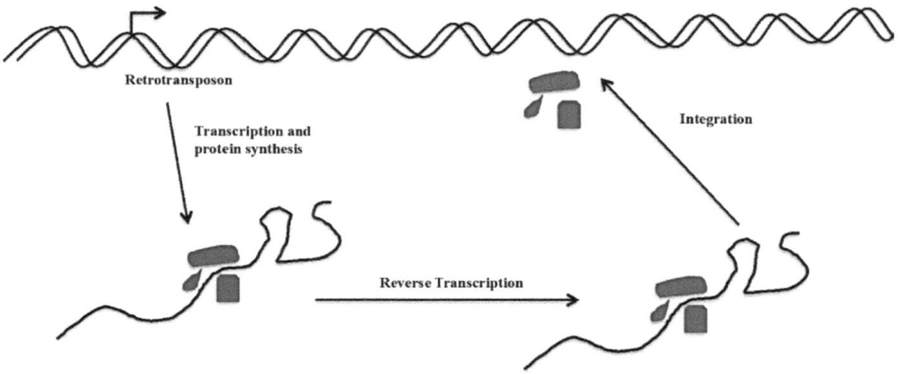

FIGURE 12.3 Schematic representation of transposition of class I TEs or retrotransposons.

12.2 TRANSPOSITION GENERATES DNA DAMAGE

The molecular mechanism of transposition involves cleavage of DNA at the sites of excision for class II and class III TEs and at the sites of integration in case of class II, class III, and class I transposons. The transposition reactions, thus, inherently cause DNA damage in genome. The excision of class II TEs generate severe double-strand breaks (DSBs) at the site of excision (Hickman and Dyda, 2016). There is no universal pattern of DSBs generated by excision of different TEs. There are TEs, for example, *Tn*5 whose excision creates a blunt-ended DSB. The excision of *P Element* and *Sleeping Beauty* create a 17 bases and 3 bases staggered cuts at their respective doner sites (Eileen and Donald, 1997; Luo et al., 1998). Several secondary structures are produced at the excision site of the TEs. A hairpin is formed at *h*AT element excision site. *Sleeping Beauty* and *Minos* generates strong heteroduplexes (Arca et al., 1997). The integration of the class I TEs at the new sites is believed to generate single-stranded DNA gaps on either side of the newly incorporated elements (Yoder and Bushman, 2000). Nevertheless, these intermediate arrangements of class I TEs may resemble and be regarded as DSBs by the cellular machinery (Gasior et al., 2006). Further, these single-stranded gaps either become an impediment for approaching replication fork or are transformed into DSBs when the incoming replication fork counters with these single-stranded breaks (Rothstein et al., 2000). It is believed that structurally different kinds of DNA lesions produced by TEs can activate different signaling mechanisms and repair pathways in the host cell for repairing the damage in DNA caused by transposition.

12.3 OVERVIEW OF DNA REPAIR

To maintain the genomic stability constantly, cells have evolved multiple pathways for repairing damaged DNA. In eukaryotic cells, the two major mechanistic pathways involved in repairing DSBs are HR and NHEJ (nonhomologous end joining). During HR, the repair of a broken DNA molecule is guided by undamaged and intact homologous DNA template. On the other hand, in NHEJ, the repair of broken DNA ends takes place without the help of homologous DNA sequences (Davis and Chen, 2013). The recent understanding about HR pathway is that five different molecular complexes are required for sensing and initiating signaling cascade in different types of damages in DNA (Harper and Elledge, 2007). The molecular complexes involved activate a diverse set of protein kinases (PIKKs) including phosphoinositide-3-kinase-related PIKK family members ATM, ATR, and DNA-PKcs. These PIKKs play a key role in transducing signaling cascade. It is well known that the presence of single-stranded DNA and stalled replication forks activate ATR (Saldivar et al., 2017). The DSBs are known to activate ATM and DNA-PKcs (Blackford and Jackson, 2017). Downstream to PIKK kinases are mediator proteins, that are important for recruitment of additional factors at damage sites. These mediators include H2AX, 53BP1, the MRN complex (Mre11/Rad50/NBS1), Mdc1, Claspin, Brit1/Mcph1, and BRCA1 (Falck et al., 2005; Harper and Elledge, 2007; Rogakou et al., 1998; Stucki and Jackson, 2006) (Fig. 12.4).

In NHEJ pathway, several factors are involved in repairing damaged DNA ends. These include Ku70/Ku80 heterodimeric complex, DNA-dependent PIKK catalytic subunit (DNA-PKcs), Artemis, and DNA ligase IV/XRCC4 complex (Drouet et al., 2006). When there is a DSB, Ku heterodimers are recruited to it, which helps in aligning the two DNA termini. The Ku complex further stimulates the recruitment of DNA-PKcs, thereby facilitating the formation of an active DNA-PK complex (Chang et al., 2017). The DNA-PKcs act as template for recruitment of Artemis, which possesses 5′-3′ exonuclease activity (Goodarzi et al., 2006). The DNA-PKcs phosphorylate Artemis and activate it. This phosphorylation of Artemis is important for stimulation of its endonucleolytic activity on 5′ and 3′ DNA overhangs, as well as on hairpins (Chang et al., 2017). Next, the MRN complex, which possesses nuclease and DNA end-bridging activity is recruited to align DNA ends in a synaptic complex. Finally, the ligation of well-aligned broken ends is carried out by DNA ligase IV/XRCC4 complex (Chen et al., 2001; Huang and Dynan, 2002) (Fig. 12.5).

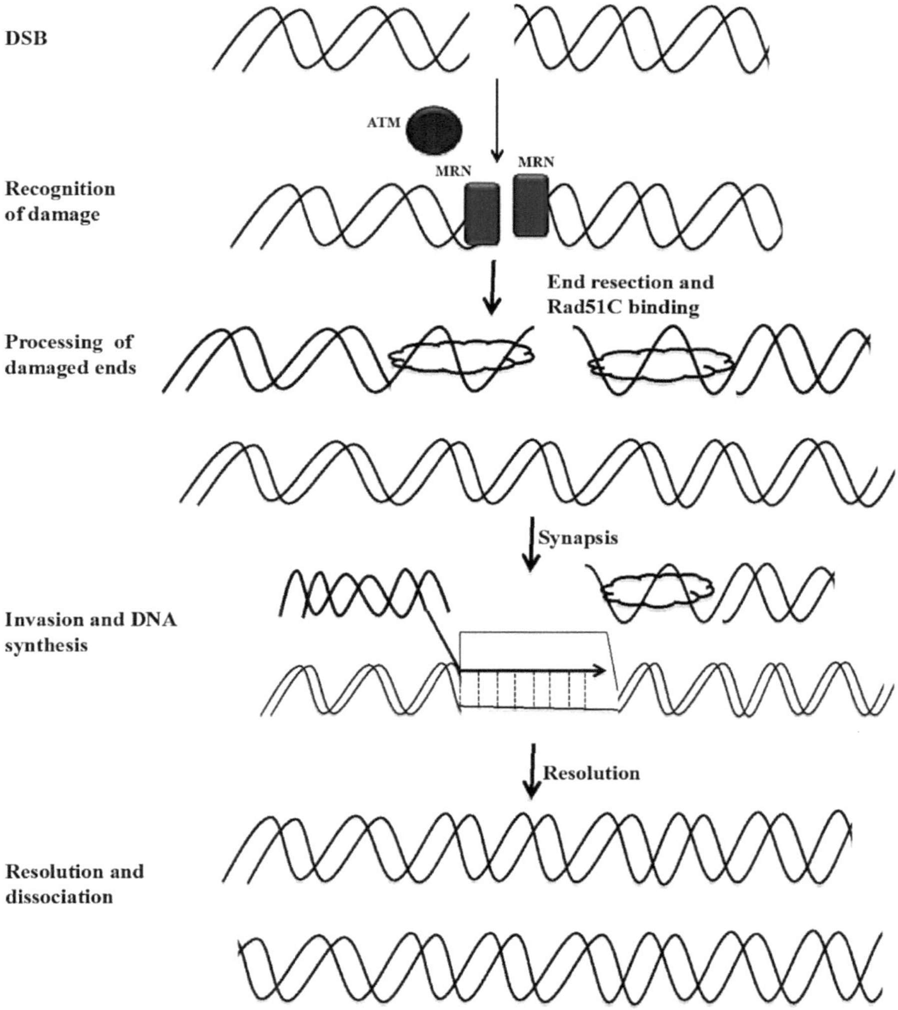

FIGURE 12.4 Homologous recombination.

12.3.1 REPAIR OF DAMAGE AT THE EXCISION SITE CAUSED BY TRANSPOSITION OF CLASS II TEs

A very prominent member in class II TEs is *Sleeping Beauty*. It belongs to *Tc*1 superfamily of TEs. This subfamily probably includes the most widespread TEs in nature. The DSB repair machinery of the host cell (both NHEJ and HR pathways) are important for efficient transposition of the *Sleeping Beauty* element. The cells that are deficient in NHEJ pathway and wild-type

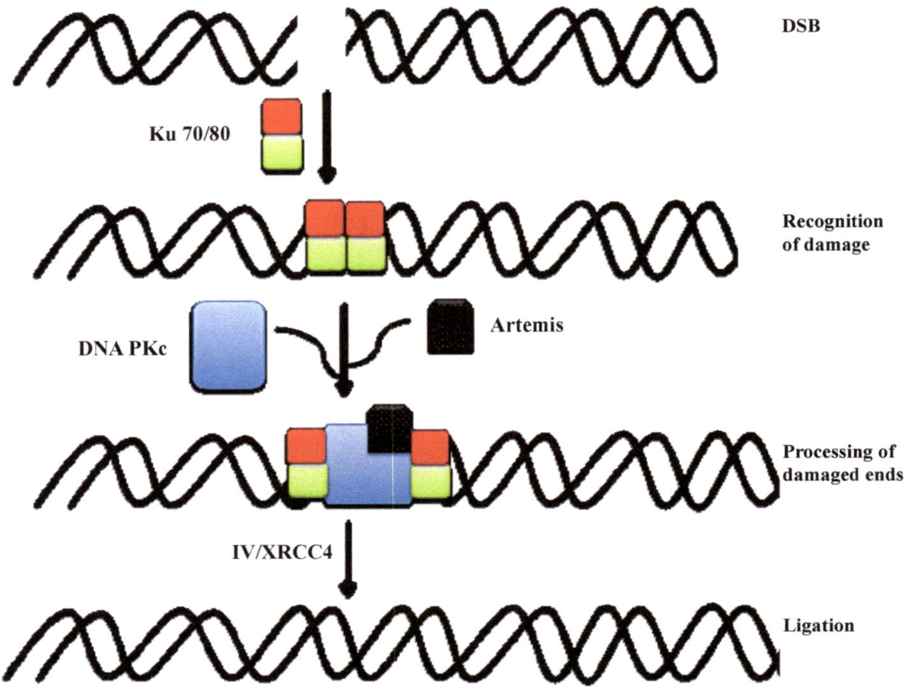

FIGURE 12.5 Nonhomologous end joining (NHEJ).

cells where expression of DNA-PKcs is defective, exhibit very low levels in *Sleeping Beauty* transposition. On the other hand, forced expression of DNA-PKcs in either mutant or wild-type cells greatly enhance the transposition efficiency of *Sleeping Beauty* element (Izsvak et al., 2004). DNA-PKcs may thus be regarded as a rate-limiting factor in transposition process. Different vertebrate cells significantly vary in the expression levels of DNA-PKcs. This variation in expression of DNA-PKcs may be a reason for the differential transpositional efficiencies of *Sleeping Beauty* in vertebrate cells (Izsvak et al., 2000; Durocher and Jackson, 2001). Ku interacts with *Sleeping Beauty* transposase. Moreover, Ku is known to associate with preintegration complexes needed for retroviral integration along with host genome (Li et al., 2001). In addition, Ku forms a complex with virus-like particles during retrotransposition of *Ty1* element in yeast (Downs and Jackson, 1999). Furthermore, a homolog of Ku has been recently purified from the phage Mu, a bacterial transposon (d'Adda di Fagagna et al., 2003). All these studies lead to the conclusions that NHEJ machinery has an active role in diverse transpositional reactions.

Several studies suggest that the process of transposition is not absolutely dependent on NHEJ. The repair of damage at doner excision sites can involve alternative pathway that requires a homology-dependent process. In NHEJ deficient cells, long fragments of transposon sequences containing microhomology at the junctions have been observed. Also, the process of repair at the excision site depends on the presence of XRCC3 (Pierce et al., 1999). The overall structural features of these repairing footprints resemble with repair synthesis by a mechanism called synthesis-dependent strand annealing (SDSA) (Nassif et al., 1994). The strand annealing in SDSA process is further followed by microhomology-directed end joining (Adams et al., 2003). The SDSA pathway engages the template provided by the homologous chromosome to repair the DSB. In addition to XRCC3, Rad51C is also involved in *Sleeping Beauty* transposition. XRCC2 and XRCC3 form separate complexes with Rad51C. Some studies suggest that Rad51C/XRCC3 complex is specifically involved in transposition process (Izsvak et al., 2004).

In *Drosophila,* both NHEJ and HR pathways are involved in the repair of DSBs produced by transposition of *P-element* (Min et al., 2004), *Tc*1 excision in *Caenorhabditis elegans* (Izsvak et al., 2004., Yant and Mark, 2003), and in integration of Ty1 (retrotransposon) in yeast (Boeke et al., 1985; Curcio et al., 2015; Sharon et al., 1994). These experimental studies suggest that NHEJ and HR may have overlapping roles in transposition of different TE elements.

12.3.2 REPAIR OF DAMAGE AT INTEGRATION SITE CAUSED BY TRANSPOSITION OF CLASS II TES

In comparison to repair events at TE excision site, not much is known about the repair of damage at integration site of TEs. It is believed that the host repair machinery is involved in the repair process at these sites. The integration of TE is presumed to produce an intermediary structure that has single-stranded gaps at ends. The repair of these single-stranded gaps reinstalls the terminal nucleotide sequences of the newly incorporated TE, and produces a characteristic duplication of the target site (target site duplication, TSD) on both sides of inserted element (Richardson and Jasin, 2000). In case there is a failure in repairing the gaps produced at the integration site, it is presumed that the damage is attenuated by cellular factors like PRR and RecQ helicase (Wu and Hickson, 2003). PRR is known to provide damage tolerance at the gapped transposition intermediates in order to dodge replication arrest (Hancks and Kazazian, 2016). There are certain transposases that interact with members of PRR. These interactions thus support the fact that PRR may have a role in neutralizing the damage caused by integration of TEs. The Pogo transposase in *Drosophila* has been found to associate with PCNA (Warbrick et al., 1998). The human counterpart of Pogo, called Tigger, synthesizes a transposase containing a conserved PCNA-interacting domain (Warbrick et al., 1998). Lemi1 that is a Pogo-like TE in *Arabidopsis* also contains a PCNA-interacting domain, suggesting that interaction of Pogo family with PCNA is a general feature in all organisms. Since PCNA has role in DSB repair, it may be assumed that DSB repair pathways also play a role in repair of damage at integration sites of TEs (Feschotte and Mouches, 2000).

12.3.3 REPAIR OF DAMAGE CAUSED BY TRANSPOSITION OF LTR RETROTRANSPOSONS

The mechanism of retrotransposition is presently not well understood. However, retrotransposon encoded enzyme integrase, which possesses an endonuclease activity, cleaves the DNA at the site of integration and produces breaks with staggered ends (Nefedova and Kim, 2017). The RNA transcript synthesized from the LTR TE interacts with the cleaved DNA ends at the site of integration. One of the DNA strands at cleaved site (at integration site) acts as the primer for reverse transcriptase to produce cDNA from RNA transcript (Boeke et al., 1985; Finnegan, 2012). How LTR-retrotransposon encoded cDNA is further converted to a double-stranded DNA element and inserted at a staggered break at the target site is not well understood. It is presumed that the DNA repair machinery of the host cell plays a critical role in the integration process.

In yeast, the transposition of the *Ty1* LTR retrotransposon is repressed by almost 30 proteins that have a role in maintaining genomic integrity. Some of these proteins like Rad3, Ssl2, nucleotide excision repair helicase (TFII), Rad18, Rad52, and Rad27/Fen1 (Lee et al., 2000; Bryk et al., 2001; Rattray et al., 2000; Sundararajan et al., 2003) have orthologs in mammals that restrict retroviral integration (Yoder et al., 2006; Lau et al., 2004). The yeast proteins like Rad3, TFIIH and Ssl2, or Rad27/Fen1 are involved in blocking *Ty1* cDNA accumulation (Lee et al., 2000). Alternatively, when these factors are silenced in yeast, the physical stability of *Ty1* cDNA is elevated, leading to increased *Ty1* retrotransposition. In humans, TFIIH complex proteins of nucleotide excision repair pathway are involved in destruction of retroviral cDNA (Yoder et al., 2006). Similarly, in mammalian cells with MMR deficiency, retroviral integration is highly compromised (Lin et al., 2001).

The genomic integration of *Ty1* requires free nucleotides present at the termini of retrotransposon cDNA. Some studies have shown that integrase deficient *Ty1* element efficiently integrates into the host genome. Although Rad52-mediated HR plays a role in inhibiting wild-type *Ty1* element transposition, the pathway is necessary for the integrase-independent retrotransposition of *Ty1*. It is possible that Rad52 and integrase mutually compete for the cDNA and when competing integrase is absent, *Ty1* cDNA is used in HR with the parental genomic *Ty1* element (Sharon et al., 1994; Rattray et al., 2000; Radford et al., 2004).

Besides HR, the NHEJ pathway plays a role in retrotransposition. Ku of the NHEJ pathway has been found to interact with *Ty1*. In the absence of Ku, the transposition frequency of *Ty1* significantly decreases. In addition, RAD6 has been shown to influence retrotransposition of Ty1 in yeast. These experimental observations suggest the role of NHEJ in *Ty1* transposition (Downs and Jackson, 1999; Freiberg et al., 2000; Ulrich, 2002).

12.3.4 REPAIR OF DAMAGE CAUSED BY TRANSPOSITION OF NON-LTR RETROTRANSPOSONS

Unlike LTR-retrotransposons, non-LTR retrotransposons use a different mechanism for their transposition, called target-site primed reverse transcription (TPRT). TPRT is dependent on endonuclease and reverse transcriptase enzyme activities of a protein encoded by non-LTR retrotransposon (Cost et al., 2002; Feng et al., 1996; Luan et al., 1993). The endonuclease enzyme cuts one strand of host DNA. The cleaved DNA end then acts as primer for reverse transcription, and reverse transcriptase copies the retrotransposon RNA template directly into the host genome. The non-LTR retrotransposons integrate 3′ end of their cDNA with the genome, whereas, the integration of 5′ end is understood to be carried out by HR machinery. The TPRT model suggests that integration of non-LTR retrotransposon needs a double strand cleavage at the integration site (Gilbert et al., 2002; Ostertag and Kazazian, 2001). Studies have demonstrated that enormous DSB signaling occurs during L1 retrotransposition (Gasior et al., 2006). There are evidences that connect NHEJ machinery to the regulation of L1 transposition. Binding sites for Ku70/80 have been discovered in mouse L1. It is intriguing to know that L1s constitute about 19% of the mouse genome, but accommodate approximately 26% of the Ku70/80 binding sites (Katz et al., 2005). The cells that lack DNA-PKcs show decreased levels of endonuclease-dependent L1 retrotransposition. On contrary, XRCC4 deficient cells have higher rates of L1 retrotransposition (Morrish et al., 2002). The deficiency in repair enzymes may impact retrotransposition of L1 via multiple pathways. The prolonged presence of unrepaired DSBs could serve as templates for insertion and increased endonuclease-independent insertion of LTR retrotransposons (Farkash and Luning, 2006).

12.3.5 REPAIR OF DAMAGE CAUSED BY TRANSPOSITION OF MITES

The TEs and their associated MITEs use "cut and paste" method for their transposition. The transposase recruits to the TIRs and is required for both excision and insertion of these small-sized TEs (Yuan and Wessler, 2011). The cutting of the genomic DNA at the integration sites produces 5′ or 3′ staggered ends with 2 or 3bp overhangs. These staggered ends create small TSDs that flank the new MITE at insertion site. The two best studied MITE families are *Tourist* and *Stowaway*. The transposase associated with *Stowaway*-like MITEs produces a 2 bp target site duplication (TSD) upon integration. Usually, excision of these MITEs leave small insertions or deletions (footprint) at their doner site (Kikuchi et al., 2003; Nakazaki et al., 2003). *Tourist*-like MITEs produce a 3 bp TSD, and leave no footprint at their doner site (Zhang et al., 2004). Experimental studies have demonstrated that *Stowaway*-like MITE, *14T32* and *Toursist*-like MITE, *mPing* elements require NHEJ pathway for repair of damage at excision and integration sites (Gilbert et al., 2015).

12.4 CONCLUDING REMARKS

TEs are best known for their mobility in the genome. Decades ago Barbara McClintock had hypothesized that transposition may cause genomic instability or conversely genomic instability may promote transposition. TEs have coexisted in prokaryotic and eukaryotic genomes for thousands of years, playing a significant role in genetic diversity and evolution. The process of transposition may be regarded as destructive because of DNA damage inflicted by movement of TEs from one place of genome to the other.

However, that is not the case as host cells have evolved strategies to repair the damage and maintain the overall integrity of genome. The involvement of host repair machinery in repairing transposition-mediated DNA damage signifies the evolutionary adaptation and tolerance for TE by host. TEs have proved both advantageous and disadvantageous to the host organisms. The TE-mediated increase in gene expression may allow certain plants and crops to adapt to changed environmental or drought-like conditions. On the other hand, TEs have been implicated in promotion of certain diseases like cancer. There is a lot of scope to understand the mechanism of transposition and the role played by TEs in regulation of different physiological pathways that in turn determine the phenotypic traits in organisms.

KEYWORDS

- **transposable elements**
- **retrotransposons**
- **MITEs**
- **DNA damage**
- **DSB**
- **homologous recombination**
- **NHEJ**

REFERENCES

Adams, M. D.; McVey, M.; Sekelsky, J. J. Drosophila BLM in Double-Strand Break Repair by Synthesis Dependent Strand Annealing. *Science* **2003,** *299* (5604), 265–267.

Arca, B.; Zabalou, S.; Loukeris, T. G.; Savakis, C. Mobilization of a Minos Transposon in Drosophila Melanogaster Chromosomes and Chromatid Repair by Heteroduplex Formation. *Genetics* **1997,** *145* (2), 267–279.

Beall, E. L.; Rio, D. C. Drosophila P-Element Transposase Is a Novel Site-Specific Endonuclease. *Genes Dev.* **1997,** *11* (16), 2137–2151.

Blackford, A. N.; Jackson, S. P. ATM, ATR, and DNA-PK: The Trinity at the Heart of the DNA Damage Response. *Mol. Cell* **2017,** *66*, 801–817.

Boeke, J. D.; Garfinkel, D. J.; Styles, C. A.; Fink, G. R. Ty Elements Transpose Through an RNA Intermediate. *Cell* **1985,** *40*(3), 491–500.

Bryk, M.; Banerjee, M.; Conte, D. Jr.; Curcio, M. J. The Sgs1 Helicase of Saccharomyces Cerevisiae Inhibits Retrotransposition of Ty1 Multimeric Arrays. *Mol. Cell Biol.* **2001,** *21* (16), 5374–5388.

Chang, H. H. Y.; Pannunzio, N. R.; Adachi, N.; Lieber, M. R. Non-Homologous DNA End Joining and Alternative Pathways to Double-Strand Break Repair. *Nat. Rev. Mol. Cell Biol.* **2017,** *18* (8), 495–506.

Chen, L.; Trujillo, K.; Ramos, W.; Sung, P.; Tomkinson, A. E. Promotion of Dnl4-Catalyzed DNA End-Joining by the Rad50/Mre11/Xrs2 and Hdf1/Hdf2 Complexes. *Mol. Cell* **2001,** *8* (5), 1105–1115.

Cost, G. J.; Feng, Q.; Jacquier, A.; Boeke, J. D. Human L1 Element Target-Primed Reverse Transcription In Vitro. *EMBO J.* **2002,** *21* (21), 5899–5910.

Curcio, M. J.; Lutz, S.; Lesage, P. The Ty1 LTR-Retrotransposon of Budding Yeast, Saccharomyces Cerevisiae. *Microbiol. Spectr.* **2015,** *3* (2), 1–35.

d'Adda di Fagagna, F.; Weller, G. R.; Doherty, A. J.; Jackson, S. P. The Gam Protein of Bacteriophage Mu Is an Orthologue of Eukaryotic Ku. *EMBO Rep.* **2003,** *4* (1), 47–52.

Davis, A. J.; Chen, D. J. DNA Double Strand Break Repair via Non-Homologous End-Joining. *Trans. Cancer Res.* **2013,** *2* (3), 130–143.

Downs, J. A.; Jackson, S. P. Involvement of DNA End-Binding Protein Ku in Ty Element Retrotransposition. *Mol. Cell Biol.* **1999,** *19* (9), 6260–6268.

Drost, H. G.; Sanchez, D. H. Becoming a Selfish Clan: Recombination Associated to Reverse-Transcription in LTR Retrotransposons. *Genome Biol. Evol.* **2019,** *11* (12), 3382–3392.

Drouet, J.; Frit, P.; Delteil, C.; Villartay, J. P.; Salles, B.; Calsou, P. Interplay Between Ku, Artemis, and the DNA-Dependent Protein Kinase Catalytic Subunit at DNA Ends. *J. Biol. Chem.* **2006,** *281* (38), 27784–27793.

Durocher, D.; Jackson, S. P. DNA-PK, ATM and ATR as Sensors of DNA Damage: Variations on a Theme? *Curr. Opin. Cell Biol.* **2001,** *13* (2), 225–231.

Farkash, E. A.; Luning, P. DNA Damage and L1 Retrotransposition. *J. Biomed. Biotechnol.* **2006,** *2006* (1), 37285.

Falck, J.; Coates, J.; Jackson, S. P. Conserved Modes of Recruitment of ATM, ATR and DNA-PKcs to Sites of DNA Damage. *Nature* **2005,** *434* (7033), 605–611.

Fattash, I.; Rooke, R.; Wong, A.; Hui, C.; Luu, T.; Bhardwaj, P.; Yang, G. Miniature Inverted-Repeat Transposable Elements: Discovery, Distribution, and Activity. *Genome* **2013,** *56* (9), 475–486.

Feng, Q.; Moran, J. V.; Kazazian, H. H. Jr., Boeke, J. D. Human L1 Retrotransposon Encodes a Conserved Endonuclease Required for Retrotransposition. *Cell* **1996,** *87* (5), 905–916.

Feschotte, C.; Mouches, C. Evidence that a Family of Miniature Inverted-Repeat Transposable Elements (MITEs) from the Arabidopsis Thaliana Genome Has Arisen from a Pogo-Like DNA Transposon. *Mol Biol Evol.* **2000,** *17*(5), 730–737.

Finnegan, D. J. Retrotransposons. *Curr. Biol.* **2012,** *22* (11), R432–R437.

Freiberg, G.; Mesecar, A. D.; Huang, H.; Hong, J. Y.; Liebman, S.W. Characterization of Novel rad6/ubc2 Ubiquitin-Conjugating Enzyme Mutants in Yeast. *Curr. Genet.* **2000,** *37*, 221–233.

Gasior, S. L.; Wakeman, T. P.; Xu, B.; Deininger, P. L. The Human LINE-1 Retrotransposon Creates DNA Double-Strand Breaks. *J. Mol. Biol.* **2006,** *357* (5), 1383–1393.

Gilbert, D. M.; Bridges, M. C.; Strother, A. E.; Burckhalter, C. E.; Burnette, J. M.; Hancock, C. N. Precise Repair of mPing Excision Sites is Facilitated by Target Site Duplication Derived Microhomology. *Mob. DNA* **2015,** 6–15.

Gilbert, N.; Lutz-Prigge, S.; Moran, J. V. Genomic Deletions Created upon LINE-1 Retrotransposition. *Cell* **2002,** *110* (3), 315–325.

Goodarzi, A. A.; Yu, Y.; Riballo, E.; Douglas, P.; Walker, S. A.; Ye, R.; Harer, C.; Marchetti, C.; Morrice, N.; Jeggo, P. A.; Lees-Miller, S. P. DNA-PK Autophosphorylation Facilitates Artemis Endonuclease Activity. *EMBO J.* **2006,** *25* (16), 3880–3889.

Hancks, D. C.; Kazazian, H. H. Roles for Retrotransposon Insertions in Human Disease. *Mob. DNA* **2016,** *7*, 9.

Han, Y., Qin, S.; Wessler, S. R. Comparison of Class 2 Transposable Elements at Superfamily Resolution Reveals Conserved and Distinct Features in Cereal Grass Genomes. *BMC Genom.* **2013,** *14*, 71.

Harper, W.; Elledge, S. J. The DNA Damage Response: Ten Years After. *Mol. Cell.* **2007,** *28* (5), 739–745.

Hickman, A. B.; Dyda, F. DNA Transposition at Work. *Chem. Rev.* **2016,** *116* (20), 12758–12784.

Huang, J.; Dynan, W. S. Reconstitution of the Mammalian DNA Double-Strand Break End-Joining Reaction Reveals a Requirement for an Mre11/Rad50/NBS1-Containing Fraction. *Nucl. Acids Res.* **2002,** *30* (3), 667–674.

Hu, J.; Zheng, Y.; Shang, X. MiteFinderII: A Novel Tool to Identify Miniature Inverted-Repeat Transposable Elements Hidden in Eukaryotic Genomes. *BMC Med. Genom.* **2018,** *11* (Suppl 5).

Izsvak, Z.; Ivics, Z.; Plasterk, R. H. Sleeping Beauty, a Wide Host-Range Transposon Vector for Genetic Transformation in Vertebrates. *J. Mol. Biol.* **2000**, *302* (1), 93–102.

Izsvak, Z.; Stuwe, E. E.; Fiedler, D.; Katzer, A.; Jeggo, P. A.; Ivics, Z. Healing the Wounds Inflicted by Sleeping Beauty Transposition by Double-Strand Break Repair in Mammalian Somatic Cells. *Mol. Cell.* **2004,** *13* (2), 279–290.

Katz, D. J.; Beer, M. A.; Levorse, J. M. Tilghman, S. M. Functional Characterization of a Novel Ku70/80 Pause Site at the H19/Igf2 Imprinting Control Region. *Mol. Cell Biol.* **2005,** *25* (10), 3855–3863.

Kikuchi, K.; Terauchi, K.; Wada, M.; Hirano, H. Y. The Plant MITE mPing is Mobilized in Anther Culture. *Nature* **2003,** *421*, 167–170.

Lau, A.; Kanaar, R.; Jackson, S. P.; O'Connor, M. J. Suppression of Retroviral Infection by the RAD52 DNA Repair Protein. *EMBO J.* **2004,** *23* (16), 3421–3429.

Lee, B. S.; Bi, L.; Garfinkel, D. J.; Bailis, A. M. Nucleotide Excision Repair/TFIIH Helicases RAD3 and SSL2 Inhibit Short-Sequence Recombination and Ty1 Retrotransposition by Similar Mechanisms. *Mol. Cell Biol.* **2000,** *20* (7), 2436–2445.

Li, L.; Olvera, J. M.; Yoder, K. E.; Mitchell, R. S.; Butler, S. L.; Lieber, M.; Martin, S. L.; Bushman, F. D. Role of the Non-Homologous DNA End Joining Pathway in the Early Steps of Retroviral Infection. *EMBO J.* **2001,** *20* (12), 3272–3281.

Lin, C. T.; Lyu, Y. L.; Xiao, H.; Lin, W. H.; Whang-Peng, J. Suppression of Gene Amplification and Chromosomal DNA Integration by the DNA Mismatch Repair System. *Nucl. Acids Res.* **2001,** *29* (16), 3304–3310.

Luan, D. D.; Korman, M. H.; Jakubczak, J. L.; Eickb, T. H. Reverse Transcription of R2Bm RNA Is Primed by a Nick at the Chromosomal Target Site: A Mechanism for Non-LTR Retrotransposition. *Cell* **1993,** *72* (4), 595–560

Luo, G.; Ivics, Z.; Izsvák, Z.; Bradley, A. Chromosomal Transposition of a Tc1/Mariner-Like Element in Mouse Embryonic Stem Cells. *Proc. Natl. Acad. Sci. U.S.A.* **1998,** *95* (18), 10769–10773.

McClintock, B. The Origin and Behaviour of Mutable Loci in Maize. *Proc. Nat. Acad. Sci. U.S.A.* **1950,** *36*, 344–355.

Min, B.; Weinert, B. T.; Rio, D. C. Interplay Between Drosophila Bloom's Syndrome Helicase and Ku Autoantigen During Nonhomologous End Joining Repair of P Element-Induced DNA Breaks. *Proc. Natl. Acad. Sci. U.S.A.* **2004,** *101* (24), 8906–89011.

Morrish, T. A.; Gilbert, N.; Myers, J. S.; Vincent, B. J.; Stamato, T. D.; Taccioli, G. E.; Batzer, M. A.; Moran, J. V. DNA Repair Mediated by Endonuclease-Independent LINE-1 Retrotransposition. *Nat. Genet.* **2002,** *31* (2), 159–165.

Nakazaki, T.; Okumoto, Y.; Horibata, A.; Yamahira, S.; Teraishi, M.; Nishida, H., Inoue, H.; Tanisaka, T. Mobilization of a Transposon in the Rice Genome. *Nature* **2003,** *421*, 170–172.

Nassif, N.; Penney, J.; Pal, S.; Engels, W. R.; Gloor, G. B. Efficient Copying of Nonhomologous Sequences from Ectopic Sites via P-Element-Induced Gap Repair. *Mol. Cell Biol.* **1994,** *14* (3), 1613–1625.

Nefedova, L.; Kim, A. Mechanisms of LTR-Retroelement Transposition: Lessons from Drosophila Melanogaster. *Viruses* **2017,** *9* (4), 81.

Orozco-Arias, S.; Isaza, G.; Guyot, R. Retrotransposons in Plant Genomes: Structure, Identification, and Classification through Bioinformatics and Machine Learning. *Int. J. Mol. Sci.* **2019,** *20* (15), 3837.

Ostertag, E. M.; Kazazian, H. H. Jr. Twin Priming: A Proposed Mechanism for the Creation of Inversions in L1 Retrotransposition. *Genome Res.* **2001,** *11* (12), 2059–2065.

Payer, L. M.; Burns, K. H. Transposable Elements in Human Genetic Disease. *Nat. Rev. Genet.* **2019,** *20* (12), 760–772.

Pierce, A. J.; Johnson, R. D.; Thompson, L. H.; Jasin, M. XRCC3 Promotes Homology-Directed Repair of DNA Damage in Mammalian Cells. *Genes Dev.* **1999,** *13* (20), 2633–2638.

Pray, L. Transposons: The Jumping Genes. *Nat. Educ.* **2008,** *1* (1), 204.

Radford, S. J.; Boyle, M. L.; Sheely. C. Increase in Ty1 cDNA Recombination in Yeast sir4 Mutant Strains at High Temperature. *Genetics* **2004,** *168* (1), 89–101.

Rattray, A. J.; Shafer, B. K.; Garfinkel, D. J. The Saccharomyces Cerevisiae DNA Recombination and Repair Functions of the RAD52 Epistasis Group Inhibit Ty1 Transposition. *Genetics* **2000,** *154* (2), 543–556.

Richardson, C.; Jasin, M. Frequent Chromosomal Translocations Induced by DNA Double-Strand Breaks. *Nature* **2000,** *405* (6787), 697–700.

Rogakou, E. P.; Pilch, D. R.; Orr, A. H.; Ivanova, V. S.; Bonner, W. M. DNA Double-Stranded Breaks Induce Histone H2AX Phosphorylation on Serine 139. *J. Biol. Chem.* **1998,** *273* (10), 5858–5868.

Rothstein, R.; Michel. B.; Gangloff, S. Replication Fork Pausing and Recombination or "Gimme a Break". *Genes Dev.* **2000,** *14* (1), 1–10.

Saldivar, J. C.; Cortez, D.; Cimprich, K. A. The Essential Kinase ATR: Ensuring Faithful Duplication of a Challenging Genome. *Nat. Rev. Mol. Cell Biol.* **2017,** *18* (10), 622–636.

Sanchez, D. H.; Gaubert, H.; Drost, H. G.; Zabet, N. R.; Paszkowski, J. High-Frequency Recombination Between Members of an LTR Retrotransposon Family During Transposition Bursts. *Nat. Commun.* **2017,** *8* (1), 1283

SanMiguel, P.; Tikhonov, A.; Jin, Y. K.; Motchoulskaia, N.; Zakharov, D.; Melake-Berhan, A.; Springer, P. S.; Edwards, K. J.; Lee, M.; Avramova, M.; Bennetzen, J. L. Nested Retrotransposons in the Intergenic Regions of the Maize Genome. *Science* **1996,** *274* (5288), 765–768.

Santiago, N.; Herráiz, C.; Goñi, J. R.; Messeguer, X.; Casacuberta, M. Genome-Wide Analysis of the Emigrant Family of MITEs of Arabidopsis Thaliana. *Mol. Biol. Evol.* **2002,** *19* (12), 2285–2293.

Sharon, G.; Burkett, T. J.; Garfinkel, D. J. Efficient Homologous Recombination of Ty1 Element cDNA When Integration Is Blocked. *Mol. Cell Biol.* **1994,** *14* (10), 6540–6551.

Stucki, M.; Jackson, S. P. Gamma H2AX and MDC1: Anchoring the DNA-Damage-Response Machinery to Broken Chromosomes. *DNA Repair (Amst).* **2006,** *5* (5), 534–543.

Sundararajan, A.; Lee, B. S.; Garfinkel, D. J. The Rad27 (Fen-1) Nuclease Inhibits Ty1 Mobility in Saccharomyces Cerevisiae. *Genetics* **2003,** *163*, 55–67.

Trizzino, M.; Park, Y.; Holsbach-Beltrame, M.; Aracena, K.; Mika, K.; Caliskan, M.; Perry, G. H.; Lynch, V. J.; Brown, C. D. Transposable Elements are the Primary Source of Novelty in Primate Gene Regulation. *Genome Res.* **2017,** *27* (10), 1623–1633.

Ulrich, H. D. Degradation or Maintenance: Actions of the Ubiquitin System on Eukaryotic Chromatin. *Eukaryot Cell.* **2002,** *1* (1), 1–10.

Warbrick, E.; Heatherington, W.; Lane, D. P.; Glover, D. M. PCNA Binding Proteins in Drosophila Melanogaster: The Analysis of a Conserved PCNA Binding Domain. *Nucl. Acids Res.* **1998,** *26*, 3925–3932.

Wicker, T.; Sabot, F.; Hua-Van, A.; Bennetzen, J.; Capy, P.; Chalhoub, B.; Flavell, A.; Leroy, P.; Morgante, M.; Panaud, O.; Paux, E.; SanMiguel, P.; Schulman, A. H. A Unified Classification System for Eukaryotic Transposable Elements. *Nat. Rev. Genet.* **2007,** *8* (12), 973–982.

Wu, L.; Hickson, I. D. The Bloom's Syndrome Helicase Suppresses Crossing over During Homologous Recombination. *Nature* **2003,** *426* (6968), 870–874.

Yang, G.; Lee, Y. H.; Jiang, Y.; Shi, X.; Kertbundit, S.; Hall, T. C. A Two-Edged Role for the Transposable Element Kiddo in the Rice Ubiquitin2 Promoter. *Plant Cell.* **2005,** *17* (5), 1559–1568.

Yant, S. R.; Mark, A. K. Nonhomologous-End-Joining Factors Regulate DNA Repair Fidelity During Sleeping Beauty Element Transposition in Mammalian Cells. *Mol. Cell. Biol.* **2003,** *23*, 8505–8518.

Yoder, K.; Sarasin, A.; Kraemer, K.; McIlhatton, M.; Bushman, F.; Fishel, R. The DNA Repair Genes XPB and XPD Defend Cells from Retroviral Infection. *Proc. Natl. Acad. Sci.* **2006,** *103*, 4622–4627.

Yoder, K. E.; Bushman, F. D. Repair of Gaps in Retroviral DNA Integration Intermediates. *J. Virol.* **2000,** *74* (23), 11191–11200.

Yuan, Y. W.; Wessler, S. R. The Catalytic Domain of All Eukaryotic Cut-and-Paste Transposase Superfamilies. *Proc. Natl. Acad. Sci. U.S.A.* **2011,** *108*(19), 7884–7889.

Zhang, X.; Jiang, N.; Feschotte, C.; Wessler, S. R. PIF- and Pong-Like Transposable Elements: Distribution, Evolution and Relationship with Tourist-Like Miniature Inverted-Repeat Transposable Elements. *Genetics* **2004,** *166* (2), 971–986.

Part IV

Transposon Applications in Genome Editing

CHAPTER 13

Role of Transposable Elements in Plants Under Abiotic Stress Response

VINCENT EZIN[1*] and RACHAEL C. SYMONDS[2*]

[1]*Faculty of Agricultural Sciences, University of Aboemy-calavi, Godomey, Benin*

[2]*School of Biological and Environmental Sciences, Liverpool John Moores University, Merseyside, UK*

**Corresponding author.*
E-mail: ishola.vincent@yahoo.com; r.c.symonds@ljmu.ac.uk

ABSTRACT

Climate change due mainly to human activities has everlasting consequences for sessile organisms on earth. Growing plants, in particular, are confronted with many challenges due to increased climatic variability, with more heat, more water stress, and more nutrient deficiency predicted. There are many adaptation measures proposed thus far in the literature to overcome these environmental challenges. The transposable elements (TEs) are an extremely promising solution to help plants withstand the adverse effects of abiotic stress. TEs' activities cause diverse changes in plants resulting in the adaptive evolution to abiotic stress. Thus, there is growing interest in the activities of TEs amongst the scientific world because they engage in crucial functions, especially in abiotic factors resistance. In the past considered as "genome genomic parasites," TEs are now labeled as genomic gold due to their huge functions in plants. In this paper, we discuss the economic importance of abiotic stress in agriculture in relation to current and future climatic conditions. We depict the different types of TEs and provide the mechanisms

Plant Transposable Elements: Biology and Biotechnology, Deepu Pandita, Anu Pandita, (Eds.)

controlling the activities of TEs. We portray the relationship between TEs activation and abiotic stress-response gene activation. We proffer the contribution of TEs to plant resistance under abiotic stress. Finally, we explain the adaptability of plants to various abiotic stress due to the presence of TEs through changes occurrence in gene expression.

13.1 INTRODUCTION

Plants due to their immovable nature are continuously constrained to acclimate to fluctuations in their surroundings that threaten their longevity and reproduction through a regulatory network of genes. Most of these environmental changes are a potential source of stress for the plant. In the context of climate change affecting our planet, many investigations aim at studying and understanding the adaptive abilities of plants to stress. Faced with the heterogeneity of the environment, plants can respond immediately with phenotypic plasticity and/or genetic adaptation (Wagner et al., 1997; Donnelly et al., 2012). The acclimation of plant to abiotic factor is due to a phenotypic character that depends on genetic and transposable elements (TEs) determinism favored by the pressure of environmental selection (Howe and Brunner, 2005). It has been demonstrated that TEs influence gene regulation and therefore phenotype. The major consequence of climate change is the increased risk of abiotic stress. Indeed, the variability of climate change and the prevalence of extreme events are a harsh reality plants are confronted with particularly due to its immobile nature. Thus, plants must develop some mechanisms to withstand water stress, salty soil, pollution, nutrient deficiency, cold, high temperatures, and flood stress for its survival, development, and reproduction. These mechanisms could be of physiological, morphological, and biochemical nature, which are under the control of genes and their components.

Apart from certain processes, such as duplications (polyploidy, segmental duplications) and genomic recombination, it is the TEs that constitute the bulk of repeated DNA (Flavell et al., 1977; Barakat et al., 1997; SanMiguel and Bennetzen, 1998). Thus, TEs constitute a key element of plant genomes. Moreover, TEs have been regarded as parasitic and junk DNA for a long due to the fact that they do not present any outward benefits. These elements, which do not appear to contain important genes to their host organism, just have the information that allows them to replicate and/or move through genomes. Thus, they have long been considered as "selfish" and "parasitic" elements only capable of disrupting genes. However, there is a paradigm shift as they play fundamental roles in plant genome and adaptation (Capy et al., 1997; Kidwell and Lisch, 2001). The determination of the nucleotides arrangements in the genomes of many plants has subsequently shown that TEs are one of the major constituents of eukaryotic genomes (Vieira et al., 2012), and particularly in plants. For instance, when they are not lethal, their evolutionary dynamics both structurally and functionally will generate genetic diversity, with various consequences for adaptation, evolution, and the diversification of plants (McClintock, 1984; Brandt et al., 2005; Wendel et al., 2016). With progress in science and more discoveries, the dogma of a genome consisting of a linear succession of perfectly stable genes was superseded by a new opinion of a complex genome comprising epigenetics, genetics, and cellular intercommunications in which many elements including TEs are implicated.

With the progress of science and the proliferation of discoveries, the dogma of a genome consisting of a linear succession of perfectly stable genes has been replaced by the view of a genome as a complex network involving genetics, epigenetics, and cellular interactions, in which TEs and other structural and functional elements intervene.

Thus, the advent of genomics has shown that these elements make up a large part of genomes and that TEs are much more dynamic entities than previously thought. Many researches have demonstrated their involvement in responses to various environmental stresses (Wessler et al., 1995; Liu and Wendel, 2000; Jiang et al., 2003). Some of these DNA sequences have their own regulatory sequences, their own coding potential and they are autonomous, while others called defective, on the contrary depend on other elements to move: this is particularly the case of small interpserced nuclear element (SINE), which use the molecular machinery encoded by the long interpserced nuclear element (LINE) to transpose. Although TEs have a replication cycle independent of the cell cycle, they still use the host's transcriptional machinery. This is why they have long been referred to as "parasitic DNA."

In this paper, we discuss the economic importance of abiotic stress in agriculture in relation to current and future climatic conditions. We depict the different types of TEs and provide the mechanisms controlling the activities of TEs. We portray the relationship between TEs activation and abiotic stress-response gene activation. We proffer the involvement of TEs in plant resistance to abiotic stress. Finally, we explain the adaptability of plants to various abiotic stresses due to the presence of TEs through changes occurrence in gene expression.

13.2 ECONOMIC IMPORTANCE OF ABIOTIC STRESS IN AGRICULTURE IN RELATION TO CURRENT AND FUTURE CLIMATIC CONDITIONS

The occurrence of a global warming is mainly caused by anthropogenic activities, including exceptional use of fossil fuels, unprecedented deforestation, and a change in land use. Our climate is getting warmer and will continue to do so due to the levels of greenhouse gases already present in the atmosphere. Climate change is the main source of increasing abiotic stress. Therefore, the average temperature will rise up and could attain 4.5°C by the year 2100, the rainfall pattern will be modified and climate variability will augment, especially the frequency of increasingly recurring extreme events such as salinity, drought, heat, cold, and flooding (IPCC, 2007). Abiotic stresses are caused by drought, strong temperature variations, frost, wind, hail, excess of water, high salinity, and flooding. These climatic variations destabilize the environment of the crops and generate stresses that impact the productivity of the plants, mainly during sensitive periods such as flowering and fruit sets. Periods of drought will be more pronounced and repeated year after year. The drought conditions will be coupled with strong variations in temperature during the daylight, which plants must face throughout their development and which can have serious consequences on crop productivity. It is estimated that 50% of yield losses are due to abiotic stresses. These climatic hazards impact the metabolism of plants, which can lead to flower abortion or irregular fruit development. Moreover, these abiotic stresses can occur at different stages of crop development and thus cause several waves of losses. Agriculture is a major user of water resources in many parts of the world and drought affects agriculture in 45% of the world's geographic area (Rockstrom, 2003). Rockstrom also reported that rainfed agriculture accounts for 80% of the world's cultivated land and contributes 60% of the world's food (Rockstrom, 2003). The low yields recorded from rainfed crops are caused by limited and unpredictable rainfall leading to drought stress. It has been estimated that drought is the most important environmental stress and accounts for 70% of yield losses in our cereal crops worldwide (Boyer, 1982). Among the abiotic stress, drought is considered as the most difficult and ravaging worldwide and causes havoc to crop production and farmers (Pennisi, 2008; Gurian-Sherman, 2012). Gurian-Sherman (2012) added that the historic 2011 drought in Texas caused a record $5.2 billion in agricultural losses, making it the most expensive drought on record. Rainfall patterns will show a decrease of more than 20–30% compared to the 1961–1990 reference level adopted by the WMO (Bigot et al., 2003). With more than

95% of African agriculture dependent on rainfall, agricultural production will be severely compromised by increased variability in rainfall combined with rising temperatures and the occurrence of devastating extreme events. These climatic events are already more frequent with serious social, economic, and ecological consequences. Heat can induce an augmentation in the permeability of the thylakoid membranes which can modify the electron flow during photosynthesis and consequently leading to starvation and death of the plants. High temperature is an abiotic factor that influences yield by affecting the growth and physiological processes of the plant. The increase of temperature beyond the limits causes the temperature stress observed in many species. Plant height and leaf elongation, flowering, fruit, and grain set maturation are among the parameters influenced by high temperature. Anther, pollination, and fertilization phases are seriously affected to temperature stress. Capiati et al. (2006) demonstrated that environmental factors such as high temperatures and inadequate moisture drastically reduces plant vigor and cause flower drop, resulting in low fruit-set and low yields. Under high temperatures or drought, the few fruits that do set often develop physiological problems such as cracking or blossom end rot and are unmarketable.

Salinity is viewed as an important abiotic stress restraining plant productivity and agricultural yield (Rozema and Flower, 2008; Abd latef, 2010). It is a serious menace to plant survival and productivity. In arid- and semi-arid agro-ecology, salinity is a result of high evaporation of water from soil, shortage and inadequate rainfall, and also agricultural irrigation. In total, 7% of cultivable lands in the world is affected by salinity which could rise to 20% in the future owing to high irrigation, bad cultural practices, and increase in change of climate (Munns and Tester, 2008). Soil salinization is regarded as "Silent killer" of natural production since it kills plants and organisms in the affected zones (Tanji, 1990). To preserve suitable balance and proliferation, plants under abiotic stress have evolved certain acclimatized responses environmental stresses through the production of enzymes. Swami et al. (2011) reported that the key events in response of plants to environmental factors are the recognition and transduction of stress signals through signaling machineries, which results in the activation of many stress-related genes. Plants in response to abiotic stress perception activate their stress signal transduction mechanisms leading to the stimulation of many genes coding for hormonal regulated stress and reprogramming the metabolism to increase molecular components enhancing plant stress tolerance (Zhu, 2003; Flower, 2004).

Forbes and Callow (1997) consider as stress any factor that affects the proper functioning of organisms. The plant is considered as tolerant to abiotic stress when the physiological answer that it develops allows its survival, in the contrary case, it is considered as sensitive. Plants tolerate stresses through morphological and biochemical adjustments and cross-tolerance to different stresses. Characteristically, stress is recognized by cellular structures on the surface of the plasma membrane of plant tissues. The signal is then transmitted to the nucleus via a cascade of reactions including secondary message carrier like calcium, inositol phosphate, and reactive oxygen species (Mahajan and Tuteja, 2005). The response to environmental factors encompasses the regulation of stress response genes. They are grouped into two classes: (i) genes that react immediately to the stress: whose induction follows very quickly the perception of stress. Their activation is also of short duration. (ii) Some genes implicated in the response to environmental factors are regarded as marker genes such as those encoding chaperone proteins or HSPs and the genes of response to oxidative stress (Scandalios, 2005; Smirnoff, 1998; Holmberg, and Bülow, 1998).

13.3 CLASSIFICATION OF TRANSPOSABLE ELEMENTS

TEs represent a significant part of all the genomes and ranging from 3% for small genomes to 80% for big genomes. They represent 50–80% of some grass genomes (SanMiguel and Bennetzen, 1998; Vicient et al., 1999) especially they comprise 80% of the maize genome. They form a major part of the genome in some organisms, particularly in plants where they can constitute more than half of the genome, as is the case for barley and maize (SanMiguel et al., 1998; Kumar and Bennetzen, 1999). Devos (2010) and Estep et al. (2003) reported more than 85% of a large part of plant genomes. The TEs play a fundamental function in the natural-selection of genomes and the acclimation of organisms (McClintock, 1984; Tenaillon et al., 2010). Although TEs appear to be of various types of sequences, it is necessary to classify them according to their structure, the similarity of their nucleotide or protein sequence, or their mode of replication. Thus, they can be classified according to several criteria. The first classification suggested by Finnegan et al. (1989) distinguishes TEs according to their transposition intermediary. Class 1 is composed of long terminal repeat (LTR) retrotransposons and LTR-free retrotransposons, essentially consisting of two major families: LINE and SINE. Their mode of transposition involves an RNA-type intermediate, transcribed produced from a copy of TEs, then retrotranscribed into DNA to form a new copy. Class 2 is composed of transposons that have a mode of transposition involving a DNA intermediate. In this case, the transposition entails the erasure of a copy of TEs and its insertion into another locus. This "cut and paste" mode of transposition is nonreplicative, since it consists of a displacement of the same copy on the chromosomes. Galindo-González et al. (2018) reported the same two classes (I and II) of TEs based on copy and paste mechanisms using RNA intermediate and on cut and paste mechanisms using DNA intermediate. According to Feschotte et al. (2002), there are various families of mobile elements that are divided into two main categories: retrotransposons (type I mobile elements), made up of the main Ty1-Copia, typ3-gypsy, LINE-like and SINE-like families, and transposons DNA (type II mobile elements) made up of the Tc1/mariner, Fot1/pogo, hAT, and miniature inverted tandem repeat (MITE) families or minitransposons (Malik and Eickbush, 2001; Feschotte and Pritham, 2007). However, the third family made its appearance with the discovery of new elements such as cryptons or polintrons. The helitrons do not require a double-strand cut of the donor site during transposition. The most recent and detailed nomenclature uses family, subfamily, superfamily, order, subclass, and class to categorize TEs according to phylogenetic classifications (Bourque et al., 2018). Another similar grouping with slight differences and details according to Kapitonov and Jurka (2008) are as follows: class I elements: retrotransposons are split into five orders, according to their transposition machinery and the evolution of their reverse transcriptase: the LTR retrotransposons, DIRS-like elements, Penelope-like elements, LINE and SINE. Class II elements include two subclasses, which differ in the number of copies of DNA generated during transposition. In the first subclass, we find the elements that transpose via a mechanism called "cut and paste," divided into two orders; the first, the ITR order (for terminal inverted repeat), where we find the best known and most studied transposons (Tc1-mariner elements), the P element, the piggyBac transposons. The second order is only composed of Cryptons elements, exotic elements little studied, identified only in fungi. In the second, subclasses have grouped the helicons and the polintons (also known as Maverick) that transpose by mechanisms totally different from the other transposons of the different from the other transposons of class II. In response to this, "universal classification" proposed by Wicker et al. (2007), two other researchers, Vladimir Kapitonov and Jerzy Jurka, published their own classification in 2008, which is based on Repbase, a computer database of eukaryotic TEs that contain more than 7600 ET and repeat sequences (Kapitonov and Jurka, 2008). This categorization was done based on structural likeness, enzymology, and rapport between DNA sequences; there are the two key classes of elements (retrotransposons and DNA transposons) further divided into five classes and then 40 superfamilies, with elements classified separately without any real justification in their article. The classification presented by Wicker and his collaborators (2007) seems simpler than the one proposed by Kapitonov and Jurka (2008), based almost exclusively on genome analysis, whereas Wicker proposes a classification based more on the transposition mechanism.

The great diversity that exists within TEs, highlighted in recent years thanks to the abundant amounts of data obtained by the various projects of sequencing, does not make it possible with the first mode of classification to position all the TE families. Some of them, like the MITE, for example, are difficult to characterize according to their simple mode of transposition. Indeed, MITEs transpose by means of a "copy and paste" machinery without going via an RNA intermediary. All known TEs are referenced in the RepBase database (http://www.girinst.org/repbase/index.html), in which developers have established a new method for their classification. By taking into account the enzymology, the structural similarities, and the relations between the sequences, Kapitonov and Jurka (2008) proposed a classification accompanied by a new nomenclature of TEs, mentioning, in particular, the name of the superfamily, a structure identifier, and an identifier of species. A similar classification was proposed by Wicker et al. (2007), and also integrates the structure and sequence of TEs. This is a unified, ranked categorization that allows all known TEs to be characterized, based on the classification established by Finnegan (1989). In this classification model, TEs are first divided into two groups, according to the presence or absence of an RNA-like transposition intermediary, and then into subclasses, orders, and families. Autonomous and nonautonomous elements have been reported in each category of TE classification (Feschotte et al., 2002).

13.4 MECHANISMS OF CONTROLLING TE ACTIVITY

Genomes are made up of genes and numerous "repeated" sequences, including TEs which, thanks to their ability to move along the DNA molecule and to insert themselves in or near genes, are responsible for a high proportion of mutations and chromosome rearrangements. Moreover, it is clearer now that these repeats impact gene expression modulation (Bhattacharyya et al., 1990; Oliver and Greene, 2009; Hua Van et al., 2011). They are, therefore, a vital source of genetic alterations. The presence of a population of TEs supposes the existence of control mechanisms that will guarantee the functional integrity of the genome. Epigenetic machinery helps to silence TEs but upon abiotic stress, they become activated again.

Many epigenetic regulatory pathways control the expression of TEs via a high level of DNA methylation on these elements. In another word, TEs are under strict epigenetic control which affects their ability to transpose under normal developmental conditions. However, various types of stresses might result in the suppression of the repression of TEs and cause their reactivation. In the model plant Arabidopsis thaliana, this targeting is partly dependent on small interfering RNAs and a strong correlation exists between this DNA hypermethylation and other epigenetic modifications affecting histones such as the methylation of lysine 9 or lysine 27 of histone H3 (Liu et al., 2000). The contiguity of the epigenetic marks is associated with the establishment of "closed" heterochromatic structures in which the highly condensed chromatin is not very conducive to transcription. However, there could be consequences of loss of epigenetic control on TEs leading to the most drastic losses of DNA methylation. These severe losses of DNA methylation are accompanied by a decondensation of chromocenters (Soppe et al., 2002). The consequence of this disruption of epigenetic control is the massive transcriptional reactivation of several hundred TEs of all types (Lippman et al., 2004; Zemach et al., 2013). This transcriptional reactivation is associated with the production of siRNAs, suggesting the endowment of post-transcriptional regulation for some of the reactivated TEs (Teixeira et al., 2009; Slotkin et al., 2009). The cause of this transcriptional reactivation remains unclear in many cases. However, certain stresses seem to be able to generate the specific expression of certain TEs due to the presence of stress response sequences in their promoters. Plant stress response pathways interact closely with epigenetic regulation. Different external or genomic stresses modulate DNA and histone methylation as well as smRNA production. The epigenetic control is associated with the response of the plant to abiotic stress. Different external or genomic stresses regulate DNA and histone methylation as well as smRNA production.

The majority of the regulation of genome activity is due to changes in chromatin. In plants, cytosines can be methylated at position five of the aromatic cycle. Although the majority of cytosines are methylated in plants, only repeat regions (including ETs) have high methylation rates compared to gene regions (Roudier et al., 2009). DNA methylation is often associated with histone methylation. The latter is done on the lysine (k) of histone H3 (at positions 9 (dimethylation) and 27 (monomethylation)) by histone methyltransferases. These two types of methylation thus play an important role in maintaining the condensed state of chromatin, inactivating the transcriptional, and thus the transpositional activity of ETs. The combinations of these different modifications confer to chromatin various properties resulting in variations of compaction levels and transcriptional activity which thus define different chromatin states (Roudier et al., 2011). In fact, many studies carried out in plants have demonstrated a transcriptional activation of TEs in response to various stresses, both biotic and abiotic (Naito et al., 2009; Yu et al., 2013).

13.5 RAPPORT AMONG TEs STIMULATION AND ENVIRONMENTAL FACTOR-RESPONSE GENE ACTIVATION

The TEs are now viewed as fundamental elements in a multitude of processes, reshaping genomes, and controlling gene activity. Genome studies will increasingly focus on repeat sequences, that is, TEs and their relationships with genes. It is in this area of analysis of gene-TE interaction networks, responsible for various phenotypes, that the most exciting discoveries are expected. The proliferation ability of TEs can cause the rise in the size of genomes in response to abiotic stress or environmental changes.

TEs, once they transpose, are powerful mutagens. They can, for example, interrupt the coding sequence of genes. Insertions of this type are for the most part particularly deleterious and are, therefore, rapidly purged by natural selection. Moreover, since TEs contain their own promoter and regulatory sequences, they can affect the regulatory sequences and the expression profiles of close genes in different ways. The results of research carried out on rice have demonstrated that miniature inverted-repeat transposable element (MITE) insertions in the range 100 bp of the transcription start site of genes can be stimulated by salinity and frost stress (Naito et al., 2009). TEs can also produce "outward reading" transcripts that extend outside the element into the adjacent gene.

The mobile elements are determinists of gene expression and the changes they induce are at the origin of genetic variability allowing plants to adapt to environmental stress. An inactivated TE by methylation leads to the inactivation of the adjacent gene while other genes are transcribed because they are not under the control of the TE. This is an indication that the immediate gene to the locus of TE is under the influence of TE under normal circumstances. It has been demonstrated that abiotic stress is related to methylation modifications in TE and nearby gene. Thus, under abiotic stress, TE is activated through a transcriptional response to stress. As a result, the nearby gene is also activated by loss of methylation in its promoter region. The newly inserted TEs may initiate stress-responsive transcription for new genes or they may result in permanent loss of function.

Some transcriptional elements can control multiple metabolic pathways and a metabolic route may involve regulation coordinated by some transcriptional elements, which a system put in place by a regulatory network of genes in plants subjected to environmental factors (Shinozaki and Dennis, 2003; Shao et al., 2006; Wu et al., 2007). Thus, the perception of the abiotic stress and its transduction activate the manifestation of stress-response genes like M MK4, VP14, DBF2, AtMEKK1, AtMPR3, AtPK19, AtPLC1, SAMK, which interact with other genes including AtCDPK1, ABI1, MP2C, AtGSK1, VP14, AtDSPTP, AtMPK4 due to the demethylation of TEs for the acclimation of plants to the environmental changes (Wu et al., 2007). At genetic level and selection, many researches are nowadays focusing on the modification of the regulatory activities of transcriptional elements so as to enhance the adaptation and resilience of plants under drought, salinity, nutrient deficiencies, flood, frost, UV-B, and other environmental factors (Casati and Walbot, 2004).

13.5.1 ABIOTIC STRESS: THE COMPLEXITIES OF TE ACTIVATION

Plants must adapt to daily fluctuations in environmental stresses alongside longer-term climatic variations, and their response is a complex process involving highly coordinated cellular and genomic mechanisms (Ma et al., 2020). Environmental stresses including high and low temperatures, and drought and flooding can be anticipated by the plant while unpredictable stresses, such as gamma radiation or other mutagens are less predictable.

The concept that TE element activity can be triggered by environmental cues was proposed in the 1930s by Barbara McClintock (1950). She developed the "Genomic Shock Model" by which TEs are part of an adaptive response to unanticipated environmental shocks (McClintock, 1950, 1984). Her work on the ruptured ends of *Zea mays* chromosomes during the 1940s led her to conclude that a traumatic event had caused the activation of previously silent "controlling elements." The initial model referred to biotic stresses rather than abiotic, nevertheless, the concept had been introduced. The model predicts the unusual responses of a genome to shocks it may receive either by errors in the cell or by external influences on the cell (McClintock, 1984).

When a TE inserts into a genome, it can increase the mutation rate which can consequently increase variation and potential for adaptation (Bennetzen,

2000). However, in normal conditions, there are epigenetic constraints that prevent damaging mutational effects of TE activation and transposition (Cheng et al., 2006). Under conditions of stress, however, an increased mutation rate might become an advantageous adaptive response.

TEs can be activated by abiotic stress (Voronova et al., 2014; Huang et al., 2017) and furthermore, some TEs have been found to contain stress-responsive regulatory sequences (Beguiristain et al., 2001). These stresses have been shown to have the potential to overcome the epigenetic restraints on TE activation and transposition (Galindo-González et al., 2018) but their role as adaptive elements is more complex. Once a TE inserts into the genome, it has the potential to have a positive, deleterious, or, as in the majority of cases, no effect on fitness at all (Lisch, 2009).

Plants must be agile in their responses to varying environmental stresses, and to rapidly adapt to this change they must obtain novel cis-regulatory sequences, which provide new patterns of gene expression (Suoniemi et al., 1996). Some TEs use cis-regulatory elements to carry out their functions and therefore they can also provide a novel source of these elements for plants when they insert or transpose in the genome (Ito et al., 2013). These are similar to the motifs used to activate stress-responsive genes (Grandbastien et al., 2005), and this way TEs can modify gene expression by inserting cis-elements into the promoter and enhancer regions (Sinzelle et al., 2009).

Upregulation of the grapevine VvTF1A gene in reproductive and vegetative tissues has been shown to correlate with the insertion of the Hatvine 1-rm a class-II TE inserted into the promoter region (Fernandez et al., 2010). This spontaneous cis-activation demonstrated a role for this class of TE in somatic as well as reproductive cell variation. Insertion of TEs can also alter gene copy number and regulation, generating further genotypic variation (Stapley et al., 2015). However, their role in abiotic stress-activated gene expression is not straightforward as TEs have been associated with both upregulation and downregulation of nearby genes (Makarevitch et al., 2015). Additionally, after initial activation in response to stress, TEs can become repressed (Secco et al., 2015).

In this way, TE insertions can provide the phenotypic variation to drive rapid responses to environmental stress. This allows plants to overcome the genetic bottlenecks that can occur as a result of speciation while rapidly adapting to new environments. Evidence for this comes from a population genomic analysis of the inbreeding *Capsella rubella*, and its sister species, the outcrossing *Coreopsis grandiflora* (Xiao et al., 2019). The promoters and downstream regions of *C. rubella* were found to be highly enriched in highly polymorphic TEs compared to *C. grandiflora* with 4.2% of the polymorphic TE insertions associated with significant changes in adjacent gene expression. This appears to be providing variation in an otherwise inbreeding species (Xiao et al., 2019). Approximately 65% or IncRNAs are derived from TEs and many of these are differentially expressed during abiotic stress (Lv et al., 2019). This is suggestive of a crucial role for TE-derived IncRNAs in moderating abiotic stresses.

Several studies have demonstrated that TEs have a positive effect on fitness (Negi et al., 2016). However, negative fitness effects in response to stress have also been observed (Mao et al., 2015; Wang et al., 2007). Overall, the literature does not show one pattern of TE regulation under abiotic stress. Several studies show TE activation, some show TE repression, and some show activation after exposure to stress. The relationship between TEs and abiotic is stress is dependent on the type of stress and the types of TEs are important for the activity under stress (Fig. 13.1).

13.5.2 PLANT GROWTH REGULATORS

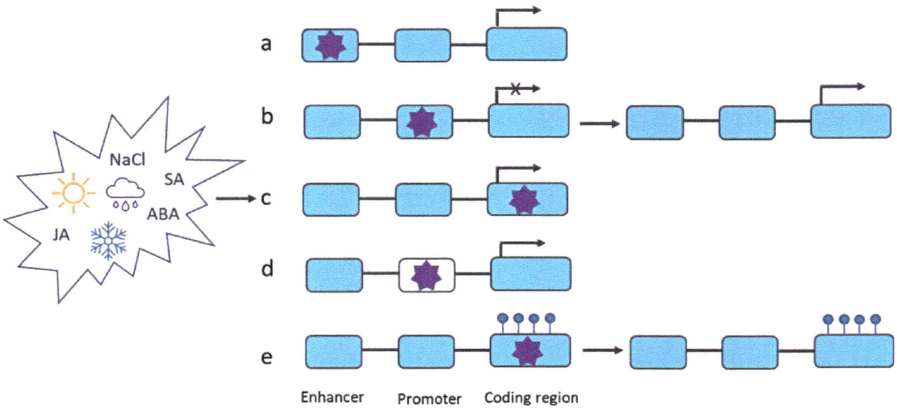

FIGURE 13.1 General overview of TE activation in response to abiotic stress. (a) TE insertion into enhancer region, (b) TE insertion into promoter depicting either enhanced or repressed gene expression. TE insertion into the promoter region may also influence the expression of nearby genes, (c) TE insertion into gene, (d) TEs may act as new cis-regulatory elements, (e) TE insertion may result in either direct DNA methylation or DNA methylation in nearby genes.

Abiotic stresses result in elevated levels of damage to DNA such as single-strand breaks and double-strand breaks (Mannuss et al., 2012). An important part of repairing this damage is DNA methylation, and DNA methylation enzymes have been shown to be an essential part of the DNA repair process in various stresses including drought (Shim et al., 2018), UV damage (Qüesta et al., 2013), and heat stress (Li et al., 2016). When TEs are activated and fall inside or close to a gene, the resulting methylation can extend to nearby genes. This change in methylation state can rapidly lower gene expression and the resulting interaction facilitates plants' responses to abiotic stress (Galindo-González et al., 2018).

TEs falling adjacent to or inside a gene can give rise to methylation that spreads to nearby genes lowering gene expression and resulting in rapid responses that facilitate plant stress adaptation (Galindo-González et al., 2018). The abiotic stress response is tightly linked to phytohormones, and abscisic acid (ABA) in particular (Sah et al., 2016), and DNA methylation is thought to be involved in ABA-dependent gene expression (Gohlke et al., 2013). The ABA deficient *Z. mays* mutant *vp10* has been used to explore the relationship between ABA induction under drought stress and DNA methylation (Sallam and Moussa, 2021). Twenty-two differentially methylated fragments were shown to have a greater than 50% sequence similarity with TE elements, the majority being gypsy and copia-like retrotransposons. TEs have been shown to exert a rapid response to environmental stress when found close to or inside a gene through epigenetic modulation (Galindo-González et al., 2018). However, more work is needed to elucidate the precise nature of the relationships between these differentially methylated fragments and ABA responses to drought stress (Sallam and Moussa, 2021). Active plant transposons contain cis-regulatory elements in their promoter regions associated with stress-responsive signal transduction pathways. For example, the barley BARE-1 copia-like retroelement contains ABA-responsive elements in the 5′ region of the LTR (Suoniemi et al., 1996).

Although ABA is often considered the major stress hormone (Tuteja, 2007), a plants response to abiotic stress is controlled by several different phytohormones in a complex pattern of crosstalk (Anderson et al., 2004). The way in which TEs respond to these signaling molecules varies with both highly specific and more general responses. Early work by Takeda et al. (1998) and Grandbastien et al. (1997) revealed that the Tto1 and Tnt1 retrotransposons of tobacco are activated by methyl jasmonate. A further study demonstrated that a 13-bp repeated motif in the promoter region of Tto1 acts as a cis-regulatory element conferring methyl jasmonate responsiveness

(Takeda et al., 1999). Variations in Tnt1 subfamily expression patterns have been correlated with nucleotide sequence differences in their U3 regions and driven by different stress-inducible cis-acting elements (Beguiristain et al., 2001). These Tnt1 subfamilies respond differently to different stresses, for example, Tnt1 A is activated in response to Jasmonic acid while Tnt1C is activated by auxin and salicylic acid (Beguiristain et al., 2001). These sequence and cis-element differences appear to allow stress responses specific to each transposon. Similarly, the oat OARE-1 Ty-1 copia LTR is highly induced by both salicylic and jasmonic acid although this pattern is not uniform with higher induction with salicylic acid (Kimura et al., 2001). The promoter region of the *Solanum chilense* TLC1.1 retrotransposon responds to different plant hormones and activates in response to multiple stresses (Salazar et al., 2007). This activity is conferred through different cis-elements within the promoter region. The PERE-boxes in the promoter integrate signals from ethylene, salicylic acid, and ABA while distinct cis-elements activate TLC1.1 in response to methyl jasmonate and auxin (Salazar et al., 2007). This ability to respond to different signaling molecules contrasts with the specific induction seen in the Tnt1 subfamilies (Beguiristain et al., 2001).

An emerging area of interest in plant responses to abiotic stress is the noncoding RNAs including the long intergenic noncoding RNAs (lincRNAs) (Jha et al., 2020). TEs that are associated with long intergenic noncoding RNAs (TE-lincRNAs) show an abiotic stress-induced expression pattern in Arabidopsis, maize, and rice, suggesting stress-responsive transcription (Wang et al., 2007). This pattern varies with different stresses including salinity, cold, and ABA stress. Two Arabidopsis TE-lincRNA mutants showed a reduced sensitivity to ABA indicating a possible role for this TE-lincRNA in the ABA—abiotic stress response (Wang et al., 2007). Given the recent finding that a novel 755 nt lincRNA has been shown to control a suite of drought-responsive genes including ABA signaling genes, this is an exciting area for further exploration (Qin et al., 2017).

13.5.3 OSMOTIC STRESSES

Drought represents a major limitation to plant productivity. Physiological responses vary across species, and in relation to specific levels of drought stress, however, all plants must adapt quickly to avoid the damaging effects of moisture deficit. An extreme example of a desiccation-tolerant species is *Craterostigma plantagineum* which can suffer extreme desiccation to just

1% water content before recovering within 24 h of rewatering (Bartels et al., 1990). However, the callus material is not dehydration tolerant unless ABA is applied exogenously (Bartels et al., 1990). This extreme example is a useful species when investigating the role of TEs in drought stress. Both desiccation and ABA treatment in *C. plantagineum* have been found to induce the same genes for ABA and dehydration responsiveness (Furini et al., 1997). A desiccation tolerance gene CDT-1 was first isolated by T-DNA activation tagging and later reclassified as a retrotransposon (Furini et al., 1997). Increased levels of dehydration correlate with elevated levels of expression, which gives the TE more opportunity to reinsert into the genome, landing in a suitable place for transcribing under stress. This provided a possible explanation for the interaction between TEs and the environment in evolutionary terms. A further study revealed siRNAs generated by CDT-1 impartthe ability to respond to desiccation in the callus (Hilbricht et al., 2008). The response to desiccation does not require translation from CDR-1, but rather, it directs siRNA synthesis to open the ABA and desiccation tolerance pathways (Hilbricht et al., 2008).

Wild species of barley make an ideal organism to study for TEs as there are a large number of accessions and ecotypes. Additionally, they contain large numbers of TEs making them useful species for the study of TE impacts on water scarcity and drought stress responses in different ecosystems (Pearce et al., 1997). The BARE-1 retrotransposon, a transcribed *copia*-like element comprises approximately 7% of the barley genome. It is a conserved, transcriptionally active element with many dispersed copies comprising a major component of Horedeum genomes (Suoniemi et al., 1996). BARE-1 copy number correlates positively with genome size but also temperature, water availability, and soil type (Vicent et al., 1999). More specifically, BARE-1 copy number correlates with ecogeographic traits such as elevation and slope aridity (Kalendar et al., 2000). This suggests an adaptive selection for an increase in genome size through retrotransposon activity. Increasing dryness correlates not only with increased BARE-1 copy number but also correlates with a lower level of solo LTRs. In barley tissues, the BARE-1 elements are transcribed from promoters within the LTR. This suggests that both the propagation and induction of these BARE-1 elements may be induced in response to stress. This is consistent with the presence of abscisic acid-response elements, within the BARE-1 promoter region, a characteristic of genes induced by drought stress (Suoniemi et al., 1996).

Another member of the *copia* superfamily is Rider which is an LTR retrotransposon ubiquitous in the tomato plant genome (Jiang et al., 2009). Rider-like elements are also present in several tomato wild relatives including

S. arcanum, *S. pennellii*, and *S. habrochaites*. Rider is constitutively transcribed, producing full-length transcripts, and is a major source of tomato phenotypic variation (Xiao et al., 2008). It is activated by both drought stress and ABA signaling pathways and versions have been discovered in diverse crops such as beetroot, rapeseed, and quinoa (Benoit et al., 2019). Both dehydration-responsive and ABA-responsive elements, and MYB recognition sequence motifs are found in *Riders*' LTRs (Benoit et al., 2019). *Rider* is controlled by both siRNA and RNA-dependent methylation pathways and inserts preferentially into genome-rich areas. This is similar to the way in which the ONSEN retrotransposon family of *A. thaliana* insert into the Arabidopsis genome (Cavrak et al., 2014). ONSEN insertions have been shown to confer heat responsiveness to neighboring genes (Ito et al., 2015). It is, therefore, possible that Rider confers drought stress in a similar way.

Members of both Gypsy and Copia TE families have been linked to derepression and high levels of expression under long-term osmotic and salt stress in maize (Forestan et al., 2016). TE activation was detectable both in wild-type plants under stress and in an epiregulator mutant involved in RNA-directed DNA methylation (RdDM) pathway siRNA biogenesis, under nonstress conditions, demonstrating the role of epigenetic mechanisms in plant responses to osmotic and drought stress.

The active MITE miniature Ping (*mPing*) was discovered in rice (Jiang et al., 2003), and studies into this transposon offer the possibility of revealing mechanisms for modifying the control of stress-induced gene expression (Yasuda et al., 2013). The *mPing* element contains putative stress-responsive cis-elements and is preferentially inserting into the 50-flanking region of genes, rendering adjacent genes stress-inducible (Naito et al., 2009). This element is highly mobile in the rice cultivar cv. Gimbozu, which makes a useful model organism for the study of this element. A screen targeting the upstream region of 17 genes selecting for high levels of stress inducibility in Gimbozu found five genes with an *mPing*-inserted promoter region (Yasuda et al., 2013). These genes were not upregulated when stress was not applied but in the majority of cases were upregulated under salt, cold, or both stress (Yasuda et al., 2013). Transgenic plants that constitutively overexpress stress tolerance genes can have reduced performance for traits such as yield and growth when stress is not applied and one solution is the use of stress-inducible promoters. This study has demonstrated the potential of harnessing an *mPing* stress-inducible promoter as a natural mutation and as part of a conventional breeding program as an alternative for producing stress tolerant plants (Yasuda et al., 2013).

Evidence has shown that stress-inducible gene upregulation can be correlated with TEs in relation to drought stress (Benoit et al., 2019) but how does this translate to the physiology? The DEEPER ROOTING 1 (DRO1) gene of rice forms part of a major QTL conferring drought avoidance (Uga et al., 2011). This gene regulates root architecture, promoting deeper roots, and allowing rice plants to avoid rather than tolerant drought stress (Kitomi et al., 2020). A INDITTO2 TE in the promoter region of DRO1 contains an auxin response element and conveys auxin-dependent transcription, resulting in enhanced drought avoidance in these plants (Zhao et al., 2021). Drought avoidance was also conferred to transgenic plants containing this element inserted into the promoter region. The mechanism has not yet been elucidated; however, this is an exciting example of a TE functioning as an autonomous auxin-responsive promoter (Zhao et al., 2021).

TE insertions into promoter regions do not always confer a positive fitness benefit. Variation in tolerance to drought in maize has been associated with the ZmNAC111 gene and the agent responsible for this variation is proposed to be a 82-bp miniature-inverted repeat (MITE) inserted into the ZmNAC111 promoter region (Mao et al., 2015). Gene expression may be being repressed via DNA and histone hypermethylation by the (RdDM) pathway. The insertion of this MITE into the maize genome is, therefore, correlated with lower expression of drought inducible genes and an increase in drought susceptibility, therefore providing an example of a negative effect on fitness.

Plants are bombarded with simultaneous multiple stresses and they must adapt using different mechanisms to overcome these stresses. Maize (*Z. mays*) has a large number of diverse mutant phenotypes making it a suitable choice for studies surrounding TE interactions with multiple abiotic stresses. Up to 85% of the maize genome is comprised of TEs, and these TEs have significant allelic variation (Baucom et al., 2009).

An extensive study by Makarevitch et al. (2015) explored the response of TEs in maize to multiple stresses including, heat and cold, salt and UV stress, and observed that approximately 20% of gene upregulated were associated with TEs. The majority of TEs responded to a single stress; however, there were examples of TEs responding to multiple stresses (Makarevitch et al., 2015). For example, genes located close to the *etug* TE associated with heat stress response while those close to the *ipiki* family were upregulated in response to salt and UV stress stresses. Overall, most TEs were correlated with up-regulation of gene expression but two TE families that were upregulated for UV stress were downregulated in response to salinity stress

(Makarevitch et al., 2015). Allelic variation in stress-responsive expression was generated by the insertion of certain TEs suggesting that TE insertions into a promoter region could be responsible for novel expression patterns. An additional and underexplored way in which TEs can modify expression patterns is by the process of TE exaptation (Joly-Lopez and Bureau, 2018). These fixed exapted or domesticated TE genes (ETEs) can potentially provide novel sources of abiotic stress response expression patterns through the co-option of TE regulatory, structural, or coding sequences. Exapted TEs have been found to play diverse roles in salinity and freezing stress tolerance, phosphate limitation, and arsenic toxicity in *A. thaliana* (Joly-Lopez et al., 2017). There is a strong link to phenotype in these ETEs in response to abiotic stress, for example, four members of the MUGA family exhibited a salinity tolerance (Joly-Lopez et al., 2017).

Flooding and waterlogging result in an inadequate supply of oxygen to the roots resulting in a major constraint to plant growth and productivity (Jackson and Colmer, 2005). Very few studies have explored TE activity in flooding stress and compared it to other abiotic stresses and the literature is very underrepresented. Given the role of TEs in other osmotic stresses, it is likely that they also play a major role in conferring flooding and waterlogging stress tolerance in plants. Some evidence come from the Sol3 transposon of tomato and potato, which has been found in the promoter region of a tomato ACC synthase gene that confers flooding tolerance (Shiu et al., 1998). Given that TEs are now playing a role in biotechnology strategies for crop improvement it is likely knowledge in this area will start to build in the near future (Springer and Schmitz, 2017; Thieme and Bucher, 2018).

13.5.4 TEMPERATURE INDUCED STRESS

Plants respond to heat stress in a variety of ways both and the physiological (Prasad et al., 2017) and molecular levels (Janni et al., 2020). Temperatures that are above optimal for that particular species can result in heat stress, with several associated consequences including elevated rates of genetic instability and somatic homologous recombination (Pecinka et al., 2009). Genome-wide modifies in TE and gene expression in *A. thaliana* in response to heat stress show many transiently expressed heterochromatic TEs and a strong correlation with chromatin reorganization (Sun et al., 2020).

A plant's reaction to high-temperature stress is not uniform and even within species, there are ecotype differences. TE activation in response to heat stress was evaluated in 10 different *A. thaliana* ecotypes which revealed a complex pattern of heat stress activation (Barah et al., 2013). The majority of TEs in seven of the ecotypes had elevated expression under stress while in the other three ecotypes the majority of TEs had lower expression.

A well-studied natural insertion of the ONSEN LTR-copia type retrotransposon (AtCOPIA78/ONSEN) in *A. thaliana* contains a heat-responsive element (HRE) (Cavrak et al., 2014). Sustained heat stress in Arabidopsis results in transient transcriptional activation of several classes of repetitive elements including ONSEN, and this is independent of heat stress signaling (Pecinka et al., 2010, 2010; Tittel-Elmer et al., 2010). Transcriptional activation often correlates with a loss of DNA methylation and in the ONSEN element this follows rather precedes activation, and therefore demethylation does not appear to be essential for activation of this element. These elements are under epigenetic silencing at ambient temperatures, supporting evidence that environmental conditions such as heat stress can override this epigenetic silencing (Pecinka et al., 2010). This TE has a complex regulation, it is normally transcriptionally silent, but its promoter region shares a cis-regulatory sequence motif with plant-derived heat stress defense factors (Cavrack et al., 2014; Ito et al., 2015), resulting in the transcription and production of full-length extra-chromosomal DNA under heat stress (Cavrak et al., 2014; Ito et al., 2016). When plants activate their heat stress defenses, extrachromosomal DNA copies are generated by the transposon which then has the ability to integrate into new genome positions. This retransposition has been observed in mutant plants impaired in siRNAs biogenesis, but not in wild-type plants indicating a key role for siRNAs in restricting retransposition (Ito et al., 2015). The insertion of this element leads to the up-regulation of genes near to the insertion site in response to heat stress (Cavrak et al., 2014; Ito et al., 2011). ONSEN appears to exploit a conserved stress response in the germline to avoid stable maintenance of DNA methylation. Its promoter region is devoid of CG and CHG sites necessary for stable DNA methylation by the host plant (Cavenak et al., 2014). To determine the effect of ONSEN insertion on heat stress tolerance, Ito et al. (2016) investigated the transcriptional regulation of ABA-responsive genes possessing the insertion. The mobilization of ONSEN created a mutation in an ABA-responsive gene, resulting in an ABA insensitive phenotype, suggesting that the ONSEN insertion does confer heat stress tolerance in *A. thaliana*. This heat-responsive TE is conserved in other brassica species including *B. oleracea* and *B. napus* raising the possibility that it is under a positive select ion (Nozawa et al., 2017).

Multiple heat-responsive COPIA families have been found in *A. thaliana* and *A. lyrata* including ONSEN, TERESTRA, COPIA37, and ROMANIAT5 (Pietzenuk et al., 2016). Heat responsiveness correlates with different HREs including variable gap and step types in all families and more specifically, 3P types in COPIA37 and ROMANIAT5, 4P types in ONSEN and TERESTRA. While gap- and step-type HREs are not sufficient to induce heat stress-related TE upregulation, 3P HREs correlate with 100-fold upregulation and the 4P type HREs up to 1000-fold increase (Pietzenuk et al., 2016). After heat stress, TEs get resilenced, a process that partly requires the Chromatin Assembly Factor 1 (CAF-1) (Pecinka et al., 2010).

Low-temperature stress can be categorized as chilling stresses in the range of 0–15°C and freezing stress below 0°C (Ruelland et al., 2009). Exposure to chilling temperatures can acclimate acclimatize plants to freezing temperatures in the process of cold acclimation (Thomashow, 1999).

Various TEs are activated by low temperatures, for example, the Tnt1 retrotransposon in tomato and tobacco is activated by freezing temperatures (Mhiri et al., 1997) and a Copia-like element is activated by cold temperatures in blood orange (Butelli et al., 2012). In blood orange, this insertion activates the Ruby MYB gene responsible for anthocyanin pigment formation under cold stress (Butelli et al., 2012). This is a striking example of transposon-induced phenotype being influenced by cold stress. An interesting example of very tightly controlled regulation is the Tam3 transposon of *Anthirrhinum majus* which is repressed at high temperatures but activated at low temperatures of 15°C (Carpenter et al., 1987). This low-temperature activation coincides with a change in the methylation state, with a reversible decrease in methylation below15°C (Hashida et al., 2003). At high temperatures, there is hypermethylation, although suppression at high temperature does not appear to be methylation dependent, as treatment with methylation inhibitors does not increase activation frequency (Hashida et al., 2005). This temperature-dependent change in the DNA methylation state occurs in developing but not fully developed tissues and is dependent on Tam3 transposase which binds to Tam3 at low temperatures (Hashida et al., 2006). Methylation proceeds rather than precedes this activation with the methylation being controlled by Tam3 activation.

The advent of transcription profiling allows large numbers of TEs to be screened, however; interpreting can be complex due to the repetitive nature of TEs (Lanciano and Cristofari, 2020). Transcriptional analysis in maize has revealed genotypic stress-specific expression in response to temperature stresses that are under the control of both genetic and epigenetic differences

(Liang et al., 2021). Expression patterns correlate with the presence of TEs in specific genotypes and TEs containing regions free of DNA methylation are much more likely to be stress responsive than methylated TEs (Liang et al., 2021).

13.5.5 ADAPTABILITY OF PLANTS TO HEAVY METAL SOIL TOXICITY

Aluminum (Al) toxicity is a key abiotic stress for crop plants, limiting crop productivity largely due to its detrimental effects on root growth (Jaiswal et al., 2018). Al is present in most soils and is more harmful in acidic soils, above pH 5. Lamounts of genotypic variation for aluminum adaptation and tolerance have been identified, reviewed in Kochian et al. (2015) and resistance can be categorized as plants that exclude Al from tissues and those with mechanisms to tolerate Al (Taylor, 1991). A key mechanism of Al tolerance in species such as barley and wheat is the efflux of several organic anions (OAs) including malate and citrate which form a stable complex with $Al3^{+}$. In plants using this mechanism, there is often a correlation between Al resistance and OA efflux genes (Delhaize et al., 2012). Various TEs have been detected inside or near OA transporters that are specifically associated with resistance to Al, and these can be classified into several distinct types (reviewed extensively by Pereira et al., 2019). The majority of these TEs are associated with increasing Al resistance in the plant by increasing OA efflux from the roots (Tovkach et al., 2013); however, some TEs are associated with decreased resistance (Collins et al., 2008) or have no detectable effect on Al resistance (Sasaki et al., 2006).

TEs have been detected near OA efflux genes in cereal crops such as barley (Kashino-Fujii et al., 2018), rice (Yokosho et al., 2016), and wheat (Tovkach et al., 2013), and are often associated with elevated levels of Al tolerance genes at the root apices. Rice tolerates Al well, and a key regulator is the aluminum resistance transcription factor1 (ART1), which regulates approximately 32 genes responsible for a diverse range of tolerance mechanisms (Yamaji et al., 2009). The rice OsFRDL4 citrate transporter gene is involved in Al-induced citrate secretion and contributes to Al tolerance (Yokosho et al., 2011). Phenotypic variation in Al resistance has been explained by a 1.2-kb insertion in the OsFRDL4 promoter region (Yokosho et al., 2016). This insertion contains nine cis-acting elements for ART1 and is responsible for elevated OsFRDL4 expression in some accessions and, therefore, genotypic variations in Al resistance. In contrast to rice, barley is extremely sensitive to high levels of Al in the soil resulting in plant death or much reduced productivity (Minella and Sorrells, 1992). There is a large genotypic variance for Al toxicity in barley; however, the situation is less complex than that of rice, with resistance being controlled by the HvAACT1 gene located on chromosome 4H, which encodes for a citrate extrusion gene (Wang et al., 2007). Higher levels of constitutive HvAACT1 expression are correlated with greater tolerance to high Al and this activity has been correlated with the presence of two independent TE insertions. In some cultivars, HvAACT1 expression at the root tips is significantly enhanced by the presence of a multiretrotransposon-like (MLR) sequence detected upstream of HvAACT1 (Kashino-Fujii et al., 2018). This results in plants with higher levels of Al resistance, but only if the promoter region of HvAACT1 has been demethylated. An independent 1023-bp CACTA-like TE located in the 5′ untranslated region upstream of the HvAACT1 gene some cultivars also correlates with increased tolerance to Al (Fujii et al., 2012) demonstrating the range of independent TE insertions that correlate with Al resistance in barley.

There is large variation for Al resistance in wheat with another Al-sensitive crop, which is related to both citrate and malate efflux (Ryan et al., 2011). The wheat gene TaMATE1B which encodes a citrate transporter is also linked to genotype-specific Al resistance. This resistance has been shown to result from a 11.1-kb transposon-like element located upstream from the start codon (Tovkach et al., 2013). The insertion extends the expression of TaMATE1B expression to the root apex, resulting in increased Al tolerance. Al resistance in wheat which is related to malate efflux at the root tip is controlled by a major gene ALMT1 (Raman et al., 2005). An analysis of genomic regions up and downstream of the ALMT1 gene revealed the presence of transposon-like insertions. Interestingly, the downstream insertion does not correlate with resistance to Al; however, the insertion located upstream contains blocks of duplicated or triplicated transposon-like insertions and the numbers of these repeats correlate with $ALMT_1$ expression and Al resistance in these plants (Sasaki et al., 2006). Transgenic studies to directly compare the promoter alleles have demonstrated that these repeated elements are linked with transcriptional regulation and elevated gene expression (Ryan et al., 2010). In comparison, the Al tolerant species *Secale cereale* L (Rye) provided a possible example of a negative effect of an insertion TE on fitness (Collins et al., 2008). Al sensitive genotypes have a splice variant of the ScALMT1 genes a result of a 227-bp MITE insertion into intron 2.

Arsenic is a nonessential metalloid toxic to plants and animals with a widespread prevalence in soils. The most abundant form of arsenic, arsenate is similar to phosphate and at levels of low soil phosphate, it can be taken up by phosphate transporters into the plant and then, in the case of crop plants, can reach the food chain (Catarecha et al., 2007). It is a particular issue of concern in rice paddies which is taken up in the rice grains (Awasthi et al., 2017). Some plants gain arsenic tolerance by restrict phosphate (Murota et al., 2012); however, the gains in arsenic tolerance are at a cost to phosphate supply. Arsenate has been shown to trigger a rapid TE burst in *A. thaliana* (Castrillo et al., 2013). The *A. thaliana* transcription factor WRKY6 rapidly represses this arsenate-induced TE burst by acting directly as a transcriptional repressor with no modification of histone marks (Castrillo et al., 2013). This mechanism of transposon silencing is similar to the heat stress response (Tittel-Elmer et al., 2010).

Iron is an essential nutrient for plants; however, high levels lead to toxicity (Stein et al., 2009). A large number of LTR retrotransposons are upregulated in response to iron stress in the rice Nipponbare cultivar (Finatto et al., 2013). Iron stress modified the transcription of 37% of all LTR retrotransposons of which the majority were up-regulated. Some cis-regulatory elements were found to be common to stress-responsive genes and LTR retrotransposons (Finatto et al., 2013).

13.5.6 TRANSPOSONS AS A STRATEGY FOR THE GENETIC IMPROVEMENT IN CROP PLANTS IN RELATION TO ABIOTIC STRESS

TEs have been a major driver in the evolution of plants and the knowledge gained can now be harnessed to drive crop improvement and development. TEs comprise a substantial component of crop genomes, for example, 85% of the genome in maize (Baucom et al., 2009), and 82% of the genome in wheat (Choulet et al., 2010) is TE derived. There is huge variation in type and number of TEs both between and within species (Lisch, 2013). These mobile elements have shaped crop evolution by providing new genes (Joly-Lopez and Bureau, 2018) and novel cis-regulatory elements (Pietzenuk et al., 2016) altered gene expression patterns (Bennetzen, 2000), and provided major structural components (Lisch, 2009). A further consequence of TE insertion into the genome is mutations such as loss of function due to insertion into introns and exons (Lisch, 2013).

Improving crop plants for abiotic stress, tolerance is challenging due to the often-complex nature of the traits and high prevalence of associated

QTLs (Reviewed in Ahmad et al., 2018), and the influence of different developmental stages on abiotic stress tolerance (Ma et al., 2020). There is evidence that TE insertions can have a positive influence on phenotype under stress conditions (Kitomi et al., 2020), which raises the possibility of novel methods of crop adaptation to different abiotic stresses. The mPing miniature inverted-repeat TE alters gene expression in rice in a stress-dependent manor but has no effect on growth or productivity under nonstress conditions suggesting that this could be a mechanism for future crop improvements (Yasuda et al., 2013). One limitation of breeding for abiotic stress tolerance is the narrow genetic diversity of the major crop plants. Natural genetic variation for stress tolerance has been linked to TE insertions (Barah et al., 2013) although how widely this occurs in crop plants is not yet known. The MITE insertion in maize is associated with drought tolerance (Mao et al., 2015) highlighting this as another target for future breeding of drought stress tolerance. An intriguing study by Long et al. (2009) looked at transposition rate of during space flight. They found that the changes in transposition rate of MITEs in rice were genotype-dependent and highlighted spaceflight as a potential mutagenic and stressful environment. With the planned manned missions to Mars, this is an existing future area of study. TEs have been shown to play an important role in maize adaptation to cooler temperatures (Lai et al., 2017). The gains in knowledge and understanding of how these TEs interact under conditions of stress are valuable in assessing their use as future crop breeding tools (Paszkowski, 2015).

13.6 CONCLUDING REMARKS

Plants are compelled to make changes at the genetic and physiological level in order to survive and reproduce under conditions of abiotic stress caused by unstable climatic environments. Barbara McClintock first postulated that TEs were "controlling agents" responsible for gene stress interactions (McClintock, 1984). Over the intervening years, an increasingly large body of research has reported on the fundamental role that TEs play in plant resistance to abiotic stress. These TEs are now viewed as fundamental elements in a large number of processes, reshaping genomes, and controlling gene activity. The mobile elements are determinists of gene expression and the changes they induce are at the origin of genetic variability allowing plants to adapt to environmental stress. Understanding the full picture of genetic and epigenetic entails full apprehension of TEs activities in plants. With the era of genomics and transcriptomics, it has become possible to acquire a fuller understanding of the different ways that different families of TEs interact with the host genome in relation to abiotic stress at the phenotypic, genetic, and epigenetic levels. As more knowledge is gained, TEs will take a central role in both the understanding of plant responses to abiotic stresses and the breeding strategies to improve them.

KEYWORDS

- **transposons**
- **abiotic stress**
- **tolerance**
- **crops**
- **epigenetics**

REFERENCES

Anderson, J. P.; Badruzsaufari, E.; Schenk, P. M.; Manners, J. M.; Desmond, O. J.; Ehlert, C.; Maclean, D. J.; Ebert, P. R.; Kazan, K. Antagonistic Interaction Between Abscisic Acid and Jasmonate Ethylene Signaling Pathways Modulates Defense Gene Expression and Disease Resistance in Arabidopsis. *Plant Cell* **2004,** *16*, 3460–3479.

Ahmad, H. M.; Ansari, M. R.; Azeem, F.; Nida, T.; Iqbal, M. QTL Mapping for Crop Improvement Against Abiotic Stresses in Cereals. *J. Anim. Plant Sci.* **2018,** *2*, 1558–1573.

Awasthi, S.; Chauhan, R.; Srivastava, S.; Tripathi, R. D. The Journey of Arsenic from Soil to Grain in Rice. *Front Plant Sci.* **2017,** *20*, 8:1007. DOI: 10.3389/fpls.2017.01007. PMID: 28676805; PMCID: PMC5476935.

Barah, P.; Jayavelu, N. D.; Mundy, J.; Bones, A. M. Genome Scale Transcriptional Response Diversity Among Ten Ecotypes of Arabidopsis Thaliana During Heat Stress. *Front Plant Sci.* **2013,** *26*, 4:532. DOI: 10.3389/fpls.2013.00532. PMID: 24409190; PMCID: PMC3872818.

Barakat, A.; Carels, N.; Bernardi, G. The Distribution of Genes in the Genomes of Gramineae. *Proc. Natl. Acad. Sci. U.S.A.* **1997,** *94*, 6857–6861.

Bartels,D.; Schneider, K.; Terstappen, G.; Piatkowski, D.; Salamini, F. Molecular Cloning of Abscisic Acid Modulated Genes Which Are Induced During Desiccation of the Resurrection Plant Craterostigma Plantagineum. *Planta* **1990**, *181*, 27–34.

Baucom, R. S.; Estill, J. C.; Chaparro, C.; Upshaw, N.; Jogi, A.; Deragon, J. M.; Westerman, R.; Sanmiguel, P.; Bennetzen, Jeffrey. Exceptional Diversity, Non-Random Distribution, and Rapid Evolution of Retroelements in the bB73 Maize Genome. *PLoS Genet* **2009,** *5*, e1000732.

Beguiristain, T.; Grandbastien, M.; Puigdomènech, P.; Casacuberta, J. Three Tnt1 Subfamilies Show Different Stress-Associated Patterns of Expression in Tobacco. Consequences for Retrotransposon Control and Evolution in Plants. *Plant physiol.* **2001**, *127*, 212–221.

Bennetzen, J. L. Transposable Element Contributions to Plant Gene and Genome Evolution. *Plant Mol. Biol.* **2000,** *42*, 251–269.

Benoit, M.; Drost, H. G.; Catoni, M.; Gouil, Q.; Lopez-Gomollon, S.; Baulcombe, D.; Paszkowski, J. Environmental and Epigenetic Regulation of Rider Retrotransposons in Tomato. *PLoS Genet.* **2019,** *16*, 15:e1008370.

Bhattacharyya, M. K.; Smith, A. M.; Ellis, T. H.; Hedley, C.; Martin, C. The Wrinkled-Seed Character of Pea Described by Mendel is Caused by a Transposon-Like Insertion in a Gene Encoding Starch Branching Enzyme. *Cell* **1990,** *60*,115–122.

Bourque, G.; Burns, K. H.; Gehring, M.; Gorbunova, V.; Seluanov, A.; Hammell, M.; Imbeault, M.; Izsvák, Z.; Levin, H. L.; Macfarlan, T. S.; Mager, D. L.; Feschotte, C. Ten Things You Should Know About Transposable Elements. *Genome Biol.* **2018,** *19*, 199. DOI: https://doi.org/10.1186/s13059-018-1577-z.

Boyer, J. S. Plant Productivity and Environment. *Science* **1982,** *218*, 443–448.

Brandt, J.; Schrauth, S.; Veith, A. M.; Froschauer, A.; Haneke, T.; Schultheis, Ch.; et al. Transposable Elements as a Source of Genetic Innovation: Expression and Evolution of a Family of Retrotransposon-Derived Neogenes in Mammals. *Gene* **2005,** *345*, 101–111.

Butelli, E.; Licciardello, C.; Zhang, Y.; Liu, J.; Mackay, S.; Bailey, P.; Reforgiato-Recupero, G.; Martin, C. Retrotransposons Control Fruit-Specific, Cold-Dependent Accumulation of Anthocyanins in Blood Oranges. *Plant Cell* **2012,** *24*, 1242–1255.

Capiati, D. A.; País, S. M.; Téllez-Iñón, M. T. Wounding Increases Salt Tolerance in Tomato Plants: Evidence on the Participation of Calmodulin-Like Activities in Cross-Tolerance Signaling. *J. Exp. Bot.* **2006,** *57*, 2391–2400.

Capy, P.; Bazin, C.; Higuet, D.; Langin, T. *Dynamics and Evolution of Transposable Elements*; Landes Company; Austin, 1997.

Carpenter, R.; Martin, C.; Coen, E. Comparison of Genetic Behaviour of the Transposable Element Tam3 at Two Unlinked Pigment Loci in *Antirrhinum majus. Mol. Gen. Genet.* **1987,** *207*, 82–89.

Catarecha, P.; Segura, M. D.; Franco-Zorrilla, J. M.; García-Ponce, B.; Lanza, M.; Solano, R.; Paz-Ares, J.; Leyva, A. A Mutant of the Arabidopsis Phosphate Transporter PHT1;1 Displaysenhanced Arsenic Accumulation. *Plant Cell* **2007,** *19*, 1123–1133.

Cavrak, V.; Lettner, N.; Jamge, S.; Kosarewicz, A.; Bayer, L.; Mittelsten, S. O. How a Retrotransposon Exploits the Plant's Heat Stress Response for Its Activation. *PLoS Genet.* **2014,** *10*, e1004115. DOI: 10.1371/journal.pgen.1004115.

Casati, P.; Walbot, V. Rapid Transcriptome Responses of Maize (*Zea mays*) to UV-B Irradiated and Shielded Tissues. *Genome Biol.* **2004,** *5*, R1628.

Castrillo, G.; Sánchez-Bermejo, E.; de Lorenzo, L.; Crevillén, P.; Fraile-Escanciano, A.; Mohan, T.C.; Mouriz, A.; Catarecha, P.; Sobrino-Plata, J.; Olsson, S.; Leo del Puerto, Y.; Mateos, I.; Rojo, E.; Hernández, L. E.; Jarillo, J. A.; Piñeiro, M.; Paz-Ares, J.; Leyva, A. WRKY6 Transcription Factor Restricts Arsenate Uptake and Transposon Activation in Arabidopsis. *Plant Cell* **2013,** *25*, 2944–2957.

Cheng, C.; Daigen, M.; Hirochika, H. Epigenetic Regulation of the Rice Retrotransposon Tos17. *Mol. Genet. Genomic.* **2006,** *276*, 378–390.

Choulet, F.; Wicker, T.; Rustenholz, C.; Paux, E.; Salse, J.; Leroy, P.; Schlub, S,; Le Paslier, M. C.; Magdelenat, G.; Gonthier, C.; Couloux, A.; Budak, H.; Breen, J.; Pumphrey, M.;

Liu, S.; Kong, X.; Jia, J.; Gut, M.; Brunel, D.; Anderson, J. A.; Gill, B. S.; Appels, R.; Keller, B.; Feuilleta, C. Megabase Level Sequencing Reveals Contrasted Organization and Evolution Patterns of the Wheat Gene and Transposable Element Spaces. *Plant Cell* **2010,** *22*, 1686–1701.

Collins, N. C.; Shirley, N. J.; Saeed, M., Pallotta, M., Gustafson, J. P. 2008. An ALMT1 Gene Cluster Controlling Aluminum Tolerance at the Alt4 Locus of Rye (Secale cereale L.). *Genetics* **2008,** *179*, 669–682.

Cowley, M.; Oakey, R. J. Transposable Elements Rewire and Fine-Tune the Transcriptome. *PLoS Genet.* **2013,** *9*, e1003234

de Souza, F. S. Franchini, L. F.; Rubinstein, M. Exaptation of Transposable Elements Into Novel Cis-Regulatory Elements: Is the Evidence Always Strong? *Mol. Biol. Evol.* **2013,** *30*, 1239–1251.

Delhaize, E.; Ma, J. F.; Ryan, P. R. Transcriptional Regulation of Aluminium Tolerance Genes. *Trends. Plant Sci.* **2012,** *17*, 341–348.

Devos, K. M. Grass Genome Organization and Evolution. *Curr. Opin. Plant Biol.* **2010,** *13*, 139–145.

Donnelly, A.; Caffarra, A.; Kelleher, C. T.; O'Neill, B. F.; Diskin, E.; Pletsers, A.; Proctor, H.; Stirnemann, R.; O'Halloran, J.; Peñuelas, J.; Hodkinson, T. R.; Sparks, T. H. Surviving in a Warmer World: Environmental and Genetic Responses. *Clim. Res.* **2012,** *53*, 245–262.

Estep, M. C.; DeBarry, J. D.;Bennetzen, J. L. The Dynamics of LTR Retrotransposon accumulation Across 25 Million Years of Panicoid Grass Evolution. *Heredity* **2003,** *110*, 194–204.

Fernandez, L.; Torregrosa, L.; Segura, V.; Bouquet, A.; Martinez-Zapater, J. M. Transposon-Induced Gene Activation as a Mechanism Generating Cluster Shape Somatic Variation in Grapevine. *Plant J.* **2010,** *61*, 545–557.

Feschotte, C.; Jiang, N.; Wessler, S. R. Plant Transposable Elements: Where Genetics Meets Genomics. *Nat. Rev. Genet.* **2002,** *3*, 329–341.

Feschotte, C.; Pritham E. J. DNA Transposons and the Evolution of Eukaryotic Genomes. *Annu. Rev. Genet.* **2007,** *41*, 331–68.

Finatto, T.; Costa de Oliveira, A.; Chaparro, C.; da Rosa Farias, D.; Woyann, L.; Mistura, C.; Soares-Bresolin, A.; Christel, L.; Panaud, O.; Picault, N. Abiotic Stress and Genome Dynamics: Specific Genes and Transposable Elements Response to Iron Excess in Rice. *Rice* **2013,** *13*, DOI: 10.1186/s12284-015-0045-6.

Flavell, R. B.; Rimpau, J.; Smith, D. B. Repeated Sequences DNA Relationships in Four Cereal Genomes. *Chromosoma* **1977,** *63*, 205–222.

Forestan, C.; Aiese, C.; Riccardo, F. S.; Lunardon, A.; Sanseverino, W.; Varotto, S. Stress-Induced and Epigenetic-Mediated Maize Transcriptome Regulation Study by Means of Transcriptome Reannotation and Differential Expression Analysis. *Sci. Rep.* **2016,** *6*, 30446. DOI: 10.1038/srep30446.

Fujii, M.; Yokosho, K.; Yamaji, N.; Saisho, D.; Yamane, M.; Takahashi, H.; Sato, K.; Nakazono, M.; Ma, J. F. Acquisition of Aluminium Tolerance by Modification of a Single Gene in Barley. *Nat. Commun.* **2012,** *3*, 713.

Fujino, K.; Hashida, S. N.; Ogawa, T.; Natsume, T.; Uchiyama, T.; Mikami, T.; Kishima, Y. Temperature Controls Nuclear Import of Tam3 Transposase in Antirrhinum. *Plant J.* **2011,** *65*, 146–155.

Furini, A.; Koncz, C.; Salamini, F.; Bartels, D. High Level Transcription of a Member of a Repeated Gene Family Confers Dehydration Tolerance to Callus Tissue of *Craterostigma plantagineum*. *EMBO J.* **1997,** *16*, 3599–3608.
Galindo-González, L.; Sarmiento, F.; Quimbaya, M. Shaping Plant Adaptability, Genome Structure and Gene Expression Through Transposable Element Epigenetic Control: Focus on Methylation. *Agronomy* **2018,** *8*, 180.
Gohlke, J.; Scholz, C. J.; Kneitz, S.; Weber, D.; Fuchs, J.; Hedrich, R.; Deeken, R. DNA Methylation Mediated Control of Gene Expression Is Critical for Development of Crown Gall Tumors *PLoS Genet* **2013,** *9*, e1003267
Grandbastien, M. A. LTR Retrotransposons, Handy Hitchhikers of Plant Regulation and Stress Response. *Biochimica et Biophysica Acta (BBA)—Gene Regul. Mech.* **2015,** *1849*, 403–416.
Grandbastien, M. A.; Audeon, C.; Bonnivard, E.; Casacuberta, J. M.; Chalhoub, B.; Costa, A. P.; Le, Q. H.; Melayah, D.; Petit, M.; Poncet, C.; Tam, S. M.; Van Sluys M. A.; Mhiri, C. Stress Activation and Genomic Impact of Tnt1 Retrotransposons in Solanaceae. Cytogenet. *Genome Res.* **2005,** *110*, 229–241.
Grandbastien, M. A.; Lucas, H.; Morel, J. B.; Mhiri, C.; Vernhettes, S.; Casacuberta, J. M. The Expression of the Tobacco Tnt1 Retrotransposon Is Linked to Plant Defense Responses. *Genetica* **1997,** *100*, 241–252.
Gurian-Sherman, D. *High and Dry, Why Genetic Engineering Is Not Solving Agriculture's Drought Problem in a Thirsty World*; Union of Concerned Scientists; UCS Publications: Cambridge, MA, 2012.
Hashida, S. N.; Kitamura, K.; Mikami, T.; Kishima, Y. Temperature Shift Coordinately Changes the Activity and the Methylation State of Transposon Tam3 in *Antirrhinum majus*. *Plant Physiol.* **2003,** *132*, 1207–1216.
Hashida, S. N.; Kishima, Y.; Mikami, T. DNA Methylation Is Not Necessary for the Inactivation of the Tam3 Transposon at Non-Permissive Temperature in Antirrhinum. *J. Plant Physiol.* **2005,** *162*, 1292–1296.
Hashida, S. N.; Uchiyama, T.; Martin, C.; Kishima, Y.; Sano, Y.; Mikami, T. The Temperature-Dependent Change in Methylation of the Antirrhinum Transposon Tam3 Is Controlled by the Activity of Its Transposase. *Plant Cell.* **2006,** *18*, 104–118.
Hilbricht, T.; Varotto, S.; Sgaramella, V.; Bartels, D.; Salamini, F.; Furini, A. Retrotransposons and siRNA Have a Role in the Evolution of Desiccation Tolerance Leading to Resurrection of the Plant *Craterostigma plantagineum*. *New Phytol.* **2008,** *179*, 877–887.
Holmberg, N.; Bülow, L. Improving Stress Tolerance in Plants by Gene Transfer. *Trends Plant Sci.* **1998,** *3*, 61–66.
Howe, G. T.; Brunner, A. M. An Evolving Approach to Understanding Plant Adaptation. *New Phytol*. **2005,** *167*, 1–5.
Huang, J.; Wang, Y.; Liu, W.; Shen, X.; Fan, Q.; Jian, S.; Tang, T. EARE-1, a Transcriptionally Active Ty1/Copia-Like Retrotransposon has Colonized the Genome of Excoecaria Agallocha Through Horizontal Transfer. *Front. Plant Sci.* **2017,** *8*, 45.
IPCC (Intergovernmental Panel on Climate Change). *Climate Change. The Physical Science Basis: Summary for Policymakers*; IPCC Secretariat: Geneva, Switzerland; 2007.
Ito, H.; Kim, J. M.; Matsunaga, W.; Saze, H.; Matsui, A.; Endo, T. A.; Harukawa,Y.; Takagi, H.; Yaegashi, H.; Masuta, Y.; Masuda, S.; Ishida, J.; Tanaka, M.; Takahashi, S.; Morosawa, T.; Toyoda, T.; Kakutani, T.; Kato, A.; Seki, M. A Stress Activated Transposon in Arabidopsis Induces Transgenerational Abscisic Acid Insensitivity. *Sci. Rep.* **2016,** *6*, 23181

Ito, H.; Gaubert, H.; Bucher, E.; Mirouze, M.; Vaillant, I.; Paszkowski, J. An siRNA Pathway Prevents Transgenerational Retrotransposition in Plants Subjected to Stress. *Nature* **2011,** *472*, 115–119.

Jackson, M. B.; Colmer, T. D. Response and Adaptation by Plants to Flooding Stress. *Ann. Bot.* **2005,** *4*, 501–505.

Jaiswal, S. K.; Naamala, J.; Dakora, F. D. Nature and Mechanisms of Aluminium Toxicity, Tolerance and Amelioration in Symbiotic Legumes and Rhizobia. *Biol. Fertil. Soils* **2018,** *54*, 309–318.

Janni, M.; Gullì, M.; Maestri, E.; Marmiroli, M.; Valliyodan, B.; Nguyen, H. T.; Marmiroli, N. Molecular and Genetic Bases of Heat Stress Responses in Crop Plants and Breeding for Increased Resilience and Productivity. *J. Exp. Bot.* **2020,** *26*, 3780–3802.

Jha, U. C.; Nayyar, H.; Jha, R.; Khurshid, M.; Zhou, M.; Mantri, N.; Siddique, K. H. M. Long Non-Coding RNAs: Emerging Players Regulating Plant Abiotic Stress Response and Adaptation. *BMC Plant Biol.* **2020,** *20*, 466.

Jiang, N.; Bao, Z.; Zhang, X.; Hirochika, H.; Eddy, S. R.; McCouch, S. R.; Wessler, S. R. An Active DNA Transposon Family in Rice. *Nature* **2003,** *421*, 163–167.

Jiang, N.; Gao, D.; Xiao, H.; van der Knaap, E. Genome Organization of the Tomato Sun Locus and Characterization of the Unusual Retrotransposon Rider. *Plant J.* **2009,** *60*, 181–193.

Joly-Lopez, Z.; Bureau, T. Exaptation of Transposable Element Coding Sequences. *Curr. Opin. Genet. Dev.* **2018,** *49*, 34–42.

Joly-Lopez, Z.; Forczek, Ewa.; Vello, E.; Hoen, D.; Tomita, A.; Bureau, T. Abiotic Stress Phenotypes Are Associated with Conserved Genes Derived from Transposable Elements. *Front. Plant Sci.* **2017,** *8*, 2027. DOI: 10.3389/fpls.2017.02027.

Kalendar, R.; Tanskanen, J.; Immonen, S.; Nevo, E.; Schulman, A. From the Cover: Genome Evolution of Wild Barley (Hordeum spontaneum) by BARE1 Retrotransposon Dynamics in Response to Sharp Microclimatic Divergence. *PNAS* **2000,** *97*, 6603–6607.

Kashino-Fujii, M.; Yokosho, K.; Yamaji, N.; Yamane, M.; Saisho, D.; Sato, K.; Ma; J. F. Retrotransposon Insertion and DNA Methylation Regulate Aluminum Tolerance in European Barley Accessions. *Plant Physiol.* **2018,** *178*, 716–727.

Kidwell, M. G.; D. R. Lisch, D. R. Perspectives: Transposable Elements, Parasitic DNA, and Genome. *Evolution* **2001,** *55*, 1–24.

Kimura, Y.; Tosa, Y.; Shimada, S.; Sogo, R.; Kusaba, M.; Sunaga, T.; Betsuyaku, S.; Eto, Y.; Nakayashiki, H.; Mayama, S. OARE-1, a Ty1-Copia Retrotransposon in Oat Activated by Abiotic and Biotic Stresses. *Plant Cell Physiol.* **2001,** *42*, 1345–1354.

Kitomi, Y.; Hanzawa, E.; Kuya, N.; Inoue, H.; Hara, N.; Kawai, S.; Kanno,N.; Endo, M.; Sugimoto, K.; Yamazaki, T.; Sakamoto, S.; Sentoku, N.; Wu, J.; Kanno, H.; Mitsuda, N.; Toriyama, K.; Sato, T.; Uga, Y. Root Angle Modifications by the DRO1 Homolog Improve Rice Yields in Saline Paddy Fields. *PNAS* **2020,** *117*, 21242–21250.

Kochian, L. V.; Piñeros, M. A.; Liu, J.; Magalhaes, J. V. Plant Adaptation to Acid Soils: The Molecular Basis for Crop Aluminum Resistance. *Annu. Rev. Plant Biol.* **2015,** *66*, 571–598.

Lai, X.; Schnable, J. C.; Liao, Z.; Xu, J.; Zhang, G.; Li, C.; Hu, E.; Rong, T.; Xu, Y.; Lu, Y. Genome-Wide Characterization of Non-Reference Transposable Element Insertion Polymorphisms Reveals Genetic Diversity in Tropical and Temperate Maize. *BMC Genom.* **2017,** *18*, 702.

Lanciano, S.; Cristofari, G. Measuring and Interpreting Transposable Element Expression. *Nat. Rev. Genet.* **2020,** *12*, 721–736.

Li, J.; Huang, Q.; Sun, M. X.; Zhang, T.; Li, H.; Chen, B.; Xu, K.; Gao, G.; Li, F.; Yan, G.; Qiao, J.; Cai, Y.; Wu, X. Global DNA methylation variations after short-term heat shock treatment in cultured microspores of *Brassica napus* cv. Topas. *Sci. Rep.* **2016,** *6*. DOI: 10.1038/srep38401.

Liang, Z.; Anderson, S.; Noshay, J.; Crisp, P.; Enders, T.; Springer, N. Genetic and Epigenetic Variation in Transposable Element Expression Responses to Abiotic Stress in Maize. *Plant Physiol.* **2021**. DOI: 10.1093/plphys/kiab073.

Lim, C.; Yang, K.; Hong, Joon.; Choi, Jin.; Yun, D. J.; Hong, J. C.; Chung, Woo.; Lee, S. Y.; Cho, Moo. Gene Expression Profiles During Heat Acclimation in Arabidopsis Thaliana Suspension-Culture Cells. *J. Plant Res.* **2006,** *119*, 373–383.

Lisch, D. Epigenetic Regulation of Transposable Elements in Plants. *Annu. Rev. Plant Biol.* **2009,** *60*, 43–66.

Lisch, D. How Important Are Transposons for Plant Evolution? *Nat. Rev. Genet.* **2013,** *14*, 49–61.

Liu, B.; Wendel, J. F. Retrotransposon Activation Followed by Rapid Repression in Introgressed Rice Plants. *Genome* **2000,** *43*, 874–880.

Liu, C.; Lu, F.; Cui, X.; Cao, X. Histone Methylation in Higher Plants. *Annu. Rev. Plant. Biol.* **2000,** *61*, 395–420.

Long, Li.; Ou, X.; Liu, J.; Lin, X.; Sheng, L.; Liu, B. The Spaceflight Environment Can Induce Transpositional Activation of Multiple Endogenous Transposable Elements in a Genotype-Dependent Manner in Rice. *J. Plant Physiol.* **2009,** *166*, 2035–2045.

Lv, Y.; Hu, F.; Zhou, Y.; Wu, F.; Gaut, B. Maize Transposable Elements Contribute to Long Non-Coding RNAs that Are Regulatory Hubs for Abiotic Stress Response. *BMC Genom.* **2019,** *20*. DOI: 10.1186/s12864-019-6245-5.

Ma, X.; Su, Z.; Ma, H. Molecular Genetic Analyses of Abiotic Stress Responses During Plant Reproductive Development. *J. Exp. Bot.* **2020,** *71*, 2870–2885.

Mahajan, S.; Tuteja, N. Cold, Salinity and Drought Stresses: An Overview. *Arch. Biochem. Biophys.* **2005,** *444*, 139–158.

Makarevitch, I.; Waters, M.; West, P.; Stitzer, M.; Hirsch, C.; Ross-Ibarra, J.; Springer, N. Transposable Elements Contribute to Activation of Maize Genes in Response to Abiotic Stress. *PLoS Genet.* **2015,** *11*, e1004915. DOI: 10.1371/journal.pgen.1004915.

Malik, H. S.; Eickbush, T. H. Phylogenetic Analysis of Ribonuclease H Domains Suggests a Late, Chimeric Origin of LTR Retrotransposable Elements and Retroviruses. *Genome Res.* **2001,** *11*, 1187–1197.

Mannuss, A.; Trapp, O.; Puchta, H. Gene Regulation in Response to DNA Damage. *Biochim. Biophys. Acta.* **2012,** *1819*, 154–165.

Mao, H.; Wang, H.; Liu, S.; Li, Z.; Yang, X.; Yan, J.; Li, J.; Tran, L. S.; Qin, F. A Transposable Element in a NAC Gene Is Associated with Drought Tolerance in Maize Seedlings. *Nat. Commun.* **2015,** *6*, 8326. DOI: 10.1038/ncomms9326.

McClintock, B. The Origin and Behavior of Mutable Loci in Maize. *PNAS U.S.A* **1950,** *36*, 344–355.

McClintock, B. The Significance of Responses of the Genome to Challenge. *Science* **1984,** *226*, 792–801.

Melayah, D.; Bonnivard, E.; Chalhoub, B.; Audeon, C.; Grandbastien, M. A. The Mobility of the Tobacco Tnt1 Retrotransposon Correlates with Its Transcriptional Activation by Fungal Factors. *Plant J.* **2001,** *28*, 159–168.

Mhiri, C.; Morel, J. B.; Vernhettes, S.; Casacuberta, J. M.; Lucas, H.; Grandbastien, M. A. The Promoter of the Tobacco Tnt1 Retrotransposon Is Induced by Wounding and by Abiotic Stress. *Plant Mol. Biol.* **1997,** *33*, 257–266.

Minella, E.; Sorrells, M. E. Aluminum Tolerance in Barley: Genetic Relationships Among Genotypes of Diverse Origin. *Crop Sci.* **1992,** *32*, 593–598.

Murota, C.; Matsumoto, H.; Fujiwara, S.; Hiruta, Y.; Miyashita, S.; Shimoya, M.; Kobayashi, I.; Hudock, M. O.; Togasaki, R. K.; Sato, N.; Tsuzuki, M. Arsenic Tolerance in a Chlamydomonas Photosynthetic Mutant Is Due to Reduced Arsenic Uptake Even in Light Conditions. *Planta* **2012,** *236*, 1395–1403.

Naito, K.; Zhang, F.; Tsukiyama, T.; Saito, H.; Hancock, C.; Richardson, A.; Okumoto, Y.; Tanisaka, T.; Wessler, S. Unexpected Consequences of a Sudden and Massive Transposon Amplification on Rice Gene Expression. *Nature* **2009,** *461*, 1130–1134.

Negi, P.; Rai, A. N.; Suprasanna, P. Moving Through the Stressed Genome: Emerging Regulatory Roles for Transposons in Plant Stress Response. *Front. Plant Sci.* **2016,** *10*, 7:1448. DOI: 10.3389/fpls.2016.01448.

Nozawa, K.; Kawagishi, Y.; Kawabe, A.; Sato, M.; Masuta, Y.; Kato, A.; Ito, H. Epigenetic Regulation of a Heat-Activated Retrotransposon in Cruciferous Vegetables. *Epigenomes* **2017,** *1*, 7.

Oliver, K. R., Greene, W. K. Transposable Elements: Powerful Facilitators of Evolution. *BioEssays News Rev. Mol. Cell. Dev. Biol.* **2009,** *31*, 703–714.

Paszkowski, J. Controlled Activation of Retrotransposition for Plant Breeding. *Curr. Opin. Biotechnol.* **2015,** *32*, 200–206.

Pearce, S. R.; Harrison, G.; Heslop-Harrison, J. S.; Flavell, A. J.; Kumar, A. Characterization and Genomic Organization of Ty1-Copia Group Retrotransposons in Rye (Secale cereale). *Genome* **1997,** *40*, 1–9.

Pecinka, A.; Dinh, H. Q.; Baubec, T.; Rosa, M.; Lettner, N.; et al. Epigenetic Regulation of Repetitive Elements Is Attenuated by Prolonged Heat Stress in Arabidopsis. *Plant Cell* **2010,** *22*, 3118–3129.

Pecinka, A.; Rosa, M.; Schikora, A.; Berlinger, M.; Hirt, H.; Luschnig, C.; Mittelsten-Scheid, O. Transgenerational Stress Memory Is Not a General Response in Arabidopsis. *PLoS ONE* **2009,** *4*, e5202.

Pennisi, E. The Blue Revolution, Drop by Drop, Gene by Gene. *Science* **2008,** *320*, 171–173.

Pietzenuk, B.; Markus, C.; Gaubert, H.; Bagwan, N.; Merotto, A.; Bucher, E.; Pecinka, A. Recurrent Evolution of Heat Responsiveness in Brassicaceae COPIA Elements. *Genome Biol.* **2016,** *17*, 209.

Prasad, P. V. V.; Bheemanahalli, R.; Jagadish, S. V. K. Field Crops and the Fear of Heat Stress Opportunities, Challenges and Future Directions. *Field Crops Res.* **2017,** *200*, 114–121.

Qin, T.; Zhao, H. Y.; Cui, P.; Albesher, N.; Xiong, LM. A Nucleus-Localized Long Non-Coding RNA Enhances Drought and Salt Stress Tolerance. *Plant Physiol.* **2017,** *175*, 1321–1236.

Qüesta, J. I.; Fina, J. P.; Casati, P. DDM1 and ROS1 Have a Role in UV-B Induced and Oxidative DNA Damage in *A. thaliana*. *Front. Plant Sci.* **2013,** *4*, 420. DOI: 10.3389/fpls.2013.00420.

Raman, H.; Zhang, K.; Cakir, M.; Appels, R.; Garvin, D. F.; Maron, L. G.; Kochian, L. V.; Moroni, J. S.; Raman, R.; Imtiaz, M.; Drake-Brockman, F.; Waters, I.; Martin, P.; Sasaki, T.; Yamamoto, Y.; Matsumoto, H.; Hebb, D. M.; Delhaize, E.; Ryan, P. R. Molecular Characterization and Mapping of ALMT1, the Aluminium-Tolerance Gene of Bread Wheat (*Triticum aestivum* L.). *Genome* **2005,** *48*, 781–791.

Ruelland, E.; Vaultier, M. N.; Zachowski, A. Hurry, V. Cold Signalling and Cold Acclimation in Plants. *Adv. Bot. Res.* **2009,** *49*, 35–150.

Ryan, P. R.; Raman, H.; Gupta, S.; Sasaki, T.; Yamamoto, Y.; Delhaize, E. The Multiple Origins of Aluminium Resistance in Hexaploid Wheat Include *Aegilops tauschii* and More Recent Cis Mutations to TaALMT1. *Plant J.* **2010,** *64*, 446–455.

Ryan, P. R.; Tyerman, S. D.; Sasaki, T.; Furuichi, T.; Yamamoto, Y.; Zhang, W. H.; Delhaize, E. The Identification of Aluminium-Resistance Genes Provides Opportunities for Enhancing Crop Production on Acid Soils. *J Exp. Bot.* **2011,** *62*, 9–20.

Sah, S.; Reddy, K.; Li, J. Abscisic Acid and Abiotic Stress Tolerance in Crop Plants. *Front. Plant Sci.* **2016,** *7*. DOI: 10.3389/fpls.2016.00571.

Sallam, N.; Moussa, M. DNA Methylation Changes Stimulated by Drought Stress in ABA-Deficient Maize Mutant vp10. *Plant Physiol. Biochem.* **2021,** *160*, 218–224.

SanMiguel, P.; Bennetzen, J. L. Evidence That a Recent Increase in Maize Genome Size Was Caused by the Massive Amplification of Intergene Retrotransposons. *Ann. Bot.* **1998,** *82*, 37–44.

Sasaki, T.; Ryan, P. R.; Delhaize, E.; Hebb, D. M.; Ogihara, Y.; Kawaura, K.; Noda, K.; Kojima, T.; Toyoda, A.; Matsumoto, H.; Yamamoto, Y. Sequence Upstream of the Wheat (*Triticum aestivum* L.) ALMT1 Gene and Its Relationship to Aluminum Resistance. *Plant Cell Physiol.* **2006,** *47*, 1343–1354.

Secco, D.; Wang, C.; Shou, H.; Schultz, M. D.; Chiarenza, S.; Nussaume, L.; Ecker, J. R.; Whelan, J.; Lister, R. Stress Induced Gene Expression Drives Transient DNA Methylation Changes at Adjacent Repetitive Elements. *Elife* **2015,** *21*, 4:e09343. DOI: 10.7554/eLife.09343.

Scandalios, J. G. Oxidative Stress: Molecular Perception and Transduction of Signals Triggering Antioxidant Gene Defenses. *Braz. J. Med. Biol. Res.* **2005,** *38*, 995–1014.

Shao, H. B.; Shao, M. A.; Liang, Z. S. Osmotic Adjustment Comparison of 10 Wheat (*Triticum aestivum* L.) Genotypes at Soil Water Deficits. *Biointerfaces* **2006,** *47*, 132139.

Shim, J. S.; Oh, N.; Chung, P. J.; Kim, Y. S.; Choi, Y. D.; Kim, J. K. Overexpression of OsNAC14 Improves Drought Tolerance in Rice. *Front. Plant Sci.* **2018,** *9*, 310.

Shinozaki, K., Dennis, E. S. Cell Signaling and Gene Regulation Global Analyses of Signal Transduction and Gene Expression Profiles. *Curr. Opin. Plant Biol.* **2003,** *6*, 405409.

Shiu, O. Y.; Oetiker, J. H.; Yip, W. K.; Yang, S. F. The Promoter of LE-ACS7, an Early Flooding-Induced 1-Aminocyclopropane-1-Carboxylate Synthase Gene of the Tomato, Is Tagged by a Sol3 Transposon. *Proc. Natl. Acad. Sci. U.S.A.* **1998,** *95*, 10334–10339.

Silva, H.; Sagardia, S.; Seguel, O.; Torres, C.; Tapia, C.; Franck, N.; Cardemil, L. Effect of Water Availability on Growth and Water Use Efficiency for Biomass and Gel Production in Aloe Vera (*Aloe barbadensis* M.). *Ind. Crop. Prod.* **2010,** *31*, 20–27.

Sinzelle, L.; Izsvak, Z.; Ivics, Z. Molecular Domestication of Transposable Elements: From Detrimental Parasites to Useful Host Genes. *Cell. Mol. Life Sci.* **2009,** *66*, 1073–1093.

Smirnoff, N. Plant Resistance to Environmental Stress. *Curr. Opin. Biotechnol.* **1998,** *9*, 214–219.

Springer, N. M.; Schmitz, R. J. Exploiting Induced and Natural Epigenetic Variation for Crop Improvement. *Nat. Rev. Genet.* **2017,** *18*, 563–575.

Stapley, J.; Santure, A. W.; Dennis, S. R. Transposable Elements as Agents of Rapid Adaptation May Explain the Genetic Paradox of Invasive Species. *Mol. Ecol.* **2015,** *24*, 2241–2252.

Stein, R. J.; Duarte, G. L.; Spohr, M. G.; Lopes, S. I. G.; Fett, J. P. Distinct Physiological Responses of Two Rice Cultivars Subjected to Iron Toxicity Under Field Conditions. *Ann. Appl. Biol.* **2009,** *154*, 269–277.

Sun, L.; Jing, Y.; Liu, X.; Li, Q.; Xue, Z.; Cheng, Z.; Wang, D.; He, H.; Qian, W. Heat Stress-Induced Transposon Activation Correlates with 3D Chromatin Organization Rearrangement in Arabidopsis. *Nat. Commun.* **2020,** *11*. DOI: 10.1038/s41467-020-15809-5.

Suoniemi, A.; Narvanto, A.; Schulman, A. H. The BARE-1 Retrotransposon Is Transcribed in Barley from an LTR Promoter Active in Transient Assays. *Plant Mol. Biol.* **1996,** *31*, 295–306.

Takeda, S.; Sugimoto, K.; Otsuki, H.; Hirochika, H. A 13-bp Cis-Regulatory Element in the LTR Promoter of the Tobacco Retrotransposon *Tto1* Is Involved in Responsiveness to Tissue Culture, Wounding, Methyl Jasmonate and Fungal Elicitors. *Plant J.* **1999,** *18*, 383–393.

Takeda, S.; Sugimoto, K.; Otsuki, H.; Hirochika, H. Transcriptional Activation of the Tobacco Retrotransposon Tto1 by Wounding and Methyl Jasmonate. *Plant Mol. Biol.* **1998,** *36*, 365–376.

Taylor, G. J. 1991. Current Views of the Aluminum Stress Response: The Physiological Basis of Tolerance. *Curr. Top. Plant Biochem. Physiol.* **1991,** *10*, 57–93.

Tenaillon, M. I.; Hollister, J. D.; Gaut. B. S. A Triptych of the Evolution of Plant Transposable Elements. *Trends Plant Sci.* **2010,** *15*, 471–478.

Thieme, M.; Bucher, E. Transposable Elements as Tool for Crop Improvement. In *Direct Plant Epigenetics Coming of Age for Breeding Applications*; Gallusci, P., Bucher, E., Mirouze, M., Eds.; Elsevier: Amsterdam, 2018; pp 165–202.

Thomashow, M. F. Plant Cold Acclimation: Freezing Tolerance Genes and Regulatory Mechanisms. *Annu. Rev. Plant Physiol. Plant Mol. Biol.* **1999,** *50*, 571–599.

Tittel-Elmer, M.; Bucher, E.; Broger, L.; Mathieu, O.; Paszkowski, J.; Vaillant, I. Stress-Induced Activation of Heterochromatic Transcription. *PLoS Genet* **2010,** *6*, e1001175.

Tovkach, A.; Ryan, P. R.; Richardson, A. E.; Lewis, D. C.; Rathjen, T. M.; Ramesh, S.; Tyerman, S. D.; Delhaize, E. Transposon-Mediated Alteration of TaMATE1B Expression in Wheat Confers Constitutive Citrate Efflux from Root Apices. *Plant Physiol.* **2013,** *161*, 880–892.

Tuteja N. Abscisic Acid and Abiotic Stress Signalling. *Plant Signal. Behav.* **2007,** *2*, 135–138.

Uga, Y.; Okuno, K.; Yano, M. Dro1, a Major QTL Involved in Deep Rooting of Rice Under Upland Field Conditions. *J. Exp. Bot*. **2011,** *62*, 2485–2494.

Van Meter, M,.; Kashyap, M.; Rezazadeh, S.; Geneva, A. J.; Morello, T. D.; Seluanov, A.; Gorbunova V. SIRT6 Represses LINE1 Retrotransposons by Ribosylating KAP1 but This Repression Fails with Stress and Age. *Nat. Commun.* **2014,** *5*, 5011.

Vernhettes, S.; Grandbastien, M. A.; Casacuberta, J. M. The Evolutionary Analysis of the Tnt1 Retrotransposon in Nicotiana Species Reveals the High Variability of Its Regulatory Sequences. *Mol. Biol. Evol.* **1998,** *15*, 827–836.

Vicient, C. M.; Suoniemi, A.; Anamthawat-Jonsson, K.; Tanskanen, J.; Beharav, A.; Nevo, E.; Schulman, A. H. Retrotransposon BARE-1 and Its Role in Genome Evolution in the Genus Hordeum. *Plant Cell* **1999,** *11*, 1769–1784.

Voronova, A.; Belevich, V.; Jansons, A.; Rungis, D. Stress-Induced Transcriptional Activation of Retrotransposon-Like Sequences in the Scots Pine (*Pinus sylvestris* L.) Genome. *Tree Genet. Genomes* **2014,** *10*, 937–951.

Wang, J.; Raman, H.; Zhou, M.; Ryan, P. R.; Delhaize, E.; Hebb, D. M.; Coombes, N.; Mendham, N. High-Resolution Mapping of the Alp Locus and Identification of a Candidate Gene HvMATE Controlling Aluminium Tolerance in Barley (Hordeum vulgare L.). *Theor. Appl. Genet.* **2007,** *115*, 265–276.

Wagner, G. P.; Booth, G.; Bagheri-Chaichian, H. A Population Genetic Theory of Canalization. *Evolution* **1997,** *51*, 329–347.

Wendel, J. F.; Jackson, S. A.; Meyers, B. C.; Wing, R. A. Evolution of Plant Genome Architecture. *Genome Biol.* **2016,** *17*, 7. DOI: 10.1186/s13059-016-0908-1.

Wessler, S. R.; Bureau, T. E; White, S. E. LTR-Retrotransposons and MITEs: Important Players in the Evolution of Plant Genomes. *Curr. Opin. Genet. Dev.* **1995,** *5*, 814–821.

Wicker, T.; Sabot, F.; Hua-Van, A.; Bennetzen, J. L.; Capy, P.; Chalhoub, B.; Flavell, A.; Leroy, P.; Morgante, M.; Panaud, O.; et al. A Unified Classification System for Eukaryotic Transposable Elements. *Nat. Rev. Genet.* **2007,** *8*, 973–982.

Wu, G.; Zhang, C.; Chu, L. Y.; Shao H. B. Responses of Higher Plants to Abiotic Stresses and Agricultural Sustainable Development. *J. Plant Interact.* **2007,** *2*, 135–147.

Xiao, H.; Jiang, N.; Schaffner, E.; Stockinger, E. J.; van der Knaap, E. A Retrotransposon-Mediated Gene Duplication Underlies Morphological Variation of Tomato Fruit. *Science* **2008,** *319*, 1527–1530.

Yamaji, N.; Huang, C. F.; Nagao, S.; Yano, M.; Sato, Y.; Nagamura, Y.; Ma, J. F. A Zinc Finger Transcription Factor ART1 Regulates Multiple Genes Implicated in Aluminum Tolerance in Rice. *Plant Cell* **2009,** *21*, 3339–3349.

Yang, A.; Dai, X.; Zhang, W. H. A R2R3-Type MYB Gene, OsMYB2, Is Involved in Salt, Cold, and Dehydration Tolerance in Rice. *J. Exp. Bot.* **2012,** *63*, 2541–2556.

Yasuda, K.; Ito, M.; Sugita, T.; Tsukiyama, T.; Saito, H.; Naito, K.; Teraishi, M.; Tanisaka, T.; Okumoto, Y. Utilization of Transposable Element *mPing* as a Novel Genetic Tool for Modification of the Stress Response in Rice. *Mol. Plant Breed.* **2013,** *32*, 505–516.

Yokosho, K.; Yamaji, N.; Fujii-Kashino, M.; Ma, J. F. Retrotransposon-Mediated Aluminum Tolerance Through Enhanced Expression of the Citrate Transporter OsFRDL4. *Plant Physiol.* **2016,** *172*, 2327–2336.

Yokosho, K.; Yamaji, N.; Ma, J. F. An Al-inducible MATE Gene Is Involved in External Detoxification of Al in Rice. *Plant J.* **2011,** *68*, 1061–1069.

Zhao, Y.; Wu, L.; Fu, Q.; Wang, D.; Li, J.; Yao, B.; Yu, S.; Jiang, L.; Qian, J.; Zhou, X.; Han, L.; Zhao, S.; Ma, C.; Zhang, Y.; Luo, C.; Dong, Q.; Li, S.; Zhang, L.; Jiang, X.; Li, Y.; Luo, H.; Li, K.; Yang, J.; Luo, Q.; Li, L.; Peng, S.; Huang, H.; Zuo, Z.; Liu, C.; Wang, L.; Li, C.; He, X.; Friml, J.; Du, Y. INDITTO2 Transposon Conveys Auxin-Mediated DRO1 Transcription for Rice Drought Avoidance. *Plant Cell Environ.* **2021**. DOI: 10.1111/pce.14029.

CHAPTER 14

Genome Engineering in Plants via Transposable Elements

POOJA SARASWAT, AMBIKA CHATURVEDI, AYUSHI SINGH, PARUL TYAGI, and RAJIV RANJAN*

Department of Botany, Dayalbagh Educational Institute, Dayalbagh, Agra, Uttar Pradesh 282005, India

**Corresponding author. E-mail: rajivranjanbt@gmail.com*

ABSTRACT

Transposable elements (TEs) are natural DNA transfer vehicles that are capable of efficient genomic insertion. DNA transposons (type II TEs) can move between the genome by cut and paste mechanism due to which they play an important role in genome evolution. The movement of DNA transposons can be controlled by providing the elements (transposase) required for transposition reaction. Thus, a DNA of interest can be cloned efficiently between the inverted repeats of transposon-based vectors for genome insertions. activator/dissociation elements, *mPing*, and *piggyBac* are utilized for gene transfer and other applications like insertional mutagenesis, tagging, etc. Therefore, DNA transposons can be applied effectively in different studies including regulatory genome, functional genomics, etc. In this chapter, we describe the different transposon systems, recent advances, and the evolving knowledge of TEs. We also highlight transposable technology in the field of genetic engineering and its applications.

14.1 INTRODUCTION

The genome of prokaryotic and eukaryotic organisms usually has a large number of repeated DNA sequences, which are mobile in nature (Kazazian, 2004). These

Plant Transposable Elements: Biology and Biotechnology, Deepu Pandita, Anu Pandita, (Eds.)

mobile elements of DNA are known as Transposable elements (TEs). TEs were first discovered by Barbara McClintock and identified as Controlling elements (McClintock, 1956). TEs are known to cover a large area of Angiosperm DNA (Biémont, 2010). TEs generally cover a large portion of the genome of any organism, which is unique to that particular species. However, TEs also show a significant role in regulation of gene expression against environmental stress. The process of mobilization depends on the intermediates formed (transposition or retrotransposition). It is found that TEs cover over 60%, 35%, and 85% of genome in human, rice, and maize genome, respectively (de Koning et al., 2011; Turcotte et al., 2001). As TEs are seen making up a large portion of genome, it is assumed that they must have been involved in the size change of genome during evolution and speciation as reported in different organisms, including plants (SanMiguel et al., 1998), primates (Locke et al., 2003), etc. TEs have identified with increase in their copy number (thousands of copies) in a genome (Khan et al., 2011) showing their immense role in evolutionary process (Muñoz-López and García-Pérez, 2010).

TEs can impact a genome either in a positive or negative way, such as their mobile nature can result in gene repression, gene activation, and can induce recombination. Genetic changes can be seen due to transposition of TEs because of insertions, duplications, or translocation at the site of integration. Say for example DNA transposons alter the gene expression by inserting themselves within introns and exons (Lerman and Feder, 2005). In addition, TEs also help in the reorganization of a genome by recombination or by mobilizing nontransposon DNA (Sayah et al., 2004). It is also seen that TEs can result in genomic DNA loss by internal deletions (Petrov and Hartl, 1997). Often, when TEs get inserted into the gene-rich regions could change the expression of other adjacent genes. Also, they can disrupt the cis-elements and thereby the interaction between cis- and trans-elements is inhibited (Martin et al., 2009). Sometimes TEs show the transcription based on the stress responses (Bucher et al., 2012). For example, the different plant's TEs get activated in response to environmental stress like Tnt1 element (tobacco) by biotic and abiotic stress (Grandbastien et al., 2005), mPing (rice) by cold and salt stress (Naito et al., 2009), and ONSEN (Arabidopsis) by heat stress (Cavrak et al., 2014). Evidences are there that some of TEs responses to environmental stress can also alter the expression of the adjacent genes. This has been proven in a study where mPing miniature inverted repeat transposable elements (MITEs) when inserted into the rice genome resulted in change of nearby genes when exposed to cold or salt stress and with no change in the normal conditions (Yasuda et al., 2013).

TEs are a good source of genetic variability in the genome and the main reason for change in gene expression that led to different forms of phenotypes. Therefore, understanding TEs and their function in the genome has become an objective for the scientist for genetic engineering. TEs are now identified as a significant genetic tool for crop improvement. The transposon systems have been used for insertional mutagenesis (May and Martienssen, 2003), as vector, activation tagging (Mori et al., 2007), etc. to apply for gene expression studies, tagging genes, markers, etc. Various computation tools and databases are present for better understanding to make use of TEs in an efficient manner (Goerner-Potvin and Bourque, 2018). All the above-mentioned applications are discussed in details (later in the chapter).

14.2 TYPES OF TES

TEs are known for their ability to change the location within the genome. TEs are classified into two classes on the basis of transposition mechanism as class I and class II (RNA transposons and DNA transposons) and further into subclasses on the basis of chromosomal integration. Each of these subclasses is subdivided into subgroups (Wicker et al., 2007).

14.2.1 CLASS I: RNA TRANSPOSONS OR RETROTRANSPOSONS

Class I element or retrotransposons, replicate through a "copy-and-paste" mechanism. The functioning of RNA transposon is based on the RNA intermediate and its reverse transcription. These can be further divided into two groups based on the presence of long terminal repeats (LTRs) at the terminal (Fig. 14.1). LTR-based retrotransposons are very much similar to the retroviruses in their functioning because integrase helps in the integration as in retroviruses (Brown et al., 1987). LTRs are generally found in all plants and animals, also in yeast (*Saccharomyces cerevisiae*). They occupy the genome in different percentages in plants (about 95%) and humans (10%). Some examples are Copia, ERV, Gypsy, etc. In case of non-LTR retrotransposons containing both long and short interspersed nuclear elements (LINEs and SINEs) (Luan et al., 1993) are present in high copy number in amniotes and about 34% in human genome (Eickbush and Malik, 2002). LINEs include R2, RTE, Jockey, L1 super families, whereas SINE consist tRNA, 7SL, and 5S (Wicker et al., 2007).

14.2.2 *CLASS 2: DNA TRANSPOSONS*

DNA transposons or class II elements generally follow the cut-and-paste mechanism to move across the genome in which the transposon is getting cut from one location and inserts itself to another place. Most of them follow the nonreplicative mechanism, with some exceptions. They are generally mobilized through a DNA intermediate (Fig. 14.1), by cutting and pasting (most common) (Rubin et al., 1982) or a peel and paste mechanism found in Helitrons (Grabundzija et al., 2016), which is a replicative mechanism that involves formation of circular DNA intermediate.

TEs consist of two flanking repeats of Terminal Inverted Repeats (TIRs) and these enables transposase to recognize and perform the cutting of the transposon, which is further integrated into a new location of the genome (see below for further details). The insertion causes duplication of target site DNA or target site duplications (TSDs), which become unique to each TE. Based on the sequence, TIRs or TSDs are further classified into subclasses. Subclass I consist of families like Tc1/mariner, P, hAT (hobo-Ac-Tam3), CACTA, crypton, merlin, piggyBac, etc. Another which belongs to the subclass II is Helitron and Maverick transposons because of their replicative mode of transposition and no double-stranded breaks during their insertion. Taking few examples, Ty3/gypsy and Ty1/copia elements (LTR retrotransposons) are found in all major groups of eukaryotes (Feschotte and Pritham, 2007). In the same way, Tc1/mariner, hAT, and MULEs (mutator-like elements) (DNA transposons) are found in the eukaryotic tree (Feschotte and Pritham, 2007).

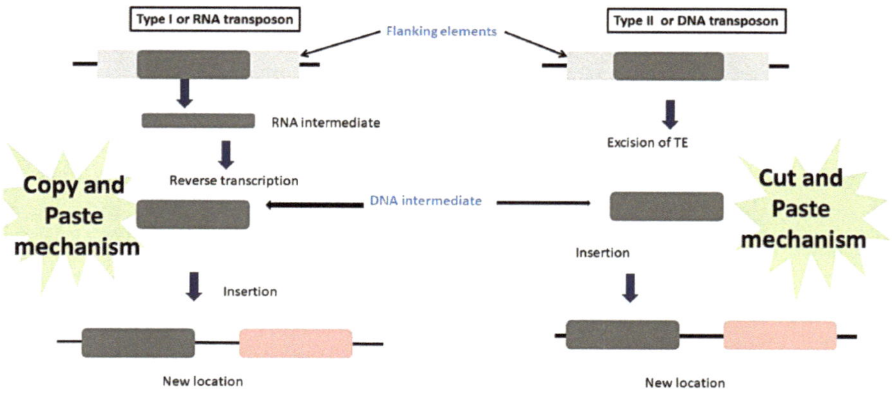

FIGURE 14.1 Types of transposon and their mechanism: Type I usually transpose through copy and paste mechanism like RNA viruses and Type II transpose via cut and paste method to insert itself into new a location in the genome.

14.3 CHARACTERISTICS OF TRANSPOSON SYSTEM

14.3.1 ACTIVATOR/DISSOCIATION (AC/DS) TES IN PLANTS

Ac elements known as the maize activator are considered as a paradigm TE belongs to hAT transposon superfamily, the member generally found in plant, fungi, and animals. Ac known to encodes the protein, which is responsible for the mobilization of autonomous *Ac* and nonautonomous *Dissociation* (*Ds*) elements, Ac/Ds become the most widely used TEs in plants for gene tagging and functional genomics approach (Lazarow et al., 2013). Ac/Ds elements are also known to generate genome rearrangements by substitute transposition reactions associated with the terminal region of closely linked transposons (Yu et al., 2012). The autonomous Ac is the simplest structure of the TE, which has ~4565 bp length and contains 11 bp TIRs (bp imperfect TIRs) with about 240-bp essential subterminal sequence at the both ends. It has been proved that the central portion of the Ac contains specific transcription unit for 3.5 kb mRNA that are responsible for the transposase protein (TPase) (Kunze et al., 1987; Coupland et al., 1988). The mRNA of Ac contains a number of transcription initiation sites; these transcription initiation sites are located between 304 and 364 nt from 5′ end and between 334 and 265 nt from the 3′-end of the element. The imperative feature for the transposition process is overlapping for an Ac promoter element with the 5′ terminal nucleotide sequence, which also includes motif for Tpase-binding (Lazarow et al., 2013).

The maize Ds elements are nonautonomous and are able to move only when the Ac autonomous element is present. There are three classes of Ds elements known for mutation, that is, Ds-del (internal deletions of the 4.6-kb *Ac* element), Ds1 (400 bp and homologous to AC), and Ds2 (variable in size with about 0.5 kb from the *Ac* termini). But Ds-l element does not require Ac for transposition (Du et al., 2011). The Ds element have amazing efficiency to cut out and then reintegrate itself anywhere within the genome, this capacity of Ds elements make Ac/Ds transposons as auspicious tool for insertional mutagenesis (Mielich et al., 2018). On the physical map of maize over 1500 elements are scattered and positioned within the genome, which start with single donor site on the chromosome 10. Particularly Ds shows a stronger priority for integration within the exon as well as in intron, whether the mutational insertions are more enhanced within the promoter region and the untranslated region among 5′ end. Initially Ac transposition model was proposed by Greenblatt and Brink (1962), which describes that during the replication Ac transposes from replicated site to replicated or unreplicated site without the restriction of receptor site's nature. The excellent spot for regional mutagenesis is present within the 2- to 3-cM zone. Ac/Ds elements transpose into sites linked genetically (Vollbrecht et al., 2010; Dooner and Belachew, 1989; Greenblatt, and Brink, 1962). Mutants can also be generated via plant transformation strategy, but the transposons tagging strategies have proven more effective for the formation of mutant population (Kondou et al., 2010). Mutant population produced via Ac/Ds system application in Arabidopsis, rice, and poplar (Kuromori et al., 2004; Fladung and Polak, 2012; Qu et al., 2008). Ac/Ds system allows the acquiesce of number of independent mutants against a single transformed population and enables the active identification of putative mutagenized sequential arrangement via the isolation of insertion flanking portion of Ds (Kuromori et al., 2004).

The particular mechanism for Ac/Ds transposition is unknown, but there are lots of works, which gives the evidence that they transpose by cut–paste mechanism (Becker and Kunze, 1996; Gorbunova and Levy, 1997, 2000). In the process of insertion, Tpase construct falter cut of 8 bp at target locus and then the transposon DNA ligated into that cut region and DNA repair mechanism play role to fill the 8 bp gap, generates 8 bp TSD flanking (Peterson and Zhang, 2013). In the excision mechanism of Ac/Ds chromosomal double-strand breaks generated, lead to the generation of excision site (Rommens et al., 1993; Rinehart et al., 1997). These excision sites are associated with the formation of footprints or minor sequence changes, at specifically on new junction most of the footprint out of them associated with the substitution as well as may or may not with deletion. If the Ac/DS integration site present with in the exon then minor sequence changes isolated from restored alleles can also be biased, because only those excision with functional ORF will detect, to conquer that complication. PCR products were analyzed and gathered from somatic Ac/Ds excision process and the most probable footprints or minor sequence changes were 1–3 bp deletion. Also, the A-T and G-transversion in which 1–3 bp deletions were quite low (Rinehart et al., 1997; Scott et al., 1996). In plants, the G-C and A-T transversion are rare; however, the short-length palindromes are quite obvious. The hairpin form by transposon-flanking DNA that open number of base pairs from the junction of flanking sequence transposon (Peterson and Zhang, 2013).

For the single gene targets chemical as well as insertional mutagenesis can be used but they generally do not generate extensive changes. Nowadays, Ac/Ds TE with combination of Cre/lox site-specific system of recombination is used as chromosome rearrangement tool rather than chemical and insertional mutagenesis as method. It involves number of steps given in Figure 14.2 but this method also has number of disadvantages (Cecchini et al., 1998; Yu et al., 2012; Osborne et al., 1995).

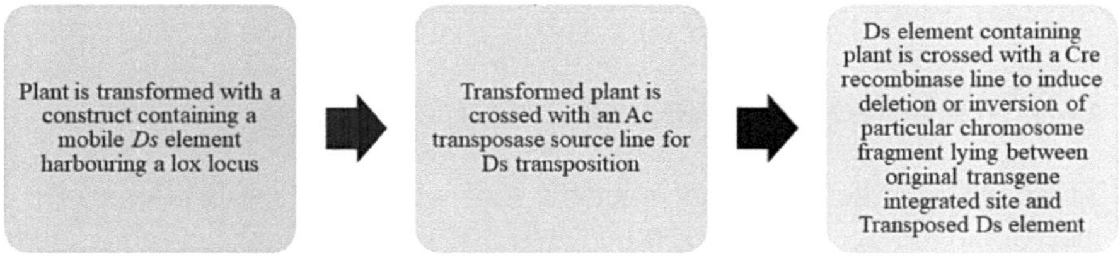

FIGURE 14.2 Cre/lox site-specific recombination system.

14.3.2 PIGGYBAC TES IN PLANTS

There is another kind of transposons that is piggyBac (PB) grab a lot of attention due to their efficiency, safety as well as stability for transposition process. Because of number of advantages provides by the PB, they are used as emerging tool for genetics (Zhao et al., 2016). PB can move comparatively larger segment of DNA about 9.1–14.3 kb (Li et al., 2013). This (PiggyBac) transposon system was initially isolate from an insect, and to transform vast diversity of organisms. Recently much progress has been made for their use in plant for gene insertion at specific site (Johnson et al., 2016). Bire and colleagues conclude in their research that the *piggyBac* transposase mRNA consider as promising path for the improvement of quality of insertion and for sustaining transposon vector expression (Bire et al., 2013). The piggyBac transposons are more rapidly used in genomic engineering approaches. It is different from other transposons; the excision site of piggyBac can repair precisely and do not leave footprints as it integrated at TTAA tetranucleotides. In Cryo-EM architecture of piggyBac transposons, synaptic complex present with in the hairpin DNA intermediates and, in the integration, process the strand get transferred. The piggyBac transposons are known for the formation of an asymmetrical dimer that is characterized by two c-terminal domains which allow the formation of a separate dimer, which contacts only to one transposon end and also two central domains known for synapse end. During the transfer of strand, the target DNA bent severely and unpaired TTAA target characterized the process (Chen et al., 2020). Nishizawa-Yokoi and colleagues conduct genome editing in plants by a targeting gene and piggyBac-mediated marker excision and provide strategy regarding modification of specific target of endogenous nucleotide fragment or gene in plants (Nishizawa-Yokoi et al., 2015). However, the use of piggyBac transposons in plants is limited and more widely used in animals.

14.4 TRANSPOSON AS GENETIC TOOL

TEs have been developed as a genetic tool with wider applications such as insertional mutagenesis, gene transfer (transformation), gene tagging, identification of genes, etc. (Lai, 1994). TEs with promoter activity can transcribe nearby intergenic DNA sequence providing a new regulatory region for the adjacent genes, which may result in novel expression (Kloeckener-Gruissem et al., 1992). Such type of changes derived from TEs is broadly used in reverse genetics. Transgenic plants need to remove their selectable marker genes, for which several methods such as transposition, co-transformation, etc. are opted (reviewed in Darbani et al., 2007; Woo et al., 2001). Among these TEs or specifically the Ac/Ds transposon system in maize is widely used in producing marker-free transgenic plants such as rice, tobacco, tomato, etc. (Goldsbrough et al., 1993; Cotsaftis et al., 2002). The maize genome is considered as an important system to study the effect of TEs on gene regulation as they have many types of TEs along with the genes. In an experiment, the gene and TE transcripts were analyzed in response to different abiotic stresses. It was concluded that most of the TE families are associated with upregulation of gene expression (22% to abiotic stress) (Makarevitch et al., 2015). The Ac/Ds elements have been used for insertional mutagenesis and functional studies in rice. The main focus of such studies is to make available all the genetic information to improve crop breeding (Kim et al., 2018). The movement of TE in the host genome may create opportunities for adaptation. The manipulation in TEs transposition process could help in modifying host function, which is possible by a new technique of genetic engineering known as clustered regulatory interspaced short palindromic repeats/CRISPR-associated protein 9 (CRISPR/Cas9). It is a RNA-guided nuclease system used for gene editing. MITEs along with the CRISPR/Cas9 system have been used to design more efficient transposon-based system (Vaschetto, 2018).

TEs are not only important in context of evolution but have also contributed in the innovation of several proteins and transcripts. Many of the genomic studies often skip them because of the related complexities. But to overcome these many software and computational tools have been developed to fill the gaps and perform all the genomic studies efficiently. All the information related to TEs is deposited in three repositories based on consensus sequence, reference genome, and polymorphism (Fig. 14.3) (Goerner-Potvin and Bourque, 2018).

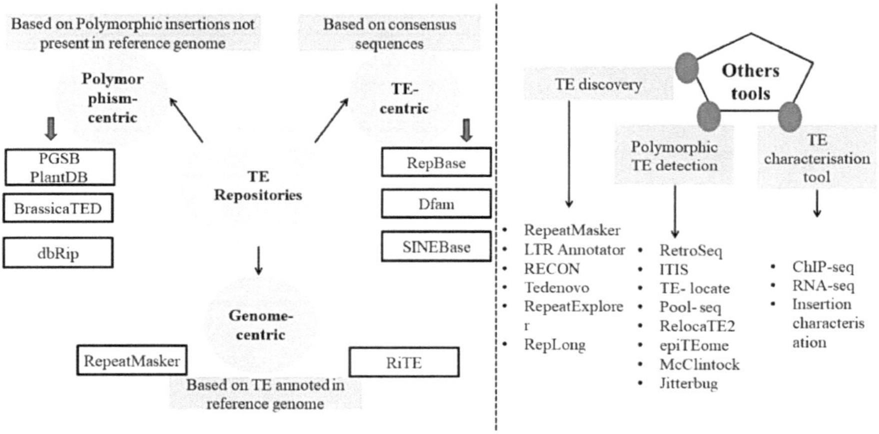

FIGURE 14.3 Computational tools for better study of TEs: Three repositories are made including TE-centric (Consensus sequence), genome-centric (reference genome), and polymorphism-centric (insertions not present in reference genome). Some other tools are also present for TE discovery, polymorphism detection, and TE characterization.

14.4.1 TRANSPOSON AS VECTORS

The development of transgenic plants has now been common for many crops. There are two methods available for the production of transgenics including naked DNA transfer (vectorless) and transformation via *Agrobacterium tumefaciens*. Both methods have their pros and cons, vectorless transfer is not restricted to specific host and is successful in various crop plants but the rearrangement of transferred DNA creates complexity of multicopy transgene which thereby limit its applications. On the other hand, *Agrobacterium*-mediated transformation show single copy or very low copy integration of foreign DNA with host-specific restrictions in most crops. A universal transposition system should fulfill some of the criteria to develop transgenics (Fig. 14.4). To overcome the limitations a new methods is developed for the gene transfer, that is, transposon based. Transposons insertional mutagenesis can be used as an alternative to T-DNA mutagenesis usually when transformation using tissue culture is feeble.

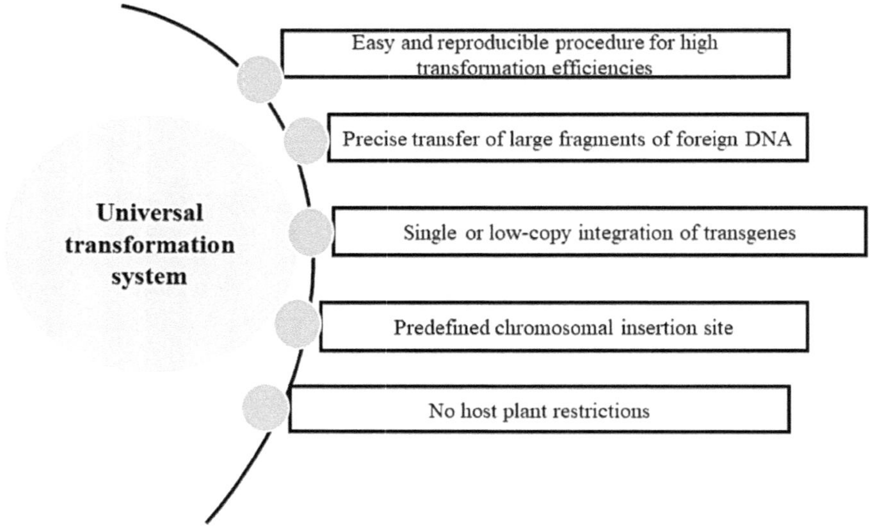

FIGURE 14.4 Characteristics of a universal system for transposition.

The Ac element that is 4.6 kb in length encodes for the very much-needed transposase for transposition and activation of defective elements. The defective elements usually do not synthesis transposase but transpose due to transactivation, are known as dissociation elements (Ds). Ac elements generally transpose through cut and paste mechanism. The Ac/Ds elements show broad range of applications in plants but still needs improvements (Fig. 14.5). Considering the needs, an experiment has been performed in which the extrachromosomal transposition (as delivery system) was used to transfer foreign gene into the chromosomes. The result revealed that the vector was efficient in delivering the long stretch of foreign DNA in the genome with intact insert of 10 kb (Lebel et al., 1995). Retrotransposons are widely distributed in plant genome and used as mutagens and vector system for delivery. Tnt1 element from tobacco (*Nicotianatabacum*) has been developed as a gene delivery system and help in producing cDNA, which can act as a donor for genetic modification (Hou et al., 2010).

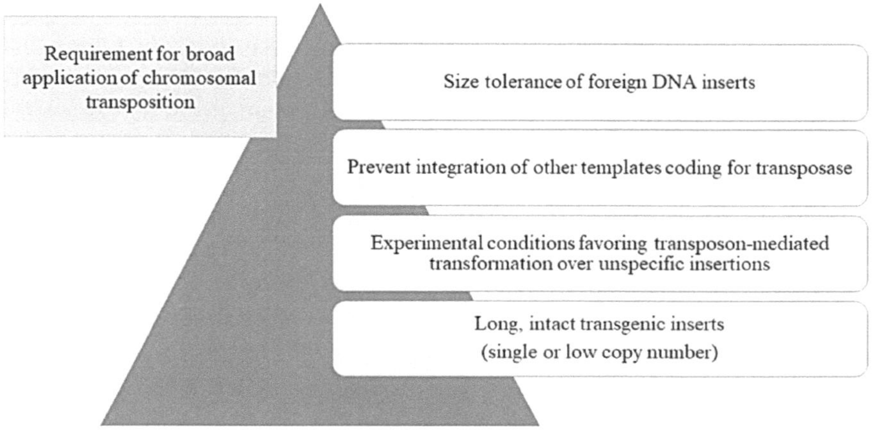

FIGURE 14.5 Requirement for broad application of chromosomal transposition via Ac/Ds elements.

14.4.2 TRANSPOSONS IN MUTAGENESIS

Mutagenesis is considered as a most efficient aspect or a tool to create genetic variations and key regulatory gene identification for crop improvement. Several techniques can be used to improve mutagenesis such as physical, chemical, and insertional mutagenesis. Technical application followed by NGS for detection of mutations in short period of time further the induce mutagenesis coupled with whole genome sequence, which provide powerful platform for forward and reverse genetic applications (Chaudhary et al., 2019). Rapid progress has been made in the genome projects of different plants subsequently large-scale transposon mutagenesis has emerged out as an important tool for functional genomics and permitting idea of functions to sequence gene using reverse genetics.

Transposons are generally used in insertional mutagenesis. On the bases of transposon properties such as their mechanism and control of transposition, transformation efficiency of host plant, etc, different approaches can be used. These approaches are successfully used in *Arabidopsis thaliana* and has made it achievable to develop reverse genetics (Ramachandran and Sundaresan, 2001). Transposons base mutagenesis provides first and also one of the most important paths for gene identification and characterization. bz1 gene was the first gene that was tagged with Ac, more than 60 genes, which are the basic component of plant development have been cloned using Suppressor-mutator (Spm) and Mutator (Mu) from maize and Tag1 from Arabidopsis (May and Martienssen, 2003). Similarly, the enhancer/suppressor (En/Spm) element from maize has been used for mutagenesis in rice. T-DNA construct was prepared with Spm-transposase and defective suppressor mutator in combination with fluorescent proteins (green fluorescent protein and *Discosoma sp.* red fluorescence protein). The results concluded that the En/Spm element can be used for functional genomic studies in rice (Kumar et al., 2005). Nowadays, most of the gene have been identified through genome sequence study but the functions of the most of the gene is unknown, there is a necessary research input in require to determine the function of gene especially those gene involve in agronomic performance. To study the function of gene, transposons tagging prove efficient in crop plants. Mathieu and colleagues established the Soybean transposon-based mutagenesis repository and concluded that soybean gene function can be disrupted by insertional mutagenesis and useful mutants can be derived (Mathieu et al., 2009). In insertional mutagenesis generally T-DNA is used, retrotransposons, Ac/Ds insertions, and transposons and create mutant libraries so the identification of tagged gene by using PCR-based techniques become easy. PCR-based techniques that are generally used are thermal asymmetric interlaced PCR (TAIL-PCR) or Inverse PCR. Insertional mutagenesis is also known to reduce the concern related to activation tagging or the gene trap approach (Springer, 2000). As we know transposon mutagenesis leads to the gene discovery using tagging that can be used for the cloning. It is seen that new genomic projects are also there and catalogs the transposon insertion sites to define all crop genes, these identified transposons insertional sites are known as "hot spot." These hot spots are used to generate new alleles of functional genomic study (Walbot, 2000).

Boucher and colleagues report transposon mutagenesis in a tomato isolate of *Pseudomonas solanacearum* by using the Tn7 and Tn5 insert in suicide conjugative plasmids in which they find that Tn5 transpose more randomly and allow the isolation of auxotrophic mutants. Screening of clones led to the 12 avirulent mutants isolation these avirulent mutants divide into two or three groups on the bases of their ability or inability to induce a collapse of tobacco leaf parenchyma (Boucher et al., 1985). Osborne and colleagues engineered new type of transposons, that is, Ds *lox*, which were made using maize transposons Ds and *lox* sites from Cre-lox site-specific recombination system and Ac transposase or Cre recombinase transcriptional fusion expression (Osborne et al., 1995). There is another example of transposon-based mutagenesis provide by Fernandes and colleagues in their study of genome-wide mutagenesis in *Zea mays L.* using *Rescue Mu* transposon. *Rescue Mu* provided the platform for the discovery of maize gene and mutant phenotype analysis of multiple plasmid recoveries and the screening led to the identification of parental insertions and insertional hotspots, which were used to produce knockout mutants (Fernandes et al., 2004). Greco and colleagues developed transposons mutation system in rice to define gene functions via forward and reverse genomic strategies. Zea mays transposons are considered as an obvious choice for the development of effective transposon-tagging systems for the knockouts of gene and their detection (Greco et al., 2001).

Nishizawa-Yokoi and coworkers (2015) performed an experiment based on animal-derived TE (piggyBac) and showed evidences of its efficiency in plant system. For this, they performed an assay to visualize the transposition of piggyBac in rice cells with emerald luciferase (Eluc) luminescence, which showed there was a high frequency of piggyBac expression in rice calli and further confirmed by other molecular techniques like PCR and southern blot. Their study concluded that efficient excision system for plants can be made using piggyBac transposon system. Another transposon known as Miniature Ping (mPing) is an active DNA transposon discovered in rice genome. These are 430 bp long DNA sequences with 15 bp LTRs which is similar to MITE family. Ping and Pong are two elements with open reading frames (ORF1 and ORF2), which provide the transposase required for the transposition activity. mPing show active transposition in some rice varieties like Gimbozu (Naito et al., 2006). It has been reported through comprehensive studies that the mPing insertion sites do not attach to the exon and instead prefer promoter regions. Therefore, show mild or moderate effect on the adjacent genes. Also, such insertion helps in expression profiling of nearby genes (Naito et al., 2014). In a study, mPing was inserted in the upstream region of a gene to evaluate the effect in response to stress. The screening results revealed that out of 17 targeted genes five of them were inserted into the promoter region. The five genes were typically seen in varying gene expression in normal and stress conditions. Such results proved that mPing insertion could be beneficial in studying gene expressions under stress condition (Yasuda et al., 2013).

A gene (See2B) responsible in coding for protease was seen mutated with the Mu (Maize mutator) insertion. The experiment was held under high and low nitrogen (N) conditions and it was found that See2B plays a significant role in nitrogen utilization under N-deficit conditions and also functional in late senescence of maize (Donnison et al., 2007). The insertion of MITE in the promoter region of NAC gene resulted in development of a natural variant of drought resistance in maize. The ZmNAC11 gene expressed in transgenic maize improved drought tolerance by the up regulation of the drought related genes (Mao et al., 2015).

Retrotransposons also proven as an important genetic tool for the study of functional genomics in crops such as Soyabean, Cui and colleagues use the Tnt1 (retrotransposon) as a mutagenesis study in *Glycine max*, which is inactive in normal plants tissues but have a capacity to become reactive by tissue culture and their research data revealed that the Tnt1 retrotransposons considered as a powerful tool for the efficient large-scale insertional mutagenesis (Cui et al., 2013). Some research proves that Hsp90 is responsible for the prevention of phenotypic variation and known for the suppressing nature of the mutagenic activity of transposons (Hummel et al., 2017).

Various types of genomic tools are generally used to identify and characterize the gene in plants. In which Ac/Ds transposons-based approaches show wide potential also for transformation in plant kingdom. In such systems, transposons are highly active resulting in mutations (by insertion, chromosomal rearrangement). The Ac/Ds system show biasness toward genic regions and also show the capacity regarding the localized transposition and provide the increase capacity of gene discovery and gene tagging especially those gene that are linked to Ds. The Ds use to map particular QTLs vicinity after which reactivation is carried out and thermal asymmetric interlaced PCR and inverse PCR help to find out new Ds location in genome of interest (Singh et al., 2012).

14.4.3 TRANSPOSON-BASED TAGGING

Transposon tagging is one of the applications of mutagenesis using TEs. The cloned TE is allowed to insert in the specific region of interest and further that inserted TE will used to clone that gene of interest. This method of tagging gene has advantages over conventional methods of mutagenesis. Firstly, the mutations developed in such a way eventually help in cloning of gene, which was more difficult with chemical methods. Also, a gene can be cloned without knowing its gene product. Although, the conventional strategies of insertional mutagenesis are important in functional genomics studies related to plants, but have a limitation in which there is a difficulty with genes whose knockout may not result in any phenotype. Activation tagging is an effective method to overcome such limitation. When the transposons use as activation tags to generate phenotypes or function without the use of transformation studies and also polyploid plant study in which gene knockout do not shows specific phenotype. In a study the introduction of a T-DNA-containing regulatory sequence of CaMV (cauliflower mosaic virus) 35S promoter in the genome of plant enhanced the expression of neighboring genes and resulted in a gain of phenotypic expression (Borevitz et al., 2000; Mori et al., 2007). Qu and colleagues developed a versatile transposon-based activation tag vector system (cis-two-element vector system) for functional genomics in cereals and other monocot plants, which eliminate crossing need and require small amount of primary transformants (Qu et al., 2008).

As the plant breeding requires the genetic information to improve crop plants. The mPing element has been modified to improve activation tagging for the expression of neighboring genes. Activation tagging in soyabean with modified mPing (mmPing20) indicated the overexpression of genes, which suggested the improvement in mutagenesis system (Johnson et al., 2021).

14.5 CONCLUSION

As discussed above in the chapter, TEs are found valuable in fulfilling the various aspects of genetic studies in terms of functional genomics and more. These are abundantly found in the most genomes not only contributing in covering a proportion of DNA but also helps in various processes like making of novel proteins and transcripts for their host. Being mutagens, TEs are more beneficial over chemical mutagens because they can generate more efficient insertion, deletions and chromosomal rearrangements in the genome. Also, they are characterized as an effective tool for tagging and cloning of genes of great value in the developmental process. Transposon tagging is seen to be very powerful in cloning genes. Different transposon systems like Ac/Ds elements, mPing, piggyBac, etc. are used in broad range of applications. It is evident that TEs can transpose from one species to another which makes them more valuable. TE-based mutagenesis and transformation could be improved by enhancing knowledge of all TE systems for its usage across different species. Also, there is need to find other possible TEs in various species to fulfill the needs for the future advances in genetic engineering.

KEYWORDS

- **transposable elements**
- **DNA transposons**
- **Ac/Ds**
- **piggyBac**
- **mPing**
- **insertional mutagenesis**

REFERENCES

Becker, H. A.; Kunze, R. Binding Sites for Maize Nuclear Proteins in the Subterminal Regions of the Transposable Element Activator. *Mol. Gen. Genet.* **1996,** *251*, 428–435.

Biémont, C. A Brief History of the Status of Transposable Elements: from Junk DNA to Major Players in Evolution. *Genetics* **2010,** *186*, 1085–1093.

Bire, S.; Ley, D.; Casteret, S.; Mermod, N.; Bigot, Y.; Rouleux-Bonnin, F. Optimization of the piggyBac Transposon Using mRNA and Insulators: Toward a More Reliable Gene Delivery System. *PloS one* **2013,** *8*, e82559.

Borevitz, J. O.; Xia, Y.; Blount, J.; Dixon, R. A.; Lamb, C. Activation Tagging Identifies a Conserved MYB Regulator of Phenylpropanoid Biosynthesis. *Plant Cell* **2000,** *12*, 2383–2394.

Boucher, C. A.; Barberis, P. A.; Demery, D. A. Transposon Mutagenesis of Pseudomonas Solanacearum: Isolation of Tn5-Induced Avirulant Mutants. *Microbiology* **1985,** *131*, 2449–2457.

Brown, P. O.; Bowerman, B.; Varmus, H. E.; Bishop, J. M. Correct Integration of Retroviral DNA In Vitro. *Cell* **1987,** *49*, 347–56.

Bucher, E.; Reinders, J.; Mirouze, M. Epigenetic Control of Transposon Transcription and Mobility in Arabidopsis. *Curr. Opin. Plant Biol.* **2012,** *15*, 503–510.

Cavrak, V. V.; Lettner, N.; Jamge, S.; Kosarewicz, A.; Bayer, L. M.; Scheid, O. M. How a Retrotransposon Exploits the Plant's Heat Stress Response for Its Activation. *PLoS Genet.* **2014,** *10*, e1004115.

Cecchini, E.; Mulligan, B. J.; Covey, S. N.; Milner, J. J. Characterization of Gamma Irradiation-Induced Deletion Mutations at a Selectable Locus in Arabidopsis. *Mutat. Res. Fund. Mol. M.* **1998,** *401*, 199–206.

Chaudhary, J.; Deshmukh, R.; Sonah, H. Mutagenesis Approaches and Their Role in Crop Improvement. *Plants* **2019,** *8*, 467.

Chen, Q.; Luo, W.; Veach, R. A.; Hickman, A. B.; Wilson, M. H.; Dyda, F. Structural Basis of Seamless Excision and Specific Targeting by piggyBactransposase. *Nat. Commun.* **2020,** *11*, 1–14.

Cotsaftis, O.; Sallaud, C.; Breitler, J.; Meynard, D.; Greco, R.; Pereira, A.; Guiderdoni, E. Transposon-Mediated Generation of T-DNA- and Marker-Free Rice Plants Expressing a Bt Endotoxin Gene. *Mol. Breed.* **2002,** *10*, 165–180.

Coupland, G.; Baker, B.; Schell, J.; Starlinger, P. Characterization of the Maize Transposable Element Ac by Internal Deletions. *EMBO J.* **1988,** *7*, 3653–3659.

Cui, Y.; Barampuram, S.; Stacey, M. G.; Hancock, C. N.; Findley, S.; Mathieu, M., Stacey, G. Tnt1 Retrotransposon Mutagenesis: A Tool for Soybean Functional Genomics. *Plant Physiol.* **2013,** *161*, 36–47.

Darbani, B.; Eimanifar, A.; Stewart, C. N; Camargo, W. N. Methods to Produce Marker-Free Transgenic Plants. *Biotechnol. J.* **2007,** *2*, 83–90.

de Koning, A. P.; Gu, W.; Castoe, T. A.; Batzer, M. A.; Pollock, D. D. Repetitive Elements May Comprise Over Two-Thirds of the Human Genome. *PLoS Genet.* **2011,** *7*, e1002384.

Donnison, I. S.; Gay, A. P.; Thomas, H.; Edwards, K. J.; Edwards, D.; James, C. L.; Ougham, H. J. Modification of Nitrogen Remobilization, Grain Fill and Leaf Senescence in Maize (Zea mays) by Transposon Insertional Mutagenesis in a Protease Gene. *New Phytol.* **2007,** *173*, 481–494.

Dooner, H. K.; Belachew, A. Transposition Pattern of the Maize Element Ac from the bz-m2 (Ac) Allele. *Genetics* **1989,** *122*, 447–457.

Du, C.; Hoffman, A.; He, L.; Caronna, J.; Dooner, H. K. The Complete Ac/Ds Transposon Family of Maize. *BMC genom.* **2011,** *12*, 1–12.

Eickbush, T. H.; Malik, H. S. Origins and Evolution of Retrotransposons. In *Mobile DNA ii*; American Society of Microbiology: Washington, DC, 2002; pp 1111–1144.

Fernandes, J.; Dong, Q.; Schneider, B.; Morrow, D. J.; Nan, G. L.; Brendel, V. Walbot, V. Genome-Wide Mutagenesis of Zea mays L. Using RescueMu Transposons. *Genome Biol.* **2004,** *5*, 1–20.

Feschotte, C.; Pritham, E. J. DNA Transposons and the Evolution of Eukaryotic Genomes. *Annu. Rev. Genet.* **2007,** *41*, 331–68.

Fladung, M.; Polak, O. Ac/Ds-Transposon Activation Tagging in Poplar: A Powerful Tool for Gene Discovery. *BMC genom.* **2012,** *13*, 1–16.

Goerner-Potvin, P.; Bourque, G. Computational Tools to Unmask Transposable Elements. *Nat. Rev. Genet.* **2018,** *19*, 688–704.

Goldsbrough, A.; Lastrella, C.; Yoder, J. Transposition Mediated Repositioning and Subsequent Elimination of Marker Genes from Transgenic Tomato. *Nat. Biotechnol.* **1993,** *11*, 1286–1292.

Gorbunova, V.; Levy, A. A. Analysis of Extrachromosomal Ac/Ds Transposable Elements. *Genetics* **2000,** *155*, 349–359.

Gorbunova, V.; Levy, A. A. Circularized Ac/Ds Transposons: Formation, Structure and Fate. *Genetics* **1997,** *145*, 1161–1169.

Grabundzija, I.; Messing, S. A.; Thomas, J.; Cosby, R. L.; Bilic, I.; Miskey, C.; Gogol-Döring, A.; Kapitonov, V.; Diem, T.; Dalda, A.; Jurka, J. A Helitron Transposon Reconstructed from Bats Reveals a Novel Mechanism of Genome Shuffling in Eukaryotes. *Nat. Commun.* **2016,** *7*, 10716.

Grandbastien, M. A.; Audeon, C.; Bonnivard, E.; Casacuberta, J. M.; Chalhoub, B. Stress Activation and Genomic Impact of Tnt1 Retrotransposons in Solanaceae. *Cytogenet. Genome Res.* **2005,** *110*, 229–241.

Greco, R.; Ouwerkerk, P. B.; Sallaud, C.; Kohli, A.; Colombo, L.; Puigdomenech, P.; Pereira, A. Transposon Insertional Mutagenesis in Rice. *Plant Physiol.* **2001,** *125*, 1175–1177.

Greenblatt, I. M. A Chromosome Replication Pattern Deduced from Pericarp Phenotypes Resulting from Movements of the Transposable Element, Modulator, in Maize. *Genetics* **1984,** *108*, 471–485.

Greenblatt, I. M.; Brink, R. A. Twin Mutations in Medium Variegated Pericarp Maize. *Genetics* **1962,** *47*, 489.

Hou, Y.; Rajagopal, J.; Irwin, P. A.; Voytas, D. F. Retrotransposon Vectors for Gene Delivery in Plants. *Mobile DNA* **2010,** *1*, 1–9.

Hummel, B.; Hansen, E. C.; Yoveva, A.; Aprile-Garcia, F.; Hussong, R.; Sawarkar, R. The Evolutionary Capacitor HSP90 Buffers the Regulatory Effects of Mammalian Endogenous Retroviruses. *Nat. Struct. Mol. Biol.* **2017,** *24*, 234.

Johnson, A.; Mcassey, E.; Diaz, S.; Reagin, J.; Redd, P. S.; Parrilla, D. R.; Nguyen, H.; Stec, A.; McDaniel, L.; Clemente, T. E.; Stupar, R. M.; Parrott, W. A.; Hancock, C. N. Development of mPing-Based Activation Tags for Crop Insertional Mutagenesis. *Plant Direct.* **2021,** *5*, e00300.

Johnson, E. T.; Owens, J. B.; Moisyadi, S. Vast Potential for Using the piggyBac Transposon to Engineer Transgenic Plants at Specific Genomic Locations. *Bioengineered* **2016,** *7*, 3–6.

Kazazian, H. H. Mobile Elements: Drivers of Genome Evolution. *Science* **2004,** *303*, 1626–1632.

Khan, M. F.; Yadav, B. S.; Ahmad, K.; Jaitly, A. K. Mapping and Analysis of the LINE and SINE Type of Repetitive Elements in Rice. *Bioinformation* **2011,** *7*, 276–279.

Kim, S. Y.; Kim, C. K.; Kang, M.; Ji, S. U.; Yoon, U. H.; Kim, Y. H.; Lee, G. S. A Gene Functional Study of Rice Using Ac/Ds Insertional Mutant Population. *Plant Breed. Biotechnol.* **2018,** *6*, 313–320.

Kloeckener-Gruissem, B.; Vogel, J. M.; Freeling, M. The TATA Box Promoter Region of Maize Adh1 Affects Its Organ-Specific Expression. *EMBO J.* **1992,** *11*, 157–166.

Kondou, Y.; Higuchi, M.; Matsui, M. High-Throughput Characterization of Plant Gene Functions by Using Gain-of-Function Technology. *Annu. Rev. Plant Biol.* **2010,** *61*, 373–393.

Kumar, C. S.; Wing, R. A.; Sundaresan, V. Efficient Insertional Mutagenesis in Rice Using the Maize En/Spm Elements. *Plant J.* **2005,** *44*, 879–892.

Kunze, R.; Stochaj, U.; Laufs, J.; Starlinger, P. Transcription of Transposable Element Activator (Ac) of Zea mays L. *EMBO J.* **1987,** *6*, 1555–1563.

Kuromori, T.; Hirayama, T.; Kiyosue, Y.; Takabe, H.; Mizukado, S.; Sakurai, T.; et al. A Collection of 11 800 Single-Copy Ds Transposon Insertion Lines in Arabidopsis. *Plant J.* **2004,** *37*, 897–905.

Lai, C. Genetic Applications of Transposable Elements in Eukaryotes. *Genome* **1994,** *37*, 519–525.

Lazarow, K.; Doll, M. L.; Kunze, R. Molecular Biology of Maize *Ac/Ds* Elements: An Overview. In *Plant Transposable Elements*; Peterson, T., Eds.; Methods in Molecular Biology (Methods and Protocols) 1057; Humana Press: Totowa, NJ, 2013.

Lebel, E. G.; Masson, J.; Bogucki, A.; Paszkowski, J. Transposable Elements as Plant Transformation Vectors for Long Stretches of Foreign DNA. *Theor. Appl. Genet.* **1995,** *91*, 899–906.

Lerman, D. N.; Feder, M. E. Naturally Occurring Transposable Elements Disrupt hsp70 Promoter Function in Drosophila Melanogaster. *Mol. Biol. Evol.* **2005,** *22*, 776–783.

Li, R.; Zhuang, Y.; Han, M.; Xu, T.; Wu, X. piggyBac as a High-Capacity Transgenesis and Gene-Therapy Vector in Human Cells and Mice. *DMM* **2013,** *6*, 828–833.

Locke, D. P.; Segraves, R.; Carbone, L.; Archidiacono, N.; Albertson, D. G.; Pinkel, D.; Eichler, E. E. Large-Scale Variation Among Human and Great Ape Genomes Determined by Array Comparative Genomic Hybridization. *Genome Res.* **2003,** *13*, 347–357.

Luan, D. D.; Korman, M. H.; Jakubczak, J. L.; Eickbush, T. H. Reverse Transcription of R2Bm RNA Is Primed by a Nick at the Chromosomal Target Site: A Mechanism for Non-LTR Retrotransposition. *Cell* **1993,** *72*, 595–605.

Makarevitch, I.; Waters, A. J.; West, P. T.; Stitzer, M.; Hirsch, C. N.; Ross-Ibarra, J.; Springer, N. M. Transposable Elements Contribute to Activation of Maize Genes in Response to Abiotic Stress. *PLoS Genet.* **2015,** *11*, e1004915.

Mao, H.; Wang, H.; Liu, S.; Li, Z.; Yang, X.; Yan, J.; et al. A Transposable Element in a NAC Gene Is Associated with Drought Tolerance in Maize Seedlings. *Nat. Commun.* **2015,** *6*, 1–13.

Martin, A.; Troadec, C.; Boualem, A.; Rajab, M.; Fernandez, R.; Morin, H.; Pitrat, M.; Dogimont, C.; Bendahmane, A. A Transposon-Induced Epigenetic Change Leads to Sex Determination in Melon. *Nature* **2009,** *461*, U1135–U1237.

Mathieu, M.; Winters, E. K.; Kong, F.; Wan, J.; Wang, S.; Eckert, H.; et al. Establishment of a Soybean (Glycine Max Merr. L) Transposon-Based Mutagenesis Repository. *Planta* **2009,** *229*, 279–289.

May, B. P.; Martienssen, R. A. Transposon Mutagenesis in the Study of Plant Development. *Crit. Rev. Plant Sci.* **2003,** *22*, 1–35.

McClintock, B. Controlling Elements and the Gene. In *Cold Spring Harbor Symposia on Quantitative Biology*; 21; Cold Spring Harbor Laboratory Press, 1956; pp 197–216.

Mielich, K.; Shtifman-Segal, E.; Golz, J. C.; Zeng, G.; Wang, Y.; Berman, J.; Kunze, R. Maize Transposable Elements Ac/Ds as Insertion Mutagenesis Tools in Candida Albicans. *G3: Genes Genom. Genet.* **2018,** *8*, 1139–1145.

Mori, M.; Tomita, C.; Sugimoto, K.; Hasegawa, M.; Hayashi, N.; Dubouzet, J. G.; Ochiai, H.; Sekimoto, H.; Hirochika, H.; Kikuchi, S. Isolation and Molecular Characterization of a Spotted Leaf 18 Mutant by Modified Activation-Tagging in Rice. *Plant Mol. Biol.* **2007,** *63*, 847–860.

Naito, K., Zhang, F., Tsukiyama, T., Saito, H., Hancock, C. N.; Richardson, A. O.; Okumoto, Y.; Tanisaka, T.; Wessler, S. R. Unexpected Consequences of a Sudden and Massive Transposon Amplification on Rice Gene Expression. *Nature* **2009,** *461*, 1130–1134.

Naito, K.; Cho, E.; Yang, G.; Campbell, M. A.; Yano, K.; Okumoto, Y.; Tanisaka, T.; Wessler, S. R. Dramatic Amplification of a Rice Transposable Element During Recent Domestication. *Proc. Natl. Acad. Sci. U.S.A.* **2006,** *103*, 17620–17625.

Naito, K.; Monden, Y.; Yasuda, K.; Saito, H.; Okumoto, Y. mPing: The Bursting Transposon. *Breed. Sci.* **2014,** *64*, 109–114.

Nishizawa-Yokoi, A.; Endo, M.; Ohtsuki, N.; Saika, H.; Toki, S. Precision Genome Editing in Plants via Gene Targeting and Piggy B Ac-Mediated Marker Excision. *Plant J.* **2015,** *81*, 160–168.

Osborne, B. I.; Wirtz, U.; Baker, B. A System for Insertional Mutagenesis and Chromosomal Rearrangement Using the Ds Transposon and Cre—lox. *Plant J.* **1995,** 7, 687–701.

Peterson, T.; Zhang, J. The Mechanism ofAc/DsTransposition. *Plant Transp. Genome Dyn. Evol.* **2013,** 41–59.

Petrov, D. A.; Hartl, D. L. Trash DNA Is What Gets Thrown Away: High Rate of DNA Loss in Drosophila. *Gene* **1997,** 205, 279–289.

Qu, S.; Desai, A.; Wing, R.; Sundaresan, V. A Versatile Transposon-Based Activation Tag Vector System for Functional Genomics in Cereals and Other Monocot Plants. *Plant Physiol.* **2008,** *146*, 189–199.

Ramachandran, S.; Sundaresan, V. Transposons as Tools for Functional Genomics. *Plant Physiol. Biochem.* **2001,** *39*, 243–252.

Rinehart, T. A.; Dean, C.; Weil, C. F. Comparative Analysis of Non-Random DNA Repair Following Ac Transposon Excision in Maize and Arabidopsis. *Plant J.* **1997,** *12*, 1419–1427.

Rommens, C. M.; Van Haaren, M. J.; Nijkamp, H. J. J.; Hille, J. Differential Repair of Excision Gaps Generated by Transposable Elements of the 'Ac Family'. *BioEssays* **1993,** *15*, 507–512.

Rubin, G. M.; Kidwell, M. G.; Bingham, P. M. The Molecular Basis of P-M Hybrid Dysgenesis: The Nature of Induced Mutations. *Cell* **1982,** *29*, 987–94.

SanMiguel, P.; Gaut, B. S.; Tikhonov, A.; Nakajima, Y.; Bennetzen, J. L. The Paleontology of Intergene Retrotransposons of Maize. *Nat. Genet.* **1998,** *20*, 43–45.

Sayah, D. M.; Sokolskaja, E.; Berthoux, L.; Luban, J. Cyclophilin A Retrotransposition into TRIM5 Explains Owl Monkey Resistance to HIV-1. *Nature* **2004,** *430*, 569–573.

Scott, L.; LaFoe, D.; Weil, C. F. Adjacent Sequences Influence DNA Repair Accompanying Transposon Excision in Maize. *Genetics* **1996,** *142*, 237–246.

Singh, S.; Tan, H. Q.; Singh, J. Mutagenesis of Barley Malting Quality QTLs with Ds Transposons. *Funct. Integ. Genom.* **2012,** *12*, 131–141.

Springer, P. S. Gene Traps: Tools for Plant Development and Genomics. *Plant Cell.* **2000,** *12*, 1007–1020.

Turcotte, K.; Srinivasan, S.; Bureau, T. Survey of Transposable Elements from Rice Genomic Sequences. *Plant J.* **2001,** *25*, 169–179.

Vaschetto, L. M. Modulating Signaling Networks by CRISPR/Cas9-Mediated Transposable Element Insertion. *Curr. Genet.* **2018,** *64*, 405–412.

Vollbrecht, E.; Duvick, J.; Schares, J. P.; Ahern, K. R.; Deewatthanawong, P.; Xu, L.; et al. Genome-Wide Distribution of Transposed Dissociation Elements in Maize. *Plant Cell* **2010,** *22*, 1667–1685.

Walbot, V. Saturation Mutagenesis Using Maize Transposons. *Curr. Opi. Plant Biol.* **2000,** *3*, 103–107.

Wicker, T.; Sabot, F.; Hua-Van, A.; Bennetzen, J. L.; Capy, P.; Chalhoub, B.; et al. A Unified Classification System for Eukaryotic Transposable Elements. *Nat. Rev. Genet.* **2007,** *8*, 973–982.

Woo, H.; Suh, S.; Cho, Y. Strategies for Developing Marker-Free Transgenic Plants. *Biotechnol. Bioproc. Eng.* **2001,** *16*, 1053–1064.

Yasuda, K.; Ito, M.; Sugita, T.; Tsukiyama, T.; Saito, H.; Naito, K.; Teraishi, M.; Tanisaka, T.; Okumoto, Y. Utilization of Transposable Element as a Novel Genetic Tool for Modification of the Stress Response in Rice. *Mol. Breed.* **2013,** *32*, 505–516.

Yu, C.; Han, F.; Zhang, J.; Birchler, J.; Peterson, T. A Transgenic System for Generation of Transposon Ac/Ds-Induced Chromosome Rearrangements in Rice. *Theor. Appl. Genet.* **2012,** *125*, 1449–1462.

Zhao, S.; Jiang, E.; Chen, S.; Gu, Y.; Shangguan, A. J.; Lv, T.; et al. PiggyBac Transposon Vectors: The Tools of the Human Gene Encoding. *Transl. Lung Cancer Res.* **2016,** *5*, 120.

CHAPTER 15

Transposon Tagging and Mutagenesis

PRAVEEN KUMAR[1,2*], MAINU HAZARIKA[1,2], LOPAMUDRA SINHA[3], GAGAN KUMAR[4], and NAMRATA[4]

[1]*Agricultural Research Station, Mandore, Agriculture University, Jodhpur 352304, Rajasthan, India*

[2]*Mayurakshi College of Agriculture, Mayurakshi Group of Institute, Jodhpur 342001, Rajasthan, India*

[3]*National Bureau of Plant Genetic Resources, Pusa Campus, New Delhi 110012, India*

[4]*Agricultural Research Station, Mandore, Agriculture University, Jodhpur 352304, India*

[*]*Corresponding author. E-mail: bkpraveen13@gmail.com*

ABSTRACT

Transposon tagging, an important tool for the sequence detection of mutant genes, is used to identify mutant genes of a particular trait, and major importance of this technique is to find out the function of unknown cloned genes and these techniques have been used in the different species, such as maize (Zea mays L.), antirrhinum, etc. When the transposons disrupt the function of wild phenotype, then it can be used for cloning and perform sequencing after amplification the cloned fragment in the polymerase chain reaction. For screening the clone from the genomic library, develop the DNA fragments of transposons elements and DNA fragments of gene of interest which is our objective of the investigation. Because of this tagging technic, we can know the detail information of unknown gene of a specific trait. Mutation can be produced by site-directed mutagenesis technique (random or specific),

Plant Transposable Elements: Biology and Biotechnology, Deepu Pandita, Anu Pandita, (Eds.)

which is highly specific. There are many other techniques that can use to create the mutation of the cell or tissue, such as transposons insertion, radiation, chemical (MMS, EMS, EtBr, etc.), eco-tilling, tilling, somaclonal variation, and double haploid, among them transposon is a potent souse for the producing the mutants of known or unknown phenotype.

15.1 BACKGROUND

These mobile elements or transposable elements jumping elements were discovered in maize (McClintock, 1948) and have now been discovered in all species of plants and animals. Transposon tagging is a tagging technique that stimulates transposons (transposable elements) inside a biological cell. Several organisms often use transposon tagging to isolate genes. The procedure can be used although the nature of the specific genes is unspecified genetic elements from maize (corn) and antirrhinum has been fully integrated into a variety of other crops by researchers, notably tobacco, aspen, and others, using transposon tagging (https://en.wikipedia.org). When movement is triggered by the involvement of a mutant phenotype, a gene responsible for an individual of the community may be cloned within a given species. Spontaneous loss of function of autonomous elements occurs as well, a phenomenon McClintock refers to as "changes of phase" (Fedoroff, 1989). "Changes in step" in Ac and Spm lines are epigenetic, reversible, and occur infrequently. The occurrence has been linked with the increase the DNA methylation in Ac and Spm at particular sites (Banks et al., 1988; Chomet et al., 1987; Cone et al., 1986; Schwartz and Dennis, 1986).

15.1.1 WHAT ARE MOBILE ELEMENTS OR TRANSPOSABLE ELEMENTS?

Mobile elements, also known as "jumping genes" or transposons, are transposable elements (TEs), these are the short DNA fragments, which could move or jump within the genome and they would be multiply themselves by copy and paste or cut and paste mechanism. Insertion sequences (IS) (Figs. 15.1 and 15.2), cassettes, and portable DNA elements are several other labels for them. Generally, these TEs are 500–10,000 nucleotide long sequence but some are larger than this range of nucleotides. Transposons have a variety of methods of relocation or transposition, which are used to categorize them into two groups, this is depending on whether or not they are RNA molecules or DNA molecules (Hua-Van et al., 2005).

I. Transposons (also called Type-II elements). These elements transpose to new sites directly as DNA, for example, IS and composite transposons.

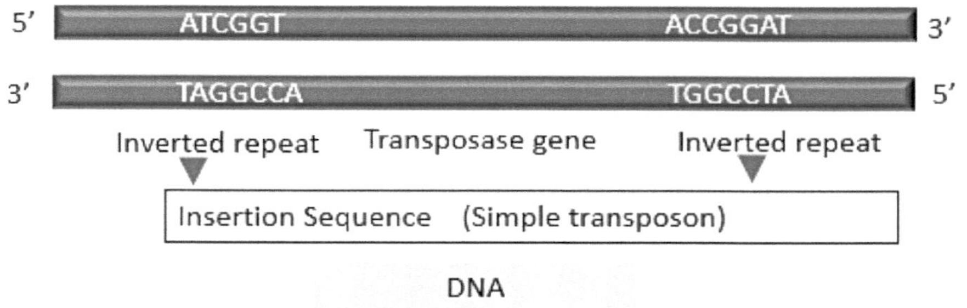

FIGURE 15.1 Insertion series has a general structure with an inverted repeat at each end.

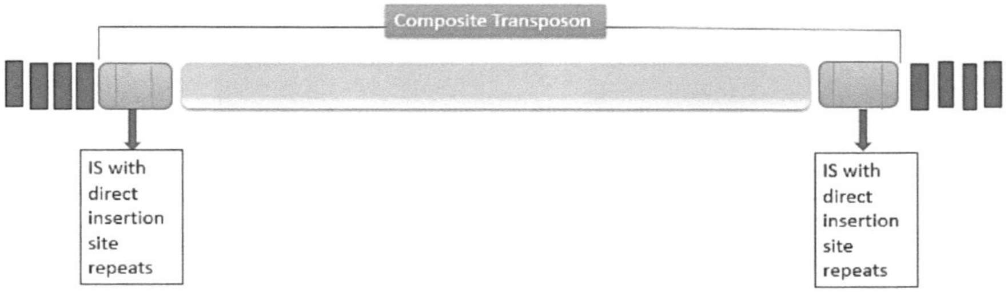

FIGURE 15.2 Composite mobile element structural organization. It is made up of two IS components that surround a central block of protein-coding genes (antibiotic resistance genes).

II. Retrotransposons are elements that are first transcribed into an RNA copy and then reverse transcribed into DNA. The presence or absence of long terminal repeats divides retrotransposons into two categories: LTR and non-LTR (LTRs).

15.1.2 DISCOVERY OF TRANSPOSABLE ELEMENT

B. Mc-Clintock recognized TEs as driving elements while investigating genetic instability in maize (Ac–Ds system) in the 1950s (Fig. 15.4). She noticed that the phenotypic effect of Ds (dissociater), a nonautonomous factor, was dependent on the presence of other elements she coined Ac (activator) (Fig. 15.3).

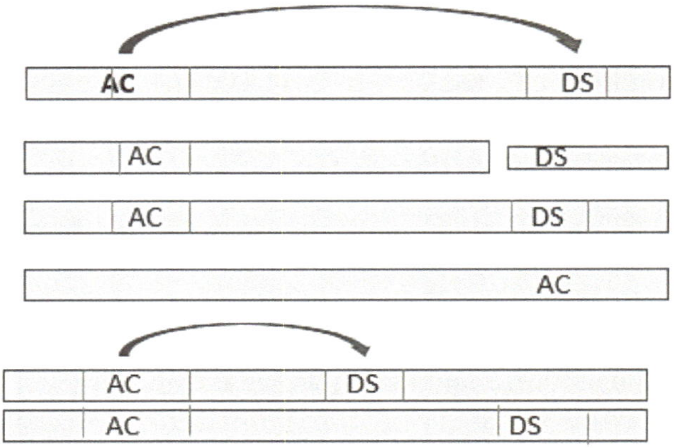

FIGURE 15.3 Maize with the Ac–Ds system. Ds is not transposable without the Ac portion, and chromosome breaks do not occur.

FIGURE 15.4 Different variegated seed colors appear in corn, these are because of the P element.

15.2 TRANSPOSON TAGGING

Maize and snapdragon were the first crops to be transposon-tagged species where mutable alleles are already present (Balcells et al., 1991). Only if the sequences of the elements are identified, are TEs useful tags. Barker et al. (1984), Chomet et al. (1991), Fedoroff et al. (1983), Pohlman et al. (1984), and Schwarz-Sommer et al. (1984) reported that the transposon trapping tool is used to clone the various jumping elements for the example activation (Ac), Ds (dissociation), En/Spm, and Mu genomic DNA sequences. In the transposon tagging approach, relocation of jumping elements, identifying the mutants, which are caused by insertion of the jumping elements, detects the jumping elements which are caused by the mutation in the genome, and finally making copies of targeted gene through to cloning method.

In bacteria and plants, the conventional transposon tagging methodology has been successfully used (Kunze et al., 1997). The use of an inserted transposon sequence as a tag to distinguish adjacent host sequences ensures that an inactivated gene can be isolated quickly. To broaden this traditional strategy into the epoch of functional genomics, more quantitative approaches are frequently used. From isolating a single gene to genome-wide quantitative mutagenesis, the goal shifted. This was accomplished by saturation mutagenesis with either TEs or T-DNA transferred to the plant from Agrobacterium tumefaciens. The cloning of genes in which TEs reside is now partly owing to the molecular isolation of TEs. The ability to clone genes whose function is unspecified is a real advantage of this system. First of all, we have identified the plant/s which is having mutation due to TEs for a specific trait/s, and it is because of TEs move itself into a gene of this specific trait and its became inactive. The cloning of this specific fragment of the plant stock (often in bacteriophage lambda) is being created. A clone is then used to screen this library for the transposon and screen out the transposon, which contains a specific portion from the clones. The mutated gene's fragment may be close to the clone's element. The original clone's nontransposable factor DNA is also used to make a subclone with gene sequences. Following that, this particular clone could be used to show a library, which is containing DNA fragment from a hepatocellular carcinoma. Following that, a genomic library containing DNA from a healthy individual plant is displayed using this clone. As a result, any chosen clone should have a complete and normal copy of the gene (Fig. 15.2).

Since this framework is so efficient, scientists have started using novel approaches to introgression the TEs from maize and antirrhinum into other plant species. These elements in the new species have been shown to be induced to migrate from one location to another. The gene which causes the phenotype can be cloned in the species if the movement associated with the appearance of the mutant phenotype. By using this transposon, tagging is now possible in plant species to detect the active TEs.

Known transposon gene sequences would be used as molecular tag for the cloning purposes and several transposons already studied through the molecular tagging. By rallying transposition and then locating transposon insertions in target genes novel transposons could be "trapped." After the

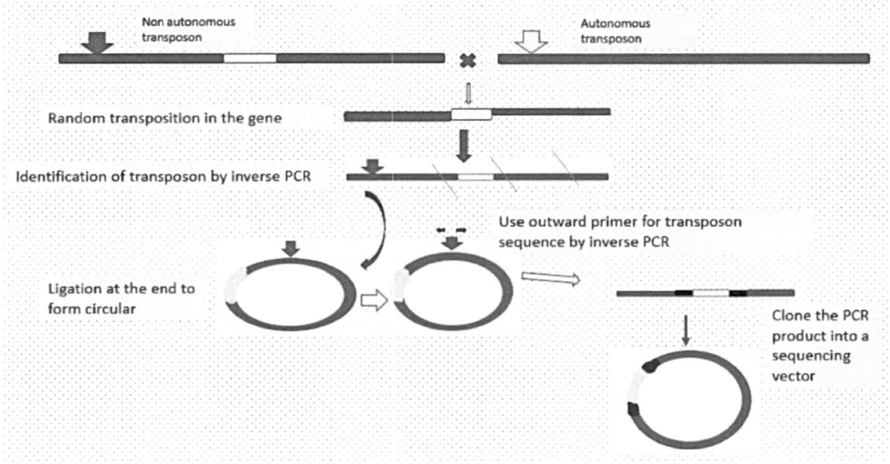

FIGURE 15.5 Nonautonomous transposons are activated.

transposition, mutant gene is used for the sequencing. The efficacy of transposon tags for a given gene is affected by multiple factors. While there can be no preference for the host genome integration of the transposon for tagging, many transposons are having some kind of preference but not all the transposons. Many transposons have a proclivity for transposing to related genomic sites. For some screen types, it may be useful, but it complicates spontaneous transposon mutagenesis. Methods for filtering for transposition to unlinked sites have been formulated in some cases.

Other aspects of transposons tagging is that whether we have to use unknown genomic sequence, which is generally applying in certain genomic background of plant species or already constructed transposons used for creating a single of few the transposons insertion per genome. The higher the level of transposition and the greater the number of mobile transposons, the more common it would be to tag a genetic variant. In the other hand, a vast number of reference transposons are used for the study to classify the particular transposon inserted into the gene of a specific trait. When creates a vast number of transposition then resultant a number of defective phenotypes will be produced and then it is become complicated to make correlation between the mutant gene and transposon. If one or a few transpositions have to do in a single genome then it is become very easy to make the correlation between the mutant gene and transposon, also speeding up the cloning process. One step to secure that the presence of the transposase, only a few elements in the genome are coordinated is to use a heterologous two-element transposon-tagging framework, which encourages the transposase to be eliminated by segregation to resist further transposition. To promote plasmid rescue cloning of flanking DNA, transposons can be arranged in such a way that transposon should have selectable marker gene and it should be the segment of *Escherichia coli*. A selectable marker gene would be linked in a such a way that transposase gene would remove from the tagged gene during the segregation in the later generation.

15.3 STRATEGIES FOR TRANSPOSON TAGGING

15.3.1 MUTAGENESIS

After insertional mutagenesis, the transposon sequence is used to recognize the flanking segments. During transposon tagging, two types of mutagenesis are used (Fig. 15.6).

15.3.2 RANDOM MUTAGENESIS

Randomly various transposon insertions have been done in the genome of individuals then potential mutant gene of interest is screened out from the mutant individuals. In the genome of nonautonomous transposons, a stable transposase is introduced by a two-element scheme, which is carried out by performing various crosses, such as P elements of drosophila, and transferring the transposase gene into the embryo, which is incorporating the nonautonomous DNA segments. Transposition is deployed in other cells as a result of advancement at low temperatures. Dominant mutants can be detected in the M1 generation and recessive mutants in the M2 generation. Transposons are people who have a transposed component in their DNA.

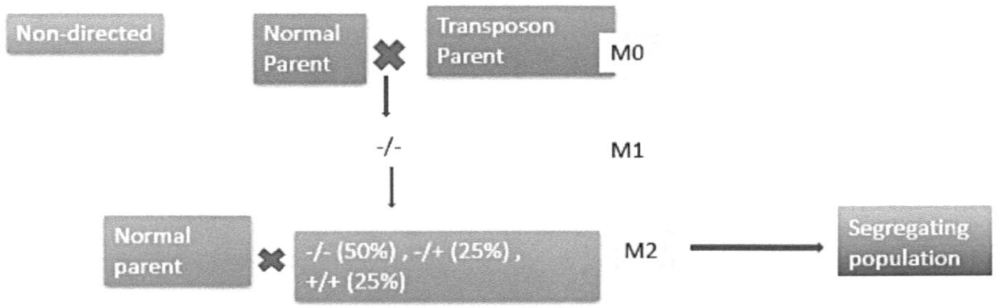

FIGURE 15.6 Mutagenesis in a population that would not be directed.

15.3.3 *DIRECTED MUTAGENESIS*

Identify the mutants of the gene of interest before starting directed mutagenesis. Crossing a wild-type human tissue with autonomous transposon filters with an organism homozygous for a previously mentioned recessive mutation for insertion into a genetic variant Otherwise, choose an individual with a stable transposase gene and a mutant gene; this individual will then cross with a healthy or wild-type individual carrying nonautonomous elements. The gene of interest is activated by a TE that has been mapped and is understood to be attached to it. The transposon is activated in the genome of healthy of wild-type individuals, and next generation is used to identify the mutants. If the transposition appeared in the same location as the gene of interest, we will compare it to the random tagging mutagenesis technique for confirmation (Fig. 15.7).

15.3.4 *ENHANCER TRAP*

Enhancer traps must be used if we want to investigate the gene expression pattern of a target gene (Peter and O'Brochta, 2009). Enhancer trap and gene trap both are similar except promotor, that is present in enhancer trap and gene trap contains a reporter gene. In the enhancer trap, a week promotor regulates the reporter gene, which is transported by the transposon. The reporter gene is expressed pattern that is determined by an enhancer when a transposon is inserted near it. In this way, genes with interesting tissue-specific or developmentally regulated patterns can be detected. This technic is very good for screening the disrupted phenotype, which usually could be done by traditional gene tagging technics. Enhancer trap is having the tissue, cell-type, and developmental markers, which is used to screen out the variety of mutant. Enhancer trap and gene trap both are similar except promotor, that is present in enhancer trap and gene trap contains reporter gene.

15.4 TAGGED GENE CLONING

In the cloning process, first thing is that with the help of southern blotting find out the transposon for the tagged gene, which was related to the mutation. The homozygous mutant individuals are detected using a complementary DNA probe, which is not used in the wild-type individual's progeny. It will become very difficult for the isolation of the mutant phenotypes, if multiple transposons will be segregating in the background.

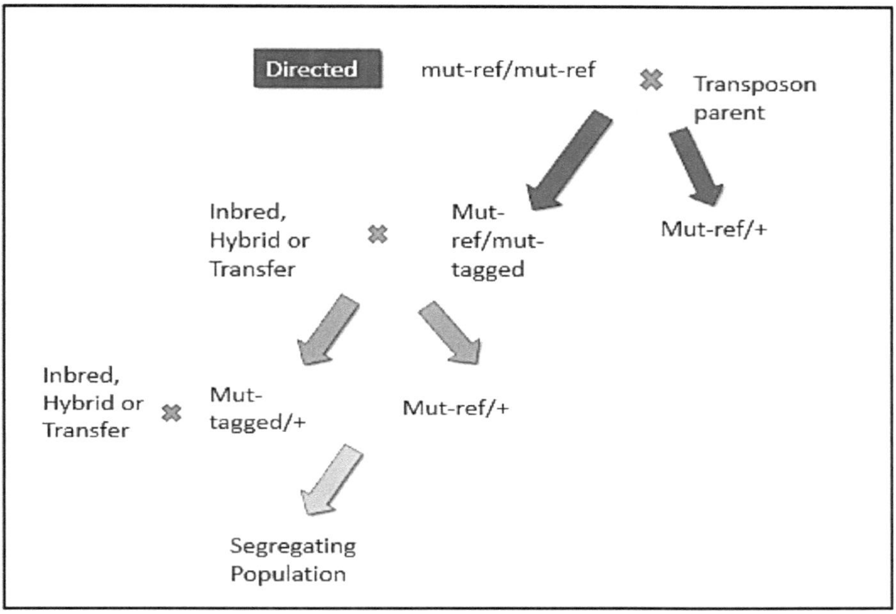

FIGURE 15.7 In a population, directed mutagenesis.

For this, we examined all of the segregating transposon mutant phenotypes, and otherwise backcrossing with individuals who did not have transposons. Once a related transposon has been recognized, targeting the gene into which the transposon has been inserted can be done in a range of methods. Here are three approaches that are widely used.

I. **Library screening:** The tagged mutant was digested with an enzyme that does not cut within the inserted transposon. Any flanking host DNA is included in the recombinant vector containing the transposon. The library is monitored with a transposon-specific probe to locate clones with the transposon and flanking DNA.

II. **Inverse polymerase chain reaction (IPCR):** The IPCR machine is used to isolate the transposon's flaking DNA sequence, and in order to do so, we must first choose the plasmid that contains the tagged gene, then digest the genomic DNA with the restriction enzyme, and then isolate the flanking DNA sequence and transposon.

III. **Plasmid recovery:** When we use a restriction enzyme to digest the genomic DNA of a tagged gene, it releases plasmid from transposons but does not digest the plasmid itself. A small fragment of DNA containing plasmid and flanking host DNA sequence is obtained as a result of digestion. We must relegate the fragments in a high dilution to ensure ligation. The ligation DNA fragment is converted into an *E. coli* plasmid and then extracted.

15.4.1 IDENTIFICATION OF TAGGING

After cloning the adjoining DNA sequence of host, it would be used to look for a band gap on hybridized DNA fragments of mutant individuals and normal individuals, which indicates transposon integration into the mutant's complementary sequences. Therefore, this is confirmation of that flanking sequence of transposon is represented by the cloning DNA rather than a cloning artifact. After that, flanking DNA sequences should be sequenced; otherwise, the probe may be used to obtain complete sequence information as well as further investigation. The flanking sequences must be used to start the chromosome walk and identify the gene of interest, this is done when mutation does not occur in the gene of interest.

Switching the cloned gene back into the mutant is the most convincing way to show that transposon suppression of the cloned gene causes a mutant phenotype. It is also likely that the correct gene has been identified if any of the mutated alleles had mutations in the cloned gene. Phenotypic revertants that show a lack of the transposon from the cloned sequence often affirm the identity of the tagged gene when investigated using sequencing or southern blotting techniques to differentiate and propose additional alleles of the gene from previously produced mutants

15.5 SITE-SELECTED MUTAGENESIS

The objective of site-directed mutagenesis is to identify DNA sequence of the null-phenotype of a gene through to incorporation the transposon. For the identification, the sequence of gene of interest we have to create a large population of transposons and divided into the different groups. Two types of a specified primer sequence id needed for the PCR reaction for identification the sequence (1) gene-based primer (2) a primer sequence of transposons and this sequence has taken from the adjacent region of transposons. For conducting the PCR reaction, mix the both primers with various groups of DNAs of transposons. If the mutation has happened in the gene of interest or its periphery then DNA polymerization will be occurred in the PRC reaction. We can rescreen groups of transposons DNA until the mutation is no longer detectable.

15.5.1 CONFIRMATION OF MUTAGENESIS

I. **Complementation test:** The mutant is transformed once more with a functional copy of the gene for which it is intended. When mutation by a tagged gene occurred, the function is restored.

II. **Revertants:** The mutant is either selfed (endogenous system) or crossed with a mutant transgenic line with a transposable portion that is autonomous (heterologous system).

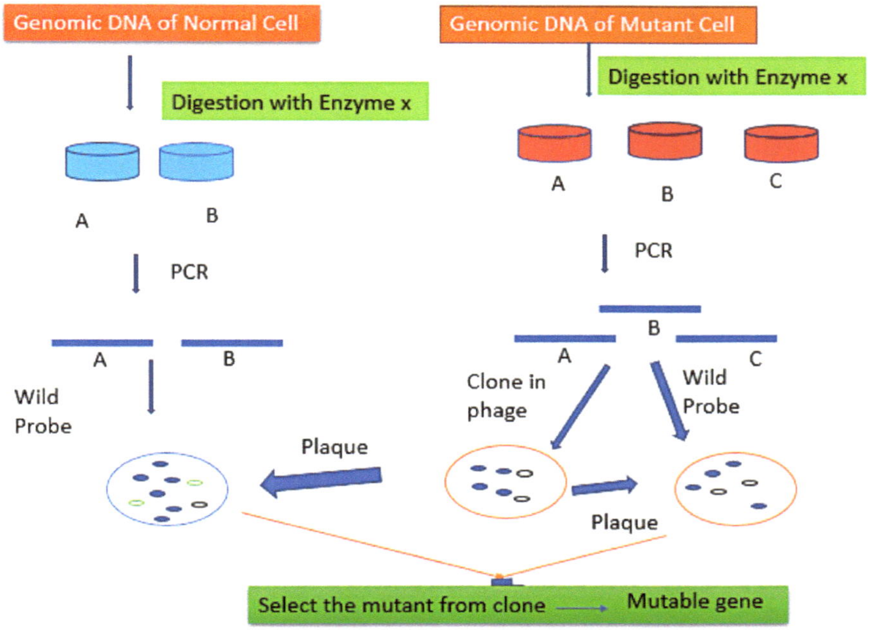

FIGURE 15.8 Endogenous system.

15.6 FUTURE PROSPECT

Methods for transposon tagging are useful for a variety of purposes. To differentiate genes based on their expression patterns, enhancer-trap and gene-trap methods are used, as well as to define and clone different genes with phenotypic proof. Gene tagging is a time saver then the tradition cloning approach, such as chromosome walking. The screening done by using site-specific mutagenesis now permits for the detection of transposon insertions into known-sequence genes. As a result of such insertions of transposons, null alleles are created, or may be produced without mutation null-genes. These null-gens are highly important for determining a gene's functional roles. They are used in genetic science for things like researching double-mutant associations and having null histories for gene mutation studies. Transposon-induced alleles may also be used to construct an allelic array by remobilizing the mobile elements and again incorporation the mobile elements in the functional sequence in the genome of individuals.

Multiple alleles of a gene may provide useful details about an individual's phenotypes. Analysis the many details information about physiological process in the developmental tissues or different types of tissues by the somatic excision. For both of these purposes, transposon labeling is and will remain an excellent method for gene cloning and gene processing.

KEYWORDS

- **transposable elements**
- **mobile elements**
- **transposon tagging**
- **Ac-Dc elements**
- **mutagenesis**
- **maize**

REFERENCES

Alonso, J. M.; Stepanova, A. N.; Leisse, T. J.; Kim, C. J.; Chen, H.; Shinn, P.; Stevenson, D. K.; Zimmerman, J.; Barajas, P.; Cheuk R. Genome-Wide Insertional Mutagenesis of Arabidopsis Thaliana. *Science* **2003**, *301*, 653–657.

Balcells, L.; Swinburne, J.; Coupland, G. Transposons as Tools for the Isolation of Plant Genes. *TIBTECH* **1991**, *9*, 31–37.

Banks, J. A.; Masson, P.; Fedoroff, N. Molecular Mechanisms in the Developmental Regulation of the Maize Suppressor-Mutator Transposable Element. *Genes Dev.* **1988**, *2*, 136480.

Barker, R. F.; Thompson, D. V.; Talbot, D. R.; Swanson, J.; Bennetzen, J. L. Nucleotide Sequence of the Maize Transposable Element *Mul. Nucleic Acids Res.* **1984**, *12*, 5955–5967.

Chomet, P.; Lisch, D.; Hardeman, K. J.; Chandler, V. L.; Freeling, M. Identification of a Regulatory Transposon that Control the *Mutator* Transposable Element System in Maize. *Genetics* **1991,** *129*, 261–270.

Chomet, P. S.; Wessler, S.; Dellaporta, S. L. Inactivation of the Maize Transposable Element Activator (Ac) Is Associated with Its DNA Modification. *EMBO J.* **1987**, *6*, 295–302.

Cone, K. c.; Burr, F. A.; Burr, B. Molecular Analysis of the Maize Anthocyanin Regulatory Locus Cl. *Proc. Natl. Acad. Sci. U.S.A.* **1986**, *83*, 9631–9633.

Cowperthwaite, M.; Park, W.; Xu, Z.; Yan, X.; Maurais, S. C.; Dooner, H. K. Use of the Transposon Ac as a Gene-Searching Engine in the Maize Genome. *Plant Cell* **2002**, *14*, 713–726.

Fedoroff, N.; Wessler, S.; Shure, M. Isolation of the Transposable Maize Controlling Elements *Ac* and *Ds*. *Cell* **1983**, *35*, 235–242

Fedoroff, N. V. Maize Transposable Elements. *See Ref.* **1989**, *16*, 375–411. https://en.wikipedia.org/wiki/Transposon_tagging

Hua-Van, A.; Le Rouzic, A.; Maisonhaute, C.; Capy, P. Abundance, Distribution and Dynamics of Retrotransposable Elements and Transposons: Similarities and Differences. *Cytogenet. Genome Res.* **2005**, *110*, 426–440.

McClintock, B. Mutable Loci in Maize. *Carnegie Inst. Wash. Year B.* **1948**, *47*, 155–169.

Peter, W. A.; O'Brochta, D. A. *Encyclopedia of Insects*; 2nd ed.; 2009.

Pohlman, R. F.; Fedoroff, N. V.; Messing, J. The Nucleotide-Sequence of the Maize Controlling Element *Activator*. *Cell* **1984**, *37*, 635–643.

Robertson, D. S. Differential Activity of the Maize Mutator Mu at Different Loci and in Different Cell Lineages. *Mol. Gen. Genet.* **1985**, *200*, 9–13.

Schwartz, D.; Dennis, E. A. Transposase Activity of the Ac Controlling Element in Maize Is Regulated by Its Degree of Methylation. *Mol. Gen. Genet.* **1986**, *205*, 476–82.

Schwarz-Sommer, Z.; Gierl, A.; Klosgen, R. B.; Wienand, U.; Peterson, P. A.; Saedler, H. The *Spm* (*En*) Transposable Element Controls the Excision of a 2-Kb DNA Insert at the *wx-m8* Allele of *Zea Mays*. *EMBO J.* **1984**, *3*, 1021–1028.

Walbot, V. Strategies for Mutagenesis and Gene Cloning Using Transposon Tagging and T-DNA Insertional Mutagenesis. *Annu. Rev. Plant Physiol. Plant Mol. Biol.* **1992**, *43* (1), 49–78.

Part V
TEs of Model Plants

CHAPTER 16

Classification, Structure, Function, and Evolution of Transposable Elements in *Oryza sativa*

SWARNAVO CHAKRABORTY and ARYADEEP ROYCHOUDHURY*

Post Graduate Department of Biotechnology, St. Xavier's College (Autonomous), 30, Mother Teresa Sarani, Kolkata, West Bengal 700016, India

**Corresponding author. E-mail: aryadeep.rc@gmail.com*

ABSTRACT

An appreciable portion of different plant genomes contain mobile genetic elements, also called as transposable elements (TEs). It has been widely accepted and demonstrated that these TEs generate a wide range of alterations in the plant genome, leading to great variations in the expression patterns and functions of many plant genes. Moreover, these mobile elements can bring about marked genetic changes including translocations, excision, and recombination at abrupt genetic locations, breakage of chromosomes and inversions, etc. that aids in the process of evolution via reproductive isolation and generation of new species. One of the most important crops in the global agricultural sector is *Oryza sativa* (rice) and thus it has been widely used as a model plant to decipher the mechanisms of transposition. In order to study the evolution of rice genome, a lucid study of the patterns of the various forms of TEs in different closely related varieties of rice is mandatory. Therefore, a clear understanding of the various rice genes and their structural and functional insights, along with their mechanisms of evolution is needed, in order to comprehend the roles of TEs in the plant genomic diversity.

Plant Transposable Elements: Biology and Biotechnology, Deepu Pandita, Anu Pandita, (Eds.)

16.1 INTRODUCTION

Transposable elements (TEs) can be defined as pieces of DNA that have the potential of getting inserted into new genomic locations and also have the power of getting duplicated into newer copies of themselves during the entire event of insertion. TEs constitute the largest portion of the genome of most eukaryotes ranging from about 50% to 80% of the genome in the members of the grass family (Feschotte et al., 2002). Barbara McClintock was the first scientist responsible for demonstrating the role of TEs in the alteration of pigmentation pattern in the mutant maize kernels (McClintock, 1951). She was also the first to genetically characterize these elements and explained the potential role of these TEs, for which she received the Nobel Prize in the year 1983.

Oryza sativa (rice) is a domesticated monocotyledonous crop plant. It serves as one of the most common model plants in the laboratory. The genomic size of rice plant is around 430 Mb, which is comparatively smaller than other members of the grass family. It is known that the size of the genome has a direct correlation with the proportion of the repeats in the genome (Turcotte et al., 2001). While comparing the genomes of the members of the grass family, plants like maize have a relatively larger size of its genome and tend to harbor a large number of nested retro-transposons in the intergenic portions of their genome. However, for rice, this number is lesser by virtue of its small genome size. Almost 40% of the genome of rice plant is comprised of sequences related to TEs (Jiang and Panaud, 2013). Therefore, these TEs act as definitive features of a particular genome. The patterns of these elements in the genome of the closely related members of different rice species have been useful in the study of the role of TEs in the evolution of the rice genome (Li et al., 2017a, 2017b). Rice being an agriculturally important crop has been widely used as a model plant to study the role of the various forms of transposons. The various TEs in rice have contributed to the betterment of growth, yield, and quality of grains in rice. Therefore, for the sake of enhanced biomass production and yield, the role of TEs in rice plants have been elaborately studied for ensuring food security throughout the globe.

16.2 CLASSIFICATION AND STRUCTURE OF TRANSPOSONS IN RICE

Majority of the eukaryotic genomes are characterized by a substantial amount of repetitive sequences, which includes TEs or transposons, often referred to as the mobile genetic elements (Turcotte et al., 2001). These elements tend to get inserted within new positions in the genome and hence called so. Moreover, the high degree of genetic variation in higher plants can be attributed to such TEs (Bennetzen and Wang, 2014). Every TE has two basic characteristics. The first property can be deciphered from its name itself. Transposons are referred to as mobile genetic elements, by virtue of their potentiality to move from one location of the genome to another. The second property is their immense ability of amplifying the copy number of a particular segment of the genome via the process of transposition (Bennetzen, 2000).

The genome of rice plants contains almost all the well-defined TEs. Its genome harbors the miniature inverted-repeat transposable elements (MITEs), the long and short interspersed elements, the miniature terminal repeat retro-transposons, and the long terminal repeat (LTR) retro-transposons (Panaud et al., 2002; Komatsu et al., 2003; Chaparro et al., 2007; Wicker et al., 2018; Tsuchimoto et al., 2008; Gao et al., 2016). In a typical eukaryotic genome, there can be a large number of inactive transposons, ranging from few hundreds to thousands. It is quite challenging to decipher whether a TE is in its active state or inactive state. Sometimes, an element can be epigenetically silenced when present at a certain location in the eukaryotic genome. Each particular group of transposons comprise autonomous elements and nonautonomous elements. The autonomous elements have ORFs (Open Reading Frames) that can encode proteins essential for the process of transposition. On the other hand, nonautonomous elements that have the capability of transposition lack the potential of coding for essential protein products (Feschotte et al., 2002). However, they tend to retain certain cis-elements that are needed for transposition. There is a duplication event associated with a short sequence of the genome, post integration of almost all TEs and these target site duplications (TSDs) tend to show a wide variability in terms of their size. TEs can be classified on the basis of their different manner of mobility (Charlesworth et al., 1994). Basically, depending on whether the transposition intermediate is DNA or RNA, they are classified as class I and class II TEs. The class I elements are commonly referred to as retro-transposons, because their movement is associated with an RNA intermediate, which undergoes the process of reverse transcription before their integration within the genome. However, after insertion, the class I elements fail to excise out (Finnegan, 1992; Grandbastien, 1992). Depending on the structure and the mechanism of transposition, the class I elements can be classified into two categories: LTR retro-transposons and the non-LTR retro-transposons. The LTR retro-transposon class I elements are commonly characterized by flanking LTRs in direct orientation that often contains several genes encoding proteins involved in the process of integration and reverse-transcription (Kubis et al., 1998; Zhang and Gao, 2017). Autonomous elements are commonly characterized by two genes named *pol* and *gag*. The *pol* gene encodes for a poly-protein that have four essential activities: protease, integrase, reverse transcriptase, and RNase H activities; while the *gag* gene generates a capsid-like protein. On the other hand, the nonautonomous elements lack almost all of these coding regions and the internal regions show a wide range of variability in length (Jin and Bennetzen, 1989; Witte et al., 2001). The non-LTR retro-transposon class I elements include SINEs (Short interspersed nuclear elements) and LINEs (Long interspersed nuclear elements), which have become widely associated with the present field of plant research (Flavell et al., 1994; Hirano et al., 1994; Leepton and Smyth, 1993; Mochizuki et al., 1992; Umeda et al., 1991). The coding regions of these nuclear elements consist of genes that encode a *gag*-like protein (ORF1), reverse transcriptase, and endonuclease. A simple sequence repeat containing a stretch of A, that is, poly As is the most striking characteristic of SINEs and LINEs and serves as a terminating element. The 5′-end of almost all SINEs contains a RNA polymerase III promoter. Irrespective of the unknown origin of the 3′- half of SINEs, LINEs and SINEs present in the common genome often tend to show homology in sequences at the 3′ terminus, thereby indicating that the transposition capability of partner LINEs can be jeopardized and parasitized by the associated SINEs (Ogiwara et al., 1999). The typical class I elements in plants are LTR-associated retro-transposons and include predominantly the Ty3-gypsy and Ty1-copia groups. The class I LTR retro-transposons constitute almost 14% of the entire genomic DNA in rice, which is the highest amongst all other forms of TEs (Jiang et al., 2003). Retro-transposons that are usually activated during callus vitro culture are *Karma, Lullaby*, and *Tos 17* (Hirochika et al., 1996; Komatsu et al., 2003; Picault et al., 2009). Transposons like *mPing* and *nDart* are usually activated when rice plants are subjected to irradiation using laser and inhibitors of methylation, respectively (Eun et al., 2012; Li et al., 2017a, 2017b).

On the other hand, the class II TEs are characterized by Terminal inverted repeats (TIRs) and exhibit variability in their lengths. They tend to have target insertion sites (TSDs), which exhibit a conserved length. It is commonly observed that the class II elements encode a protein called transposase. These elements have the dual power of excision and insertion within new locations in the genome. It is believed that by virtue of internal deletions, the autonomous elements give rise to nonautonomous elements.

The class II transposons are commonly referred to as DNA transposons (Finnegan, 1992; Flavell et al., 1994). The commonly observed elements of this type that were identified first in maize plants, includes the *Ac/Ds* (McClintock, 1984), *MuDR/Mu* (Yoshida et al., 1998; Lisch et al., 1999), and *Spm/dSpm* (Raina et al., 1998; Fedoroff, 1999). A particular family of DNA transposons can be categorized by virtue of the presence of same TIR sequences and within this family, some of the members generate a protein called transposase, which helps in the recognition of the family-specific TIR sequence and eventually aids in the process of transposition (Kunze et al., 1997). During the process of transposition involving DNA transposons, the transposase protein encoded by the TE is responsible for the recognition of its family-specific TIRs and excises the portion of the DNA between the site of its reinsertion in other genomic locations and the TIRs. The next step involves the repairing of the gap created due to the excision of the DNA segment. The repair can occur in two different ways: (i) through the process of gene conversion via recombination using the sister chromatid or the other homologous portion as the template, where no net excision takes place, and (ii) through a simple step of ligation, where a part of the DNA gets excised out, that is, net excision takes place (Bennetzen, 2000).

The most diverse with almost 1 lakh elements and comprising of around 100 different families are the MITEs. These constitute around 6% of the entire rice genome. These repeats are commonly found in the members of the grass family, but are also found in other flowering plants and even in a number of members of the animal kingdom, including humans. The MITEs are mostly located with the noncoding regions of many genes within plants (Jiang et al., 2003). These are reminiscent forms of the class II nonautonomous DNA TEs. MITEs are quite smaller, around 600 base pairs in size and the TIRs are also shorter, ranging from 10 to 30 base pairs. However, these are quite different from the other common nonautonomous DNA TEs. The MITEs are characterized by strong target site preference for a TAA trinucleotide or TA dinucleotide sequence and have a very high copy number, around 10^4 copies for each family (Feschotte et al., 2002). Depending on the similarity in the TIRs and the duplication of their TSD, the MITEs can be classified into following two types: *Stowaway* like and *Tourist* like MITEs. Moreover, these two forms of MITEs can be put into two superfamilies: Tc1/mariner and PIF/ Harbinger, respectively (Zhang et al., 2001; Jiang and Wessler, 2001; Turcotte et al., 2001; Feschotte et al., 2002). The MITEs lack any autonomous element, which has already been identified. It has been hypothesized that certain undiscovered autonomous elements encode transposases that have the specificity for particular MITEs or even the transposition can be mediated by certain trans-acting element that encodes the protein in question normally involved with some other essential cellular process. However, there are lesser evidences to prove either hypothesis (Wessler et al., 1995). In rice plants, scientists have deciphered a repetitive sequence of 430 bp and designated it as a potential MITE candidate. The sequence displayed a high degree of conservation in sequence amongst the different copies and was named as "Miniature Ping" or mPing. When subjected to cell culture media, *indica* rice varieties exhibited an increase in the mPing number, indicating active transposition to have taken place. Calculation from the entire genomes of *indica* and *japonica* rice varieties indicated the presence of 14 and 70 mPing copies, respectively (Jiang et al., 2003).

16.3 EVOLUTION AND FUNCTIONAL ROLE OF TRANSPOSONS IN RICE

TEs are ubiquitous throughout different plant genomes, thus indicating their early origin and also provide considerable evidences regarding their great diversity throughout the plant world. The mobility of TEs makes them suitable candidates for horizontal transfer. These elements are usually located on the bacterial plasmids and they get activated during bacterial mating and subsequent replication of the DNA genome (Kleckner, 1990). On the other hand, retro-viruses can efficiently transmit both within and amongst members of a particular species. Therefore, the proper mechanism and the time of origin of the transposons still remains a mystery. However, the mechanism of action of retro and DNA TEs suggests that these have originated due to independent evolutionary events. The process of natural selection will only allow the amplification of a certain sequence within the genome, until such amplification has no negative impact on the host fitness. However, these elements should arise via multiple independent events in accordance with the parasitic or selfish DNA concept (Doolittle and Sapienza, 1980; Orgel and Crick, 1980). For instance, multiple independent origin events associated with the products of different RNA polymerase III is the probable way by which a range of SINEs have originated. For instance, the *Ac/Ds* family and *Spm/dspm* family of TEs tend to have similarity in the transposase protein, the TIRs, and other related elements in different species of plants (Bonas et al., 1984; Hehl et al., 1991; Kunze et al., 1997). Therefore, it indicates that these elements should have been present in the primordial forms of the angiosperms. However, the rate of evolution of TEs is quite faster than the genomic DNA. So, it cannot be negated that plant TEs and their families share vertical evolutionary relationship and ancestral traits with those of the elements found in animals.

Initially, it was considered that transposons have a rather negative impact on the genome. These elements were earlier considered as mutagenic agents since upon insertion into coding regions of certain genes, they adversely affect normal functioning of those genes. However, current research has been successful in establishing the positive roles of transposition in various plant genomes. The transposons can either provide new cis-acting regulatory target sites or generate alternate promoters or sometimes epigenetically alter functioning of a target (Mirouze and Vitte, 2014; Hirsch and Springer, 2017). Thus, transposition can help in the development of entirely novel traits, thereby actively playing role in the generation of genetic diversity and plant evolution, including evolution of various agronomic traits in rice plants like pigmentation or color, size, shape, etc. (Xiao et al., 2008; Zhang et al., 2019). TEs are rather manageable and less comparable as entities encompassing the genome, although they have the huge potential of moving about the genome. The count of the different families of transposons and their various members are rather limited and they uncover limited genetic portions. Therefore, it is obvious to conclude that the genetic impact imposed by transposons is quite restricted. McClintock during her experiments observed novel variegating mutations in the experimental cultures at quite a huge frequency that can be associated with the perturbations in the genome due to existence of broken chromosomes. This indicated that TEs tend to remain genetically silent, but they inhabit the genome in somewhat regular fashion (McClintock, 1946; McClintock, 1978). Plants like rice tend to have greater number of copies of a given TE than can be identified as genetic elements. Moreover, most of these sequences fail to match the criterion of a complete transposon, that is, they genetically do not resemble an entire transposon. However, whether a transposon is genetically active or inactive, can be deciphered by reading its methylation patterns (Fedoroff et al., 1983; Fedoroff et al., 1984; Pereira et al., 1985; Masson et al., 1987; Banks and Fedoroff, 1989).

Transposons are considered to be the most variable and diverse genetic components of a plant system. Interestingly, one can expect a wide variety of TEs even in the members of rice plants belonging to closely related species. These elements can bring about wide range of changes in the functionality and constitution of various genes, including the evolution of sequences that code for novel proteins, insertions leading to simple mutations, or even large-scale mutations and movement of genes causing the reprogramming of pre-existing genes. Often in order to put a check on the transposition process, certain elements are silenced epigenetically by the plant system; thereby leading to genetic as well as epigenetic diversity in the genetic constitution and functioning of many genes (Feschotte et al., 2002; Lisch, 2013). These gradual accumulations of variations act as the subject of operation of natural selection and subsequent plant adaptations to the changed environment. However, the role of transposition in the process of plant-adaptive evolution is rather very restricted due to lesser availability of evidences. Currently, due to advancement in the field of phenomics and genomics, it has become quite easy to decipher and assign function to parts of the rice genome that have potential role as transposons. The various activities of different TEs in the development of genetic diversity in rice plants can be effectively deciphered through the systematic recognition and designation of an array of alleles that have crucial phenotypic effects on the plants (Lisch, 2013). Domestication of plant species has currently generated suitable models for the lucid understanding of the phenomenon. Domesticated species of rice allow proper comparison with the wild relatives due to the presence of well-characterized TEs on the basis of the possible trait variations. Moreover, domesticated plants are usually subjected to directional selection, thus making the process of comparison with their wild relatives rather easier.

McClintock first established transposons to be involved in gene and genome rearrangement (McClintock, 1946; McClintock, 1948). The rearrangement of the genome leads to alternation in structure and function of various genes. A range of genome alterations can be brought about by the transposons, which includes marked genetic changes including breaking of chromosomes, translocations, insertion, excision, amplification of certain sequences, transposition, recombination at abrupt genetic locations, breakage of chromosomes and inversions, etc., that can ultimately alter genetic structure, function, and expression thereby generating diversity in phenotype. However, TEs are commonly described in literature for the unique process of transposition. This process can even have adverse effects on the genome function, as transposition of an element of the genome into a functional gene can inactivate that genetic locus concerned (Bennetzen, 2000). Moreover, the excision of DNA TEs can often take place during gametophytic or somatic stage of the plant life cycle, and leads to generation of alleles that have immense potential to mutate and are quite unstable, which is basically an exact reversion as compared to the wild type rice plants. Sometimes, these reversions tend to leave certain sequences behind like the direct repeats that flank TEs, which were generated upon insertion. Even a small number of bases or a certain sequence segment of the TE can be generated by various repair mechanisms at the site of excision. This can occur due to illegitimate conversion or short deletions of the target location at the site of excision (Peacock et al., 1984; Doring and Starlinger, 1986; Levy and Walbot, 1991; Bennetzen, 2000). Therefore, TEs can generate mutations both due to insertion of the associated elements and subsequent cycles of excision in the following rounds, leading to alteration in the sequence. This is one of the reasons behind the huge mutability rate of the TEs as compared to genic locations associated with the same genome. Interestingly, genetic alterations like internal rearrangement and deletion of genic sequences commonly takes place during failed attempts of transposition.

TEs have the unique capacity to increase the copy number of many genes, by virtue of their power of acquiring certain segments of the genome containing genes and finally integrating those genes into new locations. However, after such integration events, the genes can get fragmented to a great extent, but these fragmented portions of the genes can give rise to novel composite genes upon assembly of the fragments. Sometimes, TEs tend to play role as introns under certain circumstances (Giroux et al., 1994).

It has been observed that rice TEs have associated with regulatory sequence elements that have the potential to alter the patterns of expression of the adjacent loci. Insertion of such an element carrying regulatory sequences into the promoter of certain genes can lead to modulation of the regulation of the gene in question under the jurisdiction of the transposon that has been inserted (Martienssen et al., 1989). Interestingly, certain DNA TEs are associated with terminal sequences that modulate the element to serve as an intron under certain conditions. Moreover, due to the binding of specific transposase to these elements, the activity of the gene in question can get suppressed via a transposase-dependent mechanism (Kim et al., 1987; Wessler et al., 1987; Wessler, 1992). Epigenetic silencing of plant transgenes is a commonly encountered phenomenon. However, the exact mechanism associated with the epigenetic silencing of plant TEs, especially in rice plants is still a mystery. Scientists have reported similarities of this phenomenon with the process of homology-based silencing of genes. It can be speculated that plants might have evolved this mechanism in the form of a secondary outcome of the process of silencing of plant transgenes, in order to inactivate and disrupt the activities of certain TEs and some deadly plant viruses that affect normal plant growth, development and yield of rice (Matzke et al., 1989; Matzke and Matzke, 1998).

16.4 CONCLUSION AND FUTURE PERSPECTIVES

The mobile genetic elements or transposons constitute a large portion of the genome of the members of the grass family of angiosperms, including rice. These mobile elements are known for their hyperactivity in alteration of the structure of plant genes and the genome. Almost 40% of the rice genomic DNA contains stretches of repetitive DNA having no defined and assigned function. Moreover, these repetitive parts of the rice genome have not been found to encode any functional protein. The class I LTR retro-transposons and the MITEs encompass the largest portions of the rice genomic DNA. Unfortunately, the role of TEs in the creation of phenotypic diversity and genetic variation in plants has always been overlooked. Currently, with the advancement in rice research, rice genomic resources are available in the scientific databases. A lucid understanding of these basic aspects of the rice genetic information and their functions will help the researchers to unravel newer strategies and formulate newer technologies to accelerate the process of improvement of rice as a commercially important crop. Improvisation of Genome-Wide Association Studies can aid in the discovery of many hidden agronomic characters and unpredictable resources in rice plants. Moreover, the use of TEs for the improvement of essential traits in rice has to be implemented in large scale. It is known that the magnitude of morphological differences between two related plant species can never be directly correlated with the extent of chromosomal or genetic differences they harbor. However, the role of transposons in the evolution of plant genome is indispensable. Thus, researchers must try to fish out the extent of genomic and chromosomal alterations that can potentially aid the process of evolution of new species, along with the role of transposable elements as a whole in the entire process of species evolution.

ACKNOWLEDGMENTS

Financial assistance from Science and Engineering Research Board, Government of India through the grant [EMR/2016/004799] and Department of Higher Education, Science and Technology and Biotechnology, Government of West Bengal, through the grant [264(Sanc.)/ST/P/S&T/1G-80/2017] to Dr. Aryadeep Roychoudhury is gratefully acknowledged.

KEYWORDS

- **transposable elements**
- ***Oryza sativa***
- **translocations**
- **excision**
- **recombination**
- **inversions**
- **genomic diversity**

REFERENCES

Banks, J. A.; Fedoroff, N. Patterns of Developmental and Heritable Change in Methylation of the *Suppressor-mutator* Transposable Element. *Dev. Genet.* **1989,** *10*, 425–437.

Bennetzen, J. L. Transposable Element Contributions to Plant Gene and Genome Evolution. *Plant Mol. Biol.* **2000,** *42*, 251–269.

Bennetzen, J. L.; Wang, H. The Contributions of Transposable Elements to the Structure, Function, and Evolution of Plant Genomes. *Annu. Rev. Plant Biol.* **2014,** *65*, 505–530.

Bonas, U.; Sommer, H.; Saedler, H. The 17-kb *Tam1* element of *Antirrhinum majus*induces a 3-bp Duplication upon Integration into the Chalcon Synthase Gene. *EMBO J.* **1984,** *3*, 1015–1019.

Chaparro, C.; Guyot, R.; Zuccolo, A.; Piegu, B.; Panaud, O. RetrOryza: A Database of the Rice LTR-Retrotransposons. *Nucleic Acids Res.* **2007,** *35*, D66–D70.

Charlesworth, B.; Sniegowski, P.; Stephan, W. The Evolutionary Dynamics of Repetitive DNA in Eukaryotes. *Nature* **1994,** *371*, 215–220.

Doolittle, W. F.; Sapienza, C. Selfish Genes, the Phenotype Paradigm and Genome Evolution. *Nature* **1980,** *284*, 601–603.

Doring, J. -P.; Starlinger, P. Molecular Genetics of Transposable Elements in Plants. *Annu. Rev. Genet.* **1986,** *20*, 175–200.

Eun, C. -H.; Takagi, K.; Park, K. -I.; Maekawa, M.; Iida, S.; Tsugane, K. Activation and Epigenetic Regulation of DNA Transposon *nDart1* in Rice. *Plant Cell Physiol.* **2012,** *53*, 857–868.

Fedoroff, N.; Wessler, S.; Shure M. Isoltion of the Transposable Maize Controlling Elements Ac and Ds. *Cell* **1983,** *35*, 243–251.

Fedoroff, N. V. The Suppressor-Mutator Element and the Evolutionary Riddle of Transposons. *Genes Cells* **1999,** *4*, 11–19.

Fedoroff, N., Furtek, D.; Nelson, O. Cloning of the Bronze Locus in Maize by a Simple and Generalizable Procedure Using the Transposable Controlling Element Activator (Ac). *Proc. Natl. Acad. Sci. U.S.A.* **1984,** *81*, 3825–3829.

Feschotte, C.; Jiang, N.; Wessler, R. S. Plant Transposable Elements: Where Genetics Meets Genomics. *Nat. Rev. Genet.* **2002,** *3*, 329–341.

Feschotte, C.; Wessler, S. R. Mariner-Like Transposases Are Widespread and Diverse in Flowering Plants. *Proc. Natl. Acad. Sci. U.S.A.* **2002,** *99*, 280–285.

Finnegan, D. J. Transposable Elements. *Curr. Opin. Genet. Dev.* **1992,** *2*, 861–867.

Flavell, A. J.; Pearce, S. R.; Kumar, A. Plant Transposable Elements and the Genome. *Curr. Opin. Genet. Dev.* **1994,** *4*, 838–844.

Gao, D.; Li, Y.; Kim, K. D.; Abernathy, B.; Jackson, S. A. Landscape and Evolutionary Dynamics of Terminal Repeat Retrotransposons in Miniature in Plant Genomes. *Genome Biol.* **2016,** *17*, 7.

Giroux, M. J.; Clancy, M.; Baier, J.; Ingham, L.; McCarty, D.; Hannah, L. C. De Novo Synthesis of an Intron by the Maize Transposable Element *Dissociation. Proc. Natl. Acad. Sci. U.S.A.* **1994,** *91*, 12150–12154.

Grandbastien, M. A. Retroelements in Higher Plants. *Trend. Genet.* **1992,** *8*, 103–108.

Hehl, R.; Nacken, W.; Krause, A.; Saedler, H.; Sommer, H. Structural Analysis of *Tam*3, a Transposable Element from *Antirrhinum majus*, Reveals Homologies to the *Ac* Element from Maize. *Plant Mol. Biol.* **1991,** *16*, 369–371.

Hirano, H. Y.; Mochizuki, K.; Umeda, M.; Ohtsubo, H.; Ohtsubo, E.; Sano, Y. Retrotransposition of a Plant SINE into the wx Locus During Evolution of Rice. *J. Mol. Evol.* **1994,** *38*, 132–137.

Hirochika, H.; Sugimoto, K.; Otsuki, Y.; Tsugawa, H.; Kanda, M. Retrotransposons of Rice Involved in Mutations Induced by Tissue Culture. *Proc. Natl. Acad. Sci. U.S.A.* **1996,** *93*, 7783–7788.

Hirsch, C. D.; Springer, N. M. Transposable Element Influences on Gene Expression in Plants. *Biochimica et Biophysica Acta (BBA)—Gene Regul. Mech.* **2017,** *1860*, 157–165.

Jiang, N.; Bao, Z.; Zhang, X.; Hirochika, H.; Eddy, S. R.; McCouch, S. R.; Wessler, S. R. An Active DNA Transposon Family in Rice. *Nature* **2003,** *421*, 163–167.

Jiang, N.; Panaud, O. Transposable Element Dynamics in Rice and Its Wild Relatives. In *Plant Genetics and Genomics: Crops and Models*; Zhang, Q., Wing, R. A., Eds.; Genetics and Genomics of Rice: New York, USA, 2013; pp 55–69.

Jiang, N.; Wessler, S. R. Insertion Preference of Maize and Rice Miniature Inverted Repeat Transposable Elements as Revealed by the Analysis of Nested Elements. *Plant Cell* **2001,** *13*, 2553–2564.

Jin, Y. K.; Bennetzen, J. L. Structure and Coding Properties of *Bs1*, a Maize Retrovirus-Like Transposon. *Proc. Natl. Acad. Sci. U.S.A.* **1989,** *86*, 6235–6239.

Kim, H. -Y.; Schiefelbein, J. W.; Raboy, V.; Furtek, D. B.; Jr. Nelson, O. E. RNA Splicing Permits Expression of a Maize Gene with a Defective *Suppressor-mutator* Transposable Element Insertion in an Exon. *Proc. Natl. Acad. Sci. U.S.A.* **1987,** *84*, 5863–5867.

Kleckner, N. Regulation of Transposition in Bacteria. *Annu. Rev. Cell Dev. Biol.* **1990,** *6*, 297–327.

Komatsu, M.; Shimamoto, K.; Kyozuka, J. Two-Step Regulation and Continuous Retrotransposition of the Rice LINE-Type Retrotransposon Karma. *Plant Cell* **2003,** *15*, 1934–1944.

Kubis, S.; Schmidt, T.; Heslop-Harrison, J. S. Repetitive DNA Elements as a Major Component of Plant Genomes. *Ann. Bot.* **1998,** *82*, 45–55.

Kunze, R.; Saedler, H.; Lonnig, W. E. Plant Transposable Elements. *Adv. Bot. Res.* **1997,** *27*, 331–470.

Leepton, P. J.; Smyth, D. R. An Abundant LINE-Like Element Amplified in the Genome of *Lilium speciosum*. *Mol. Genet. Genom.* **1993,** *237*, 97–104.

Levy, A. A.; Walbot, V. Molecular Analysis of the Loss of Somatic Instability in the *bz2:mu1* Allele of Maize. *Mol. Genom. Genet.* **1991,** *229*, 147–151.

Li, S.; Xia, Q.; Wang, F.; Yu, X.; Ma, J.; Kou, H.; Lin, X.; Gao, X.; Liu, B. Laser Irradiation-Induced DNA Methylation Changes Are Heritable and Accompanied with Transpositional Activation of *mPing* in Rice. *Front. Plant Sci.* **2017a,** *8*, 363.

Li, X.; Guo, K.; Zhu, X.; Chen, P.; Li, Y.; Xie, G.; Wang, L.; Wang, Y.; Persson, S. Domestication of Rice has Reduced the Occurrence of Transposable Elements Within Gene Coding Regions. *BMC Genom.* **2017b,** *18*, 55–67.

Lisch, D. How Important Are Transposons for Plant Evolution? *Nat. Rev. Genet.* **2013,** *14*, 49–61.

Lisch, D.; Girard, L.; Donlin, M.; Freeling, M. Functional Analysis of Deletion Derivatives of the Maize Transposon MuDR Delineates Roles for the MURA and MURB Proteins. *Genetics* **1999,** *151*, 331–341.

Martienssen, R. A.; Barkan, A.; Taylor, W. C.; Freeling, M. Somatically Heritable Switches in the DNA Modification of *Mu* Transposable Elements Monitored with a Suppressible Mutant in Maize. *Genes Dev.* **1989,** *4*, 331–343.

Masson, P.; Surosky, R.; Kingsbury, J. A.; Fedoroff, N. V. Genetic and Molecular Analysis of the *Spm*-Dependent *a-m2* Alleles of the Maize *a* Locus. *Genetics* **1987,** *117*, 117–137.

Matzke, M. A.; Matzke, A. J. Epigenetic Silencing of Plant Transgenes as a Consequence of Diverse Cellular Defence Responses. *Cell Mol. Life Sci.* **1998,** *54*, 94–103.

Matzke, M. A.; Primig, M.; Trnovsky, J.; Matzke, A. J. Reversible Methylation and Inactivation of Marker Genes in Sequentially Transformed Tobacco Plants. *EMBO J.* **1989,** *8*, 643–649.

McClintock, B. Maize Genetics. *Year B. Carnegie Inst. Wash.* **1946,** *45*, 176–186.

McClintock, B. Mechanisms that Rapidly Reorganise the Genome: (Genome Reorganisations, Transposable Gene Control, Restriction Enzymes, Stabilising Mechanisms). *Stadler Genet. Symp.* **1978,** *10*, 25–48.

McClintock, B. The Significance of Responses of the Genome to Challenges. *Science* **1984,** *2261*, 792–801.

McClintock, B. Mutable Loci in Maize. *Carnegie Inst. Wash. Year B.* **1948,** *47*, 155–169.

McClintock, B. Chromosome Organization and Genic Expression. *Cold Spring Harb. Symp. Quant. Biol.* **1951,** *16*, 13–47.

Mirouze, M.; Vitte, C. Transposable Elements, a Treasure Trove to Decipher Epigenetic Variation: Insights from *Arabidopsis* and Crop Epigenomes. *J. Exp. Bot.* **2014,** *65*, 2801–2812.

Mochizuki, K.; Umeda, M.; Ohtsubo, H.; Ohtsubo, E. Characterization of Plant SINE, p-SINE1, in Rice Genomes. *Jpn. J. Genet.* **1992,** *57*, 155–166.

Ogiwara, I.; Miya, M.; Ohshima, K.; Okada, N. Retropositional Parasitism of SINEs on LINEs: Identification of SINEs and LINEs in Elasmobranchs. *Mol. Biol. Evolut.* **1999,** *16*, 1238–1250.

Orgel, L. E.; Crick, F. H. C. Selfish DNA: the Ultimate Parasite. *Nature* **1980,** *284*, 604–607.

Panaud, O.; Vitte, C.; Hivert, J.; Muzlak, S.; Talag, J.; Brar, D.; Sarr, A. Characterization of Transposable Elements in the Genome of Rice (*Oryza sativa* L.) Using Representational Difference Analysis (RDA). *Mol. Genet. Genom.* **2002,** *268*, 113–121.

Peacock, W. J.; Dennis, E. S.;Gerlach, W. L.; Sachs, M. M.; Schwartz, D. Insertion and Excision of *Ds* Controlling Elements in Maize. *Cold Spring Harb. Symp. Quant. Biol.* **1984,** 49, 347–354.

Pereira, A.; Schwarz-Sommer, Z.; Gierl, A.; Bertram, I.; Peterson, P. A.; Saedler, H. Genetic and Molecular Analysis of the Enhancer (En) Transposable Element System of *Zea mays*. *EMBO J.* **1985,** *4*, 17–23.

Picault, N.; Chaparro, C.; Piegu, B.; Stenger, W.; Formey, D.; Llauro, C.; Descombin, J.; Sabot, F.; Lasserre, E.; Meynard, D. Identification of an Active LTR Retrotransposon in Rice. *Plant J.* **2009,** *58*, 754–765.

Raina, R.; Schlappi, M.; Karunanandaa, B.; Elhofy, A.; Fedoroff, N. Concerted Formation of Macromolecular Suppressor Mutator Transposition Complexes. *Proc. Natl. Acad. Sci. U.S.A.* **1998,** *95*, 8526–8531.

Tsuchimoto, S.; Hirao, Y.; Ohtsubo, E.; Ohtsubo, H. New SINE Families from Rice, OsSN, with Poly (A) at the 3′ Ends. *Genes Genet. Syst.* **2008,** *83*, 227–236.

Turcotte, K.; Srinivasan, S.; Bureau, T. Survey of Transposable Elements from Rice Genomic Sequences. *Plant J.* **2001,** *25*, 169–179.

Umeda, M.; Ohtsubo, H.; Ohtsubo, E. Diversification of the Rice *Waxy* Gene by Insertion of Mobile Elements into Introns. *Jpn. J. Genet.* **1991,** *66*, 569–586.

Wessler, S. R. The Splicing of Transposable Elements and Its Role in Intron Evolution. *Genetica* **1992,** *86*, 295–305.

Wessler, S. R.; Baran, G.; Varagona, M. The Maize Transposable Element *Ds* Is Spliced from RNA. *Science* **1987,** *237*, 916–918.

Wessler, S. R.; Bureau, T. E.; White, S. E. LTR-Retrotransposons and MITEs: Important Players in the Evolution of Plant Genomes. *Curr. Opin. Genet. Dev.* **1995,** *5*, 814–821.

Wicker, T.; Gundlach, H.; Spannagl, M.; Uauy, C.; Borrill, P.; Ramırez-Gonzalez, R. H.; DeOliveira, R.; Mayer, K. F. X.; Paux, E.; Choulet, F. Impact of Transposable Elements on Genome Structure and Evolution in Bread Wheat. *Genome Biol.* **2018,** *19*, 103.

Witte, C. P.; Le, Q. H.; Bureau, T.; Kumar, A. Terminal Repeat Retrotransposons in Miniature (TRIM) Are Involved in Restructuring Plant Genomes. *Proc. Natl. Acad. Sci. U.S.A.* **2001,** *98*, 13778–13783.

Xiao, H.; Jiang, N.; Schaffner, E.; Stockinger, E. J.; van der Knaap, E. A Retrotransposon-Mediated Gene Duplication Underlies Morphological Variation of Tomato Fruit. *Science* **2008,** *319*, 1527–1530.

Yoshida, S.; Tamaki, K.; Watanabe, K.; Fujino, M.; Nakamura, C. A Maize MuDR-Like Element Expressed in Rice Callus Subcultured with Proline. *Hereditas* **1998,** *1291*, 95–99.

Zhang, L.; Hu, J.; Han, X.; Li, J.; Gao, Y.; Richards, C. M.; Zhang, C.; Tian, Y.; Liu, G.; Gul, H. A High-Quality Apple Genome Assembly Reveals the Association of a Retrotransposon and Red Fruit Colour. *Nat. Commun.* **2019,** *10*, 1494.

Zhang, Q. J.; Gao, L. Z. Rapid and Recent Evolution of LTR Retrotransposons Drives Rice Genome Evolution During the Speciation of AA-Genome *Oryza* Species. *G3 (Bethesda)* **2017,** *7*, 1875–1885.

Zhang, X.; Feschotte, C.; Zhang, Q.; Jiang, N.; Eggleston, W. B.; Wessler, S. R. P Instability Factor: an Active Maize Transposon System Associated with the Amplification of Tourist-Like MITEs and a New Superfamily of Transposases. *Proc. Natl. Acad. Sci. U.S.A.* **2001,** *98*, 12572–12577.

Index

E

F

G

H

I

J

K

L

U

Z